Special Functions
of Mathematics
for Engineers

Special Functions of Mathematics for Engineers

Larry C. Andrews
University of Central Florida

Second Edition

McGraw-Hill, Inc.

New York St. Louis San Francisco Auckland Bogotá
Caracas Lisbon London Madrid Mexico Milan
Montreal New Delhi Paris San Juan São Paulo
Singapore Sydney Tokyo Toronto

Library of Congress Cataloging-in-Publication Data

Andrews, Larry C.
    Special functions of mathematics for engineers / Larry C. Andrews.
    —2nd ed.
          p.   cm.
    Rev. ed. of :   Special functions for engineers and applied
mathematicians. New York : Macmillan, c1985.
    Includes bibliographical references and index.
    ISBN 0-07-001848-0
    1. Functions, Special.   I. Andrews, Larry C.   Special functions
for engineers and applied mathematicians.   II. Title.
QA351.A75   1992
515.5—dc20   91-16093
                                                                      CIP

For Louise

First Edition published by Macmiiian © 1985 as *Special Functions for
Engineers and Applied Mathematicians.*

1 2 3 4 5 6 7 8 9 0   DOC/DOC   9 7 6 5 4 3 2 1

ISBN   0-07-001848-0

*The sponsoring editor for this book was Robert Hauserman, the editing
supervisor was Peggy Lamb, and the production supervisor was Donald
F. Schmidt. This book was set in Century Schoolbook by The
Universities Press (Belfast) Ltd.*

Printed and bound by R. R. Donnelley & Sons Company.

Preface to the Second Edition

The primary changes in this second edition include the introduction
of many more applications, chosen from a variety of fields such as
statics, dynamics, statistical communication theory, fiber optics, heat
conduction in solids, vibration phenomena, and fluid mechanics,
among others. In many cases these applications appear in the
chapter in which the particular special function is introduced.
However, because applications involving Bessel functions and
hypergeometric-type functions are far more extensive than those of
the other functions, they carry over to separate chapters devoted
entirely to applications (Chaps. 8 and 12).

As in the first edition, the text is suitable for use either as a
classroom text in various courses dealing with higher mathematical
functions or as a reference text for practicing engineers and scientists. To this end I have tried to preserve the readability of the first
edition, improving it where I could by the addition of further
examples or clearer exposition. For instance, I have rearranged the
order of topics in Chap. 1 so that asymptotic formulas follow the
discussion of improper integrals, and in addition to the chapter on
applications, the discussion of Bessel functions has been expanded to
two chapters—one chapter devoted entirely to Bessel functions of the
first and second kinds (Chap. 6) and one devoted to Bessel functions
of other kinds, such as modified Bessel functions and spherical Bessel
functions (Chap. 7). These discussions on Bessel functions also
include some new material such as the introduction of addition
formulas, Kelvin's functions, and Struve functions.

I am grateful to a number of students and colleagues for their
helpful suggestions concerning this second edition. In particular, I
wish to thank B. K. Shivamoggi, K. Vajravelu, and M. Belkerdid for
their input concerning the choice of certain applications. I am further
indebted to B. K. Shivamoggi for reading most of the new material

and offering many useful suggestions. Finally, I wish to thank the entire production staff of McGraw-Hill and, in particular, acknowledge my editor, Robert Hauserman, for his continued support of this project.

L. C. Andrews

Preface to the First Edition

Modern engineering and physics applications demand a more thorough knowledge of applied mathematics than ever before. In particular, it is important to have a good understanding of the basic properties of *special functions*. These functions commonly arise in such areas of application as heat conduction, communication systems, electro-optics, nonlinear wave propagation, electromagnetic theory, quantum mechanics, approximation theory, probability theory, and electric circuit theory, among others. Special functions are sometimes discussed in certain engineering and physics courses, and mathematics courses such as partial differential equations, but the treatment of special functions in such courses is usually too brief to focus on many of the important aspects, such as the interconnecting relations between various special functions and elementary functions. This book is an attempt to present, at the elementary level, a more comprehensive treatment of special functions than can ordinarily be done within the context of another course. It provides a systematic introduction to most of the important special functions that commonly arise in practice and explores many of their salient properties. I have tried to present the special functions in a broader sense than is often done by not introducing them as simply solutions of certain differential equations. Many special functions are introduced by the generating-function method, and the governing differential equation is then obtained as one of the important properties associated with the particular function.

In addition to discussing special functions, I have injected throughout the text by way of examples and exercises some of the techniques of applied analysis that are useful in the evaluation of nonelementary integrals, summing series, and so on. All too often in practice a problem is labeled "intractable" simply because the practitioner has not been exposed to the "bag of tricks" that helps the applied analyst deal with formidable-looking mathematical expressions.

During the last 10 years or so at the University of Central Florida we have offered an introductory course in special functions to a mix of advanced undergraduates and first-year graduate students in mathematics, engineering, and physics. A set of lecture notes developed for that course has finally led to this textbook. The prerequisites for our course are the basic calculus sequence and a first course in differential equations. Although complex variable theory is often utilized in studying special functions, knowledge of complex variables beyond some simple algebra and Euler's formulas is not required here. By not developing special functions in the language of complex variables, the text should be accessible to a wider audience. Naturally, some of the beauty of the subject is lost by this omission.

The text is not intended to be an exhaustive treatment of special functions. It concentrates heavily on a few functions, using them as illustrative examples, rather than attempting to give equal treatment to all. For instance, an entire chapter is devoted to the Legendre polynomials (and related functions), while the other orthogonal polynomial sets, including Hermite, Laguerre, Chebyshev, Gegenbauer, and Jacobi polynomials, are all lumped together in a single separate chapter. However, once the student is familiar with Legendre polynomials (which are perhaps the simplest set) and their properties, it is easy to extend these properties to other polynomial sets. Some applications occur throughout the text, often in the exercises, and Chap. 7 is devoted entirely to applications involving boundary-value problems. Other interesting applications which lead to special functions have been omitted, since they generally presuppose knowledge beyond the stated prerequisites.

Because of the close association of infinite series and improper integrals with the special functions, a brief review of these important topics is presented in the first chapter. In addition to reviewing some familiar concepts from calculus, this first chapter contains material that is probably new to the student, such as the Cauchy product, index manipulation, asymptotic series, Fourier trigonometric series, and infinite products. Of course, our discussion of such topics is necessarily brief.

I owe a debt of gratitude to the many students who took my course on special functions over the years while this manuscript was being developed. Their patience, understanding, and helpful suggestions are greatly appreciated. I want to thank my colleague and friend, Patrick J. O'Hara, who graciously agreed on several occasions to teach from the lecture notes in their early rough form, and who made several helpful suggestions for improving the final version of the manuscript. Finally, I wish to express my appreciation to Ken Werner, Senior Editor of Scientific and Technical Books Department, for his continued faith in this project and efforts in getting it published.

Notation for Special Functions

Notation	*Name of function*
Ai (x), Bi (x)	Airy functions of the first and second kinds
bei (x), ber (x), bei$_p$ (x), ber$_p$ (x)	Kelvin's functions
$B(x, y)$	Beta function
$B_x(p, q)$	Incomplete beta function
$b_n(x)$	Bessel polynomial
$C(x)$, $C_1(x)$, $C_2(x)$	Fresnel cosine integrals
$C_n^\lambda(x)$	Gegenbauer polynomial
Ci (x)	Cosine integral
cn u, dn u	Jacobian elliptic functions
$D_n(x)$	Parabolic cylinder function
Ei (x), $E_1(x)$	Exponential integral
$E(m)$	Complete elliptic integral of the second kind
$E(m, \phi)$	Elliptic integral of the second kind
$E(a_p; c_q; x)$	MacRobert E function
$\mathbf{E}_p(x)$	Weber function
erf x, erfc x	Error functions
$\zeta(x)$	Riemann zeta function
$F(a, b, c; x) = {}_2F_1(a, b; c; x)$	Hypergeometric function
$F(m, \phi)$	Elliptic integral of the first kind
${}_pF_q(a_p; c_q; x)$	Generalized hypergeometric function
$\Gamma(x)$	Gamma function
$\gamma(a, x)$, $\Gamma(a, x)$	Incomplete gamma functions
$G(a, b; c; x)$	Hypergeometric function of the second kind
$G_{p,q}^{m,n}(x \,|_{c_q}^{a_p})$	Meijer G function

Notation	Name of function
$H_n(x), H_\nu(x)$	Hermite polynomial, Hermite function
$\mathbf{H}_p(x)$	Struve function of the first kind
$H_p^{(1)}(x), H_p^{(2)}(x)$	Hankel functions of the first and second kinds
$h_n^{(1)}(x), h_n^{(2)}(x)$	Spherical Hankel functions of the first and second kinds
$i_n(x)$	Modified spherical Bessel function of the first kind
$I_p(x)$	Modified Bessel function of the first kind
$\mathrm{Ji}_p(x)$	Integral Bessel function
$j_n(x)$	Spherical Bessel function of the first kind
$J_p(x)$	Bessel function of the first kind
$\mathbf{J}_p(x)$	Anger function
$\mathrm{kei}\,(x), \mathrm{ker}\,(x)$	Kelvin's functions
$K(m)$	Complete elliptic integral of the first kind
$k_n(x)$	Modified spherical Bessel function of the second kind
$K_p(x)$	Modified Bessel function of the second kind
$\mathrm{li}\,(x)$	Logarithmic integral
$L_n(x)$	Laguerre polynomial
$L_n^{(m)}(x), L_\nu^{(a)}(x)$	Associated Laguerre polynomial, associated Laguerre function
$\mathbf{L}_p(x)$	Modified Struve function
$M(a;c;x) = {}_1F_1(a;c;x)$	Confluent hypergeometric function
$M_{k,m}(x)$	Whittaker function of the first kind
$P_n(x), P_\nu(x)$	Legendre polynomial, Legendre function
$P_n^m(x)$	Associated Legendre function of the first kind
$P_n^{(a,b)}(x)$	Jacobi polynomial
$\Pi(m, a)$	Complete elliptic integral of the third kind
$\Pi(m, \phi, a)$	Elliptic integral of the third kind
$\psi(x)$	Digamma or psi function
$\psi^{(m)}(x)$	Polygamma function
$Q_n(x)$	Legendre function of the second kind
$Q_n^m(x)$	Associated Legendre function of the second kind
$\mathrm{Si}\,(x), \mathrm{si}\,(x)$	Sine integrals
$S(x), S_1(x), S_2(x)$	Fresnel sine integrals

Notation	*Name of function*
sn u	Jacobian elliptic function
$T_n(x)$	Chebyshev polynomial of the first kind
$U_n(x)$	Chebyshev polynomial of the second kind
$U(a;c;x)$	Confluent hypergeometric function of the second kind
$W_{k,m}(x)$	Whittaker function of the second kind
$y_n(r)$	Spherical Bessel function of the second kind
$Y_p(x)$	Bessel function of the second kind

**Special Functions
of Mathematics
for Engineers**

Infinite Series, Improper Integrals, and Infinite Products

1.1 Introduction

Because of the close relation of *infinite series* and *improper integrals* to the special functions, it can be useful to first review some basic concepts of series and integrals. *Infinite products,* which are generally less well known, are introduced here mostly for the sake of completeness, but in some instances they are also useful.

Infinite series are important in almost all areas of pure and applied mathematics. In addition to numerous other uses, they are used to define certain functions and to calculate accurate numerical estimates of the values of these functions. In calculus the primary problem is deciding whether a given series converges or diverges. In practice, however, the more crucial problem may actually be summing the series. If a convergent series converges too slowly, the series may be worthless for computational purposes. On the other hand, the first few terms of a divergent series in some instances may give excellent results. Improper integrals and infinite products are used in much the same fashion as infinite series, and, in fact, their basic theory closely parallels that of infinite series.

In the application of mathematics frequently two or more limiting processes have to be performed successively. For example, we often find the derivative (or integral) of an infinite sum of functions by taking the sum of derivatives (or integrals) of the individual terms of the series. However, in many cases of interest, performing two limit operations in one order may yield an answer different from that obtained using the other order. That is, the order in which the limiting processes are carried out may be critical. It is of utmost importance, therefore, to know the conditions under which such interchanges are permitted, and that is one of the considerations of this chapter.

Because we assume the reader already has some familiarity with series and improper integrals, our treatment here is intentionally cursory. In this regard, we state only the most relevant theorems, generally without proof. For a deeper discussion of the subject matter, the reader is advised to consult one of the standard texts on advanced calculus.

1.2   Infinite Series of Constants

If to each positive integer n we can associate a number S_n, then the ordered arrangement

$$S_1, S_2, \ldots, S_n, \ldots \tag{1.1}$$

is called an **infinite sequence**. We call S_n the *general term* of the sequence. Should it happen that

$$\lim_{n \to \infty} S_n = S \tag{1.2}$$

where S is finite, then sequence (1.1) is said to **converge** to S; and otherwise it is said to **diverge**.

An **infinite series** results when an infinite sequence of numbers $u_1, u_2, \ldots, u_k, \ldots$ is summed, that is,

$$u_1 + u_2 + \cdots + u_k + \cdots = \sum_{k=1}^{\infty} u_k \tag{1.3}$$

In this case the number u_k is called the *general term* of the series. Closely associated with the infinite series (1.3) is a particular sequence

$$
\begin{aligned}
S_1 &= u_1 \\
S_2 &= u_1 + u_2 \\
&\ \ \vdots \\
S_n &= u_1 + u_2 + \cdots + u_n = \sum_{k=1}^{n} u_k \\
&\ \ \vdots
\end{aligned}
\tag{1.4}
$$

called the **sequence of partial sums**. The relation between the convergence of the partial sums and that of the series is contained in the following definition.

Definition 1.1. If the sequence of partial sums $\{S_n\}$ converges to a finite limit S, that is, if

$$S_n = \sum_{k=1}^{n} u_k \quad \text{and} \quad \lim_{n \to \infty} S_n = S$$

we say the infinite series

$$\sum_{k=1}^{\infty} u_k$$

converges, or sums, to the value S. In this case, we write

$$S = \sum_{k=1}^{\infty} u_k$$

The series **diverges** when the limit of partial sums fails to exist, i.e., fails to approach a unique finite value.

Example 1: Determine whether the following series converges or diverges:

$$\sum_{k=1}^{\infty} \frac{1}{k(k+1)}$$

Solution: To show convergence or divergence, we need to find the sum S_n of the first n terms and examine its limit as $n \to \infty$. To do so, first we observe that

$$S_n = \sum_{k=1}^{n} \frac{1}{k(k+1)} = \sum_{k=1}^{n} \left(\frac{1}{k} - \frac{1}{k+1} \right)$$

$$= \left(1 - \frac{1}{2} \right) + \left(\frac{1}{2} - \frac{1}{3} \right) + \cdots + \left(\frac{1}{n} - \frac{1}{n+1} \right)$$

$$= 1 - \frac{1}{n+1}$$

where only the first and last terms do not cancel. Thus,

$$\lim_{n \to \infty} S_n = \lim_{n \to \infty} \left(1 - \frac{1}{n+1} \right) = 1$$

and we conclude that the series *converges* and, in particular, converges to the value unity.

The key to solving Example 1 was obtaining a formula for the nth partial sum S_n. In cases like this where S_n can be readily found, it is a relatively easy matter to then take its limit to decide the convergence or divergence of the series. Situations for which S_n cannot be easily found will be discussed shortly.

When a series diverges, it may do so for different reasons. For example, let us first consider the infinite series

$$\sum_{k=1}^{\infty} k = 1 + 2 + 3 + \cdots + k + \cdots \tag{1.5}$$

Making use of the well-known formula from calculus

$$S_n = \sum_{k=1}^{n} k = \tfrac{1}{2}n(n+1)$$

it is clear that $\lim_{n\to\infty} S_n = \infty$, and therefore the infinite series $\sum k$ diverges.* In other instances the partial sums may not approach any particular limit, as for the series

$$\sum_{k=1}^{\infty} (-1)^{k-1} = 1 - 1 + 1 - 1 + \cdots + (-1)^{k-1} + \cdots \tag{1.6}$$

Here we find the partial sums $S_1 = 1$, $S_2 = 0$, $S_3 = 1, \ldots$ so that in general $S_n = 1$ for *odd n* and $S_n = 0$ for *even n*. Hence, because a unique limit of S_n does not exist, we say that $\sum (-1)^{k-1}$ diverges.

1.2.1 The geometric series

The special series

$$1 + r + r^2 + \cdots + r^k + \cdots = \sum_{k=0}^{\infty} r^k \tag{1.7}$$

is called a **geometric series**. The value r is called the *common ratio* since it is the ratio between the $(k+1)$st term and the kth term. This series is important because it has a wide variety of applications, but also because it can be summed exactly in those cases for which it converges.

From elementary algebra we know that the sum of the first n terms

---

*We will occasionally find it convenient to use the symbol $\sum u_k$ to denote $\sum_{k=1}^{\infty} u_k$.

of (1.7) is given by (see problem 1 in Exercises 1.2)

$$S_n = \sum_{k=0}^{n-1} r^k = \frac{1-r^n}{1-r} \qquad r \neq 1 \tag{1.8}$$

where we stop the summation at $n-1$ since the series begins at $k=0$. Since (1.8) requires that $r \neq 1$, let us consider that case separately. Substituting $r=1$ into (1.7), we find that it reduces to the series

$$\sum_{k=0}^{\infty} 1 = 1 + 1 + 1 + \cdots$$

which clearly *diverges* (Why?). Also the case $r = -1$ leads to the divergent series (1.6). For other values of r we simply take the limit of (1.8) as $n \to \infty$ to find

$$\lim_{n \to \infty} S_n = \begin{cases} \dfrac{1}{1-r} & |r| < 1 \\ \text{no limit} & |r| > 1 \end{cases} \tag{1.9}$$

where we are using the fact that $r^n \to 0$ for increasing n when $|r| < 1$. We deduce, therefore, that the geometric series (1.7) *converges* for $|r| < 1$ and *diverges* for $|r| \geq 1$.

Based on the above results, we have established the important formula

$$\sum_{k=0}^{\infty} r^k = \frac{1}{1-r} \qquad |r| < 1 \tag{1.10}$$

which, in addition to identifying the values of r for which the series converges, provides the actual *sum* of the series.

Example 2: Test the series

$$3 - 2 + \frac{4}{3} - \frac{8}{9} + \cdots + 3\left(-\frac{2}{3}\right)^k + \cdots$$

for convergence.

Solution: By writing the series in the form

$$3 - 2 + \frac{4}{3} - \frac{8}{9} + \cdots + 3\left(-\frac{2}{3}\right)^k + \cdots$$

$$= 3\left[1 - \frac{2}{3} + \frac{4}{9} - \frac{8}{27} + \cdots + \left(-\frac{2}{3}\right)^k + \cdots \right]$$

$$= \sum_{k=0}^{\infty} \left(-\frac{2}{3}\right)^k$$

we recognize it as a geometric series (multiplied by 3) with $r = -\frac{2}{3}$. Since r is less than unity in absolute value, we deduce that the series *converges* and, moreover, converges to the value

$$3 \sum_{k=0}^{\infty} \left(-\frac{2}{3}\right)^k = \frac{3}{1 - (-\frac{2}{3})} = \frac{9}{5}$$

1.2.2 Summary of convergence tests

Generally speaking, the series that are most useful in practice are those that converge. For that reason we attach a great deal of importance to the task of deciding whether a particular series converges. In the case of the geometric series we were able to get the nth partial sum S_n into a "closed form" and examine its limit directly as $n \to \infty$. By doing so, we not only answered the question of convergence or divergence, but also summed the series. That is the real power of Definition 1.1. Unfortunately, the geometric series is one of the rare examples for which we are able to directly apply Definition 1.1, since we generally cannot obtain S_n for most other series. What is required in these cases, then, is a "handful" of *tests* that can be applied to the series in question from which its convergence or divergence can be established independently of Definition 1.1. A great many such convergence tests have been developed over the years, some simple to apply and others quite sophisticated. In virtually all cases, however, these tests tell us only whether the series converges or diverges—they do not actually provide the sum of the series as in Definition 1.1.

The development of various convergence tests is generally taken up in calculus courses (both elementary and advanced). Our intention here is to simply recall some of the more elementary tests for reference purposes.

Remark: Any index such as k or n can always be used in writing a series, finite or infinite, depending on which index may be convenient. That is,

$$\sum_{k=1}^{\infty} u_k = \sum_{n=1}^{\infty} u_n$$

As a general rule, however, we will ordinarily use the index n when the series is infinite and k when it is finite.

Let us first observe that if $\sum u_n = S$, where S is finite, then $S_n \to S$

and $S_{n-1} \to S$ as $n \to \infty$. Hence, it follows that

$$\lim_{n \to \infty} (S_n - S_{n-1}) = S - S = 0$$

But since

$$S_n - S_{n-1} = \sum_{k=1}^{n} u_k - \sum_{k=1}^{n-1} u_k = u_n$$

we see that a *necessary condition* (but not a sufficient condition) for the series $\sum u_n$ to converge is that $\lim_{n \to \infty} u_n = 0$. Thus, we have the following theorem.

Theorem 1.1. If $\sum u_n$ converges, then $\lim_{n \to \infty} u_n = 0$. On the other hand, if $\lim_{n \to \infty} u_n \neq 0$, then the series $\sum u_n$ diverges.

Based on Theorem 1.1, the series $\sum_{n=1}^{\infty} n/(n+1)$ diverges since

$$\lim_{n \to \infty} u_n = \lim_{n \to \infty} \frac{n}{n+1} = 1 \neq 0$$

However, testing the general term of the series $\sum_{n=1}^{\infty} 1/(n+1)$, we have

$$\lim_{n \to \infty} u_n = \lim_{n \to \infty} \frac{1}{n+1} = 0$$

so we cannot use Theorem 1.1 to draw a conclusion about the convergence or divergence of this series.*

A series is called **positive** if the terms of the series are either all positive or all negative. In general, the terms of a series may vary in sign, some terms positive and others negative. If the consecutive terms have opposite signs, we call the series an **alternating series**. Series containing terms both positive and negative converge more rapidly (when they converge) than do positive series, due to the partial cancellation of the negative terms with the positive terms. Because of these distinctions, we introduce notions of different kinds of convergence.

---

*This series in fact diverges, as can be shown by the integral test (Theorem 1.2).

Definition 1.2. The series $\sum u_n$ is said to **converge absolutely** if the associated series of positive terms $\sum |u_n|$ converges.

Definition 1.3. If the series $\sum u_n$ converges but the related series of positive terms $\sum |u_n|$ diverges, then $\sum u_n$ is said to **converge conditionally.**

If a positive series converges, it necessarily converges absolutely. Hence, the term *conditional convergence* applies only to those series which have both positive and negative terms. When the general term of a positive series is an integrable function of n, the following *integral test* of convergence can be useful.

Theorem 1.2 (Integral test). Let $f(n)$ denote the general term of the series $\sum u_n$. If the function $f(x)$ is positive, continuous, and nonincreasing for $x \geq a$, then the positive series $\sum u_n$ converges or diverges according to the convergence or divergence of the improper integral* $\int_a^\infty f(x)\, dx$.

To illustrate the use of Theorem 1.2, let us consider the **p series**†

$$\sum_{n=1}^{\infty} \frac{1}{n^p} = 1 + \frac{1}{2^p} + \frac{1}{3^p} + \cdots + \frac{1}{n^p} + \cdots \tag{1.11}$$

Here we take $f(x) = 1/x^p$, and for any constant $a > 0$, we find that

$$\int_a^\infty \frac{1}{x^p}\, dx = \begin{cases} \left. \dfrac{x^{1-p}}{1-p} \right|_a^\infty & p \neq 1 \\ \ln x |_a^\infty & p = 1 \end{cases}$$

from which we deduce that the series *converges* for $p > 1$ and *diverges* for $p \leq 1$.‡ The special value $p = 1$ leads to

$$\sum_{n=1}^{\infty} \frac{1}{n} = 1 + \frac{1}{2} + \frac{1}{3} + \cdots + \frac{1}{n} + \cdots \tag{1.12}$$

called the **harmonic series**. It plays an important role in the use of comparison tests (see Theorem 1.3 below), and although it diverges,

---

* The convergence of improper integrals is reviewed in Sec. 1.5.

† For $p > 1$, the p series is also called the *Riemann zeta function* (see Sec. 2.5.4).

‡ We follow the convention that $\log x$ means $\log_{10} x$ and $\ln x$ means $\log_e x$ ($\ln x$ is known as the *natural logarithm*).

it does so at a very slow rate. For example, the total of the first million terms is only slightly larger than 14.

In addition to the integral test, we have the following useful tests for deciding convergence and divergence of a series.

Theorem 1.3 (Comparison test). A positive series $\sum u_n$ **converges absolutely** if each term (after a finite number) is less than or equal to the corresponding term of a known convergent positive series $\sum a_n$, that is, if

$$u_n \leq a_n \qquad n > N$$

The positive series $\sum u_n$ **diverges** if each term (after a finite number) is greater than or equal to the corresponding term of a known divergent positive series $\sum b_m$, that is, if

$$u_n \geq b_n \qquad n > N$$

Theorem 1.4.   If $\lim_{n \to \infty} n^p u_n = A$, then

(*a*) $\sum u_n$ converges if $p > 1$ and A is finite.

(*b*) $\sum u_n$ diverges if $p \leq 1$ and $A \neq 0$ (A may be infinite).

Example 3: Test the following series for convergence:

(*a*) $\displaystyle\sum_{n=2}^{\infty} \frac{1}{n-1}$ (*b*) $\displaystyle\sum_{n=1}^{\infty} \frac{\log n}{\sqrt{n+1}}$

Solution

(*a*) By observing that $\sum (1/n)$ diverges and

$$\frac{1}{n-1} \geq \frac{1}{n} \qquad n > 2$$

it follows from Theorem 1.3 that the given series diverges.

(*b*) Here we choose $p = \frac{1}{2}$ and consider the limit

$$\lim_{n \to \infty} n^{1/2} \frac{\log n}{\sqrt{n+1}} = \infty$$

Based on Theorem 1.4, we deduce that the series diverges.

If the series in question is an alternating series, then we usually start with the following test.

Theorem 1.5 (Alternating-series test). If $u_n > 0$, $n = 1, 2, 3, \ldots$ and $\lim_{n \to \infty} u_n = 0$, then the alternating series $\sum (-1)^n u_n$ converges (conditionally at least). Also the sum of a convergent alternating series always lies between the partial sums S_n and S_{n+1} for each n.

Unlike in Theorem 1.1, we can conclude (conditional) convergence for an alternating series by showing that $u_n \to 0$ as $n \to \infty$. However, if we show that an alternating series converges by the alternating-series test, we must further investigate the series to determine whether it also converges absolutely. This we do by applying another test to the related series of positive terms.

Example 4: Test for convergence:

$$\sum_{n=1}^{\infty} \frac{(-1)^{n-1}}{n}$$

Solution: Clearly, the general term of the series approaches zero as $n \to \infty$. Hence, by the alternating-series test, the series converges. However, since the absolute value of the general term is $|u_n| = 1/n$, the related series of positive terms is $\sum (1/n)$, which is the divergent harmonic series. We conclude, therefore, that the given series, called the **alternating harmonic series**, converges *conditionally* but not absolutely.

Theorem 1.6 (Ratio test). Let $\sum u_n$ denote any series for which

$$\lim_{n \to \infty} \left| \frac{u_{n+1}}{u_n} \right| = L$$

(a) If $L < 1$, the series $\sum u_n$ converges absolutely.

(b) If $L > 1$, the series $\sum u_n$ diverges.

(c) If $L = 1$, the test fails (no conclusion).

The *ratio test* is probably the most widely used test of convergence, and it is a particularly useful test for those series involving factorials or exponentials. However, it fails in those cases where the general term is a rational function of n.

Example 5: Test the series for convergence:

$$\sum_{n=1}^{\infty} \frac{n^2}{2^n}$$

Solution: Forming the ratio

$$\left| \frac{u_{n+1}}{u_n} \right| = \frac{(n+1)^2}{2^{n+1}} \cdot \frac{2^n}{n^2}$$

$$= \frac{1}{2} \left(\frac{n+1}{n} \right)^2$$

we see that its limit is

$$L = \lim_{n \to \infty} \left| \frac{u_{n+1}}{u_n} \right| = \frac{1}{2}$$

Hence, since $L < 1$, the given series converges.

1.2.3 Operations with series

In many applications the need arises to combine series by such operations as addition (and subtraction) and multiplication. Although the division of one series by another is sometimes required, we do not address it here. To perform addition and multiplication, it is usually important to establish the absolute convergence of all series involved in the process, since such operations can then be performed by the familiar rules of algebra or arithmetic. Specifically, we have the following:

1. The sum of an absolutely convergent series is independent of the order in which terms are added.

2. Two absolutely convergent series may be added termwise, and the resulting series will converge absolutely.

3. Two absolutely convergent series may be multiplied (Cauchy product), and the resulting series will also converge absolutely.

   Remark: It is also possible to sum or multiply series under less restrictive conditions than those listed above. However, in this review chapter we usually do not state results in their most general forms.

The significance of property 1 above can be best realized by considering what can happen if the series we wish to sum is *not* absolutely convergent. The classic example of a series converging conditionally, but not absolutely, is the alternating harmonic series $\sum (-1)^{n-1}/n$ (recall Example 4). Let us assume that the sum of the series is some finite value S and write*

$$\sum_{n=1}^{\infty} \frac{(-1)^{n-1}}{n} = 1 - \frac{1}{2} + \frac{1}{3} - \frac{1}{4} + \frac{1}{5} - \frac{1}{6} + \cdots = S \qquad (1.13)$$

However, if we rearrange the terms of the series according to

$$1 - \frac{1}{2} + \frac{1}{3} - \frac{1}{4} + \cdots = \left(1 - \frac{1}{2}\right) - \frac{1}{4} + \left(\frac{1}{3} - \frac{1}{6}\right) - \frac{1}{8} + \left(\frac{1}{5} - \frac{1}{10}\right)$$

$$- \frac{1}{12} + \left(\frac{1}{7} - \frac{1}{14}\right) - \frac{1}{16} + \cdots$$

$$= \frac{1}{2} - \frac{1}{4} + \frac{1}{6} - \frac{1}{8} + \frac{1}{10} - \frac{1}{12} + \cdots$$

$$= \frac{1}{2}\left(1 - \frac{1}{2} + \frac{1}{3} - \frac{1}{4} + \frac{1}{5} - \frac{1}{6} + \cdots\right)$$

we may conclude that the sum of the series is $\frac{1}{2}S$. We arrive at this conclusion not because we have cleverly omitted some terms of the series. Indeed, each term of series (1.13) does eventually appear exactly once, but in the final arrangement the whole series appears multiplied by the factor $\frac{1}{2}$.

What is being illustrated here is that by rearranging the terms of a conditionally convergent series, that series may be made to converge to any desired numerical value or even to diverge! Thus, it is clear that conditionally convergent series must be handled very carefully if they are to be meaningful.

Also if the series is *not positive* and *diverges,* we can sometimes produce what appears to be a convergent series from it by rearranging or regrouping the terms. For example, if we write the divergent series $\sum (-1)^{n-1}$ in the form

$$\sum_{n=1}^{\infty} (-1)^{n-1} = (1 - 1) + (1 - 1) + \cdots$$

$$= 0 + 0 + 0 + \cdots$$

we might conclude that the series converges to the value zero. On the other hand, by writing

$$\sum_{n=1}^{\infty} (-1)^{n-1} = 1 + (-1 + 1) + (-1 + 1) + \cdots$$

$$= 1 + 0 + 0 + \cdots$$

---

* We will show in Sec. 1.3 that $S = \ln 2$.

we might conclude that the series converges to unity. However, if a series is *positive* and *diverges,* it cannot be made to converge by any rearrangement of its terms.

If two series are absolutely convergent, no rearrangement of their terms will alter the sum or difference of the two series. But again, if both of the series forming the sum or difference are divergent, it is not clear what will happen. For instance, by writing

$$\sum_{n=1}^{\infty} \frac{1}{n} - \sum_{n=1}^{\infty} \frac{1}{n+1} = \sum_{n=1}^{\infty} \left(\frac{1}{n} - \frac{1}{n+1} \right) = \sum_{n=1}^{\infty} \frac{1}{n(n+1)}$$

we can interpret the series on the right as the difference of two divergent series (both series on the left are harmonic series). Although we may be tempted to say that the series on the right-hand side diverges, because of its relation to the divergent series on the left-hand side, we have actually shown that the series on the right-hand side converges (see Example 1) and, in fact, *converges absolutely.*

In forming the product of two series, we are led to **double infinite series** of the form*

$$\sum_{m=0}^{\infty} a_m \sum_{k=0}^{\infty} b_k = \sum_{m=0}^{\infty} \sum_{k=0}^{\infty} A_{m,k} \qquad (1.14)$$

where the summand $A_{m,k} = a_m b_k$ can be treated as a function of two variables. We find that by making a change of index in (1.14), the resulting double sum can often be simplified or even partially summed. For example, suppose we introduce the change of index

$$m = n - k$$

or, equivalently, $n = m + k$. Now since $m \geq 0$, the index k must satisfy the condition $n - k \geq 0$, or $k \leq n$. Also the new index n runs over the range of values $0 \leq n < \infty$, since both m and k are nonnegative and m can be infinite. Hence, for absolutely convergent series, we deduce that

$$\sum_{m=0}^{\infty} \sum_{k=0}^{\infty} A_{m,k} = \sum_{n=0}^{\infty} \sum_{k=0}^{n} A_{n-k,k} \qquad (1.15)$$

Equation (1.15) illustrates that all absolutely convergent double infinite series can be replaced by a single infinite series of finite

---

*Most of the infinite series that we encounter later on will start with the index value zero. Also, double series are not always the result of a product.

sums. This property is particularly useful in numerical computations. If the two series forming the product are each absolutely convergent, we can also interchange the order of the infinite sums and then apply (1.15).

Equation (1.15) is related to the *Cauchy product* of two series, for which we have the following theorem.

Theorem 1.7 (Cauchy product). If $\sum a_n$ and $\sum b_n$ are both absolutely convergent series, then so is their Cauchy product, defined by

$$\sum_{n=0}^{\infty} a_n \cdot \sum_{n=0}^{\infty} b_n = \sum_{n=0}^{\infty} c_n$$

where
$$c_n = \sum_{k=0}^{n} a_k b_{n-k} = \sum_{k=0}^{n} a_{n-k} b_k$$

Remark: Although we may use the same index n to illustrate that we wish to form a product, as in Theorem 1.7, it is best in practice to use two different indices as in (1.15) to avoid confusion.

The Cauchy product may still lead to a convergent series when only one of the series converges absolutely and the other converges conditionally. Also it is possible for both $\sum a_n$ and $\sum b_n$ to converge (conditionally) while the product series $\sum c_n$ diverges.

Example 6: Use the Cauchy product to find ST, where*

$$S = \sum_{n=0}^{\infty} \frac{(-1)^n}{2^n} \qquad T = \sum_{n=0}^{\infty} \frac{(-1)^n}{3^n}$$

Solution: Using Theorem 1.7, we have

$$ST = \sum_{n=0}^{\infty} \frac{(-1)^n}{2^n} \cdot \sum_{n=0}^{\infty} \frac{(-1)^n}{3^n} = \sum_{n=0}^{\infty} c_n$$

where c_n is the finite geometric series

$$c_n = \sum_{k=0}^{n} \frac{(-1)^{n-k}}{2^{n-k}} \frac{(-1)^k}{3^k} = \frac{(-1)^n}{2^n} \sum_{k=0}^{n} \left(\frac{2}{3}\right)^k$$

---

* Notice that both S and T are geometric series which could be summed exactly. We use them simply for illustrative purposes.

or, upon summing the series,

$$c_n = \frac{(-1)^n}{2^n}\, 3\left[1 - \left(\frac{2}{3}\right)^{n+1}\right]$$

Hence,

$$ST - 3\sum_{n=0}^{\infty} \frac{(-1)^n}{2^n}\left[1 - \left(\frac{2}{3}\right)^{n+1}\right] = \frac{1}{2}$$

the last step of which we leave to the reader.

On rare occasions we find it necessary to make a change of index in the double sum (1.14) different from that leading to (1.15). For example, if we set $m = n - 2k$ in (1.14), or equivalently $n = m + 2k$, it follows that $k \leq n/2$. But since $n/2$ is not always an integer, it is conventional to introduce the bracket notation

$$\left[\frac{n}{2}\right] = \begin{cases} \dfrac{n}{2} & n \text{ even} \\[2mm] \dfrac{n-1}{2} & n \text{ odd} \end{cases} \qquad (1.16)$$

Hence, with this index change we deduce that

$$\sum_{m=0}^{\infty}\sum_{k=0}^{m} A_{m,k} = \sum_{n=0}^{\infty}\sum_{k=0}^{[n/2]} A_{n-2k,k} \qquad (1.17)$$

and upon combining (1.15) and (1.17), it also follows that

$$\sum_{n=0}^{\infty}\sum_{k=0}^{n} A_{n-k,k} = \sum_{n=0}^{\infty}\sum_{k=0}^{[n/2]} A_{n-2k,k} \qquad (1.18)$$

1.2.4 Factorials and binomial coefficients

Products of consecutive integers arise quite naturally in the study of series. For a positive integer n, the product of all positive integers from 1 to n inclusive is called **factorial n** and denoted by $n!$. That is,

$$n! = 1 \cdot 2 \cdot 3 \cdots n \qquad n = 1, 2, 3, \ldots \qquad (1.19)$$

The most fundamental property of factorials is given by

$$n! = n(n-1)! \qquad n = 2, 3, 4, \ldots \qquad (1.20)$$

If we want (1.20) to hold also for $n = 1$, this would require $1! = 1(0!)$, suggesting that we define

$$0! = 1 \qquad (1.21)$$

A particular combination of factorials that appears quite frequently in practice is the *binomial coefficient*. To introduce it, let us consider the expanded products $(a + b)^n$ for integral $n \geq 1$. The first few products lead to

$$(a + b)^1 = a + b$$

$$(a + b)^2 = a^2 + 2ab + b^2$$

$$(a + b)^2 = a^3 + 3a^2b + 3ab^2 + b^3$$

whereas in general

$$(a + b)^n = a^n + na^{n-1}b + \frac{n(n-1)}{2!} a^{n-2}b^2 + \cdots$$

$$+ \frac{n(n-1)\cdots(n-k+1)}{k!} a^{n-k}b^k + \cdots + b^n \qquad (1.22)$$

We call (1.22) the *binomial formula*, or the *finite binomial series*.

It is customary to introduce a special symbol for the coefficient of the general term in (1.22), i.e.,

$$\binom{n}{k} = \frac{n(n-1)\cdots(n-k+1)}{k!} \qquad k = 1, 2, \ldots, n \qquad (1.23)$$

However, by the use of factorials we can write this also in the form

$$\binom{n}{k} = \frac{n!}{k!\,(n-k)!} \qquad n = 0, 1, 2, \ldots; \qquad k = 0, 1, \ldots, n \qquad (1.24)$$

which we obtain by multipyling the numerator and denominator of the right-hand side of (1.23) by $(n - k)!$ and using Eq. (1.20). Note that (1.24) is valid for $0 \leq k \leq n$ while (1.23) is restricted to $1 \leq k \leq n$. Adopting this notation, we can now write (1.22) more compactly as

$$(a + b)^n = \sum_{k=0}^{n} \binom{n}{k} a^{n-k}b^k \qquad (1.25)$$

The symbol $\binom{n}{k}$ is what we call a **binomial coefficient.** Besides its connection in (1.25) with the expansion of $(a + b)^n$, the binomial coefficient also occurs in combinatorial problems, probability theory, and algorithm development. In these other applications, the upper index is often not an integer or even a positive number. For such situations we cannot use (1.24) to define the binomial coefficient; rather we resort to the form given by (1.23). Hence, for general r we define

$$\binom{r}{0} = 1 \qquad \binom{r}{k} = \frac{r(r-1)\cdots(r-k+1)}{k!} \qquad k = 1, 2, 3, \ldots \quad (1.26)$$

Remark: Notice that although r is unrestricted in (1.26), parameter k is a nonnegative integer. In all relations given below involving the binomial coefficient, it is understood that the parameter occurring in the position of k in (1.26) must be so restricted.

As simple consequences of the definitions of binomial coefficient and factorial, we have the following useful relations:

$$\binom{n}{0} = \binom{n}{n} = 1 \tag{1.27}$$

$$\binom{n}{1} = \binom{n}{n-1} = n \tag{1.28}$$

$$\binom{n}{k} = \binom{n}{n-k} \tag{1.29}$$

$$\binom{n+1}{k+1} = \binom{n}{k+1} + \binom{n}{k} \qquad 0 \le k \le n-1 \tag{1.30}$$

$$\binom{-r}{k} = (-1)^k \binom{r+k-1}{k} \tag{1.31}$$

Verifying Eqs. (1.27) to (1.30) is left to the exercises, while the proof of Eq. (1.31) is given in Example 7.

Example 7: Show that

$$\binom{-r}{k} = (-1)^k \binom{r+k-1}{k} \qquad k = 1, 2, 3, \ldots$$

Solution: From (1.26), we have

$$\binom{-r}{k} = \frac{-r(-r-1)\cdots(-r-k+1)}{k!}$$

$$= (-1)^k \frac{r(r+1)\cdots(r+k-1)}{k!}$$

$$= (-1)^k \frac{(r+k-1)(r+k-2)\cdots(r+1)r}{k!}$$

where in the last step we have reversed the order of terms in the numerator. Using (1.26) once again, we obtain our result

$$\binom{-r}{k} = (-1)^k \binom{r+k-1}{k} \qquad k = 1, 2, 3, \ldots$$

There are literally thousands of identities involving binomial coefficients that have been discovered over the years. Fortunately, only a few are required in most routine applications. In addition to Eqs. (1.27) to (1.31), the following summation formulas are very useful:

$$\sum_{k=0}^{n} \binom{n}{k} = 2^n \tag{1.32}$$

$$\sum_{k=0}^{n} \binom{r+k}{k} = \binom{r+n+1}{n} \tag{1.33}$$

$$\sum_{k=0}^{n} \binom{k}{m} = \binom{n+1}{m+1} \qquad m = 0, 1, 2, \ldots \tag{1.34}$$

$$\sum_{k=0}^{n} \binom{r}{k}\binom{s}{n-k} = \binom{r+s}{n} \tag{1.35}$$

$$\sum_{k=0}^{n} (-1)^k \binom{n}{k}\binom{s+k}{m} = (-1)^n \binom{s}{m-n} \qquad m \geq n \tag{1.36}$$

Equation (1.32) follows directly from (1.25) with $a = b = 1$. Except for (1.35), which is verified below (see Example 8), all others are left to the exercises. However, we point out that one way to prove (1.33) requires application of (1.30), whereas (1.34) follows from (1.33) with two applications of (1.29).

Example 8: Show that

$$\sum_{k=0}^{n} \binom{r}{k}\binom{s}{n-k} = \binom{r+s}{n}$$

Solution: Although the identity is valid for all values of r and s, we will restrict r and s to positive integers in our proof.
Starting with the obvious identity

$$(1+x)^r (1+x)^s = (1+x)^{r+s} \qquad r, s = 1, 2, 3, \ldots$$

let us replace each factor by its binomial series, which yields

$$\sum_{m=0}^{r} \binom{r}{m} x^m \cdot \sum_{k=0}^{s} \binom{s}{k} x^k = \sum_{n=0}^{r+s} \binom{r+s}{n} x^n$$

The left-hand side is simply the product of two (finite) series, and hence we can rewrite it by using Eq. (1.15). By making the change of index $m = n - k$, the new indices k and n are restricted to $0 \le k \le n$ and $0 \le n \le r + s$. Therefore, we obtain

$$\sum_{n=0}^{r+s} \sum_{k=0}^{n} \binom{r}{k} \binom{s}{n-k} x^n = \sum_{n=0}^{r+s} \binom{r+s}{n} x^n$$

both sides of which describe a polynomial in x of degree $r + s$. Clearly, two polynomials are equal if and only if the corresponding terms of each are identical. Hence, by comparing coefficients of like terms of x^n, we get the desired result

$$\sum_{k=0}^{n} \binom{r}{k} \binom{s}{n-k} = \binom{r+s}{n}$$

Remark: In Sec. 1.3.2 we will show that when α is not a positive integer, the binomial series for $(1+x)^{\alpha}$ becomes an infinite series, and in this case x is restricted to the interval $|x| < 1$. Hence, each of the series appearing in the proof given in Example 8 is replaced by an infinite series in the general case of arbitrary r and s, but the proof remains essentially the same.

Exercises 1.2

1. Show that the nth partial sum of a geometric series satisfies

$$1 + r + r^2 + \cdots + r^{n-1} = \frac{1-r^n}{1-r} \qquad r \ne 1$$

Hint: Observe that

$$S_n = 1 + r + r^2 + \cdots + r^{n-1}$$

$$r S_n = r + r^2 + \cdots + r^{n-1} + r^n$$

and subtract termwise.

In problems 2 through 5, find the sum of the geometric series.

2. $\displaystyle\sum_{k=0}^{10} 2^k$

4. $\displaystyle\sum_{k=2}^{10} \left(\frac{1}{2}\right)^k$

3. $\displaystyle\sum_{k=0}^{100} (-1)^k$

5. $\displaystyle\sum_{n=0}^{\infty} \sin^{2n} x, \ |x| < \frac{\pi}{2}$

In problems 6 through 12, use an appropriate test of convergence to determine whether the given series *converges absolutely, converges conditionally,* or *diverges.*

6. $\displaystyle\sum_{n=2}^{\infty} \frac{1}{n \ln n}$

10. $\displaystyle\sum_{n=1}^{\infty} \frac{(-1)^n}{2n - 1}$

7. $\displaystyle\sum_{n=1}^{\infty} \frac{1}{(n + 2)^{3/2}}$

11. $\displaystyle\sum_{n=1}^{\infty} \frac{\ln n}{n^p}$

8. $\displaystyle\sum_{n=0}^{\infty} \frac{(2n)!}{(n!)^2}$

12. $\displaystyle\sum_{n=3}^{\infty} \frac{(-1)^n}{\sqrt{n} \ln (\ln n)}$

9. $\displaystyle\sum_{n=1}^{\infty} (-1)^n \left(1 + \frac{1}{n^4}\right)$

13. By using the Cauchy product, verify the identity $e^a e^b = e^{a+b}$.
Hint: Recall that

$$e^a = \sum_{n=0}^{\infty} \frac{a^n}{n!}$$

14. Use the identity $2 \sin a \cos a = \sin 2a$ and the Cauchy product to deduce that

$$\sum_{k=0}^{n} \binom{2n + 1}{2k + 1} = 2^{2n}$$

Hint: Recall that

$$\sin a = \sum_{n=0}^{\infty} \frac{(-1)^n}{(2n + 1)!} a^{2n+1} \quad \text{and} \quad \cos a = \sum_{n=0}^{\infty} \frac{(-1)^n}{(2n)!} a^{2n}$$

15. Show that

(a) $\displaystyle\binom{n}{0} = \binom{n}{n} = 1$

(b) $\displaystyle\binom{n}{1} = \binom{n}{n - 1} = n$

16. Show that

(a) $\dbinom{n}{k} = \dbinom{n}{n-k}$

(b) $\dbinom{n+1}{k+1} = \dbinom{n}{k+1} + \dbinom{n}{k}$, $0 \le k \le n-1$

17. Show that

(a) $\dbinom{-1/2}{n} = \dfrac{(-1)^n (2n)!}{2^{2n}(n!)^2}$

(b) $\dbinom{-2k-1}{n} = (-1)^n \dfrac{(n+2k)!}{(2k)!\,n!}$, $k = 0, 1, 2, \ldots$

In problems 18 through 20, verify the given formula.

18. $\displaystyle\sum_{k=0}^{n} \dbinom{r+k}{k} = \dbinom{r+n+1}{n}$

Hint: Use Eq. (1.30).

19. $\displaystyle\sum_{k=0}^{n} \dbinom{k}{m} = \dbinom{n+1}{m+1}$, $m = 0, 1, 2, \ldots$

Hint: Use problem 18 and Eq. (1.29).

20. $\displaystyle\sum_{k=0}^{n} (-1)^k \dbinom{n}{k}\dbinom{s+k}{m} = (-1)^n \dbinom{s}{m-n}$, $m \ge n$

1.3 Infinite Series of Functions

Of special importance to us are those series that result when the general term u_n is a function of x, that is, $u_n = u_n(x)$. The nth partial sum then defines the function

$$S_n(x) = \sum_{k=1}^{n} u_k(x) \tag{1.37}$$

The question of concern here is whether there exists any value of x for which the limit of partial sums exists as $n \to \infty$. If, for a fixed value of x, we find that

$$\lim_{n \to \infty} S_n(x) = f(x) \tag{1.38}$$

we say the series $\sum u_n(x)$ **converges pointwise** to $f(x)$. The *domain* of the function f is then the totality of all such values of x for which the series converges.

Example 9: Test the series for convergence:

$$x + \sum_{n=1}^{\infty} (x^{n+1} - x^n) \qquad 0 \leq x \leq 1$$

Solution: The sum of the first n terms leads to

$$S_n(x) = x + \sum_{k=1}^{n-1} (x^{k+1} - x^k)$$

$$= x + (x^2 - x) + (x^3 - x^2) + \cdots + (x^n - x^{n-1})$$

$$= x^n$$

where all other terms cancel. Hence, we see that

$$\lim_{n \to \infty} S_n(x) = \begin{cases} 0 & 0 \leq x < 1 \\ 1 & x = 1 \end{cases}$$

The given series converges pointwise, therefore, for every x in the interval $0 \leq x \leq 1$.

Pointwise convergence treats convergence of a series at individual points in an interval. Although it is an important type of convergence in some theoretical studies, it is inadequate for others. For example, pointwise convergence is not adequate to permit termwise differentiation or even termwise integration of a series. The simplest notion of convergence that in turn implies pointwise convergence throughout a closed interval is called *uniform convergence* (see Fig. 1.1).

Definition 1.4. If, given some $\epsilon > 0$, there exists a number $N = N(\epsilon)$, independent of x in the interval $a \leq x \leq b$, such that

$$|f(x) - S_n(x)| < \epsilon \qquad a \leq x \leq b$$

for all $n > N$, we say that $S_n(x)$ **converges uniformly** to $f(x)$ in the interval $a \leq x \leq b$ as $n \to \infty$.

The usual test for establishing uniform convergence of an infinite series of functions is the *Weierstrass M test*. There are other tests, of course, but we do not discuss them.

Theorem 1.8 (Weierstrass M test). If $|u_n(x)| \leq M_n$ ($n = 1, 2, 3, \ldots$) for all x in the interval $a \leq x \leq b$ and $\sum M_n$ is a convergent series of

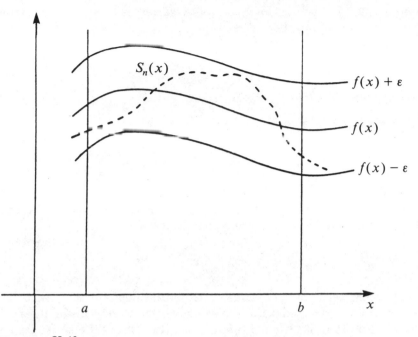

Figure 1.1 Uniform convergence.

positive constants, then the series $\sum u_n(x)$ converges uniformly on the interval $a \leq x \leq b$.

Example 10: Show that the series $\sum_{n=1}^{\infty} (\sin n^2 x)/n^2$ converges uniformly on the interval $0 \leq x \leq \pi$.

Solution: We first observe the inequality

$$\left| \frac{\sin n^2 x}{n^2} \right| \leq \frac{1}{n^2} \qquad 0 \leq x \leq \pi$$

Hence, if we choose $M_n = 1/n^2$, it follows from the p series that $\sum (1/n^2)$ converges; consequently, the given series converges uniformly on $0 \leq x \leq \pi$ by virtue of the M test. Moreover, because the above inequality holds for *all* x, the given series converges uniformly on *every* closed interval.

1.3.1 Properties of uniformly convergent series

The most basic properties of a uniformly convergent series are given in Theorems 1.9 and 1.10.

Theorem 1.9. If each term $u_n(x)$ of a series is continuous in the interval $a \leq x < b$ and the series

$$f(x) = \sum_{n=1}^{\infty} u_n(x)$$

converges uniformly in $a \leq x \leq b$, then

(a) f is a continuous function in the same interval

(b) $\displaystyle\int_a^b f(x)\, dx = \sum_{n=1}^{\infty} \int_a^b u_n(x)\, dx$

A series of continuous functions which converges to a function $f(x)$, but does not converge uniformly, need not converge to a continuous function. To show this, let us consider the series

$$f(x) = x + \sum_{n=1}^{\infty} (x^{n+1} - x^n) \qquad 0 \leq x \leq 1$$

Clearly each term of the series is a continuous function. Also we showed in Example 9 that the series converges to zero in the interval $0 \leq x < 1$ and to unity when $x = 1$. Hence, the sum of the series is the function

$$f(x) = \begin{cases} 0 & 0 \leq x < 1 \\ 1 & x = 1 \end{cases}$$

which is *not* continuous on the closed interval $0 \leq x \leq 1$.

Theorem 1.9b is particularly important in applications. In some cases, however, it may be difficult to establish the conditions of the theorem before integrating the series termwise. We might therefore assume that the conditions are satisfied and proceed formally, but then it becomes essential to attempt justification of the derived result by some independent means.

The conditions stated in Theorem 1.9 are satisfied by many of the series that commonly arise in practice, permitting termwise integration of the series. However, these conditions by themselves are not adequate to permit termwise differentiation of the series. For differentiation under the summation sign, we need the following theorem.

Theorem 1.10. Let both $u_n(x)$ and $u_n'(x)$ be continuous functions in the interval $a \leq x \leq b$ for each n. Then if

$$f(x) = \sum_{n=1}^{\infty} u_n(x)$$

converges (pointwise) in $a \leq x \leq b$ and the series $\sum u_n'(x)$ converges uniformly in $a \leq x \leq b$, we have

$$f'(x) = \sum_{n=1}^{\infty} u_n'(x)$$

Based on Theorem 1.10, we see that any series of functions $\sum u_n(x)$ satisfying the given hypotheses can be differentiated termwise and that the resulting series $\sum u_n'(x)$ represents the derivative of the function represented by $\sum u_n(x)$. The basic requirement for this to be true is *uniform convergence of the differentiated series*. To illustrate its necessity, consider the series

$$f(x) = \sum_{n=1}^{\infty} \frac{\sin n^2 x}{n^2}$$

which *converges uniformly* in every finite interval (recall Example 10). However, if we simply differentiate the given series termwise, we obtain the *divergent* series (for all x)

$$f'(x) = \sum_{n=1}^{\infty} \cos n^2 x$$

which therefore cannot represent $f'(x)$.

1.3.2 Power series

A **power series** is an infinite series of the form

$$c_0 + c_1(x - a) + \cdots + c_n(x - a)^n + \cdots = \sum_{n=0}^{\infty} c_n(x - a)^n \quad (1.39)$$

where $c_0, c_1, \ldots, c_n, \ldots$ are called the *coefficients* of the series and a is the *center* of the series.

Theorem 1.11. To every power series $\sum c_n(x - a)^n$ is a number ρ, $0 \leq \rho < \infty$, called the **radius of convergence,** with the property that the series converges absolutely for $|x - a| < \rho$ and diverges for $|x - a| > \rho$.

Notice that Theorem 1.11 does not include the values of x for which $|x - a| = \rho$. Such points must be tested separately for convergence, e.g., by one of the tests given in Sec. 1.2.

If $\rho > 0$, then for every ρ_1 such that $0 \leq \rho_1 < \rho$, it can be shown that the power series $\sum c_n(x - a)^n$ converges *uniformly* on the interval

$|x - a| \le \rho_1$. In such cases the results of Theorems 1.9 and 1.10 are applicable. That is, *a power series converges to a continuous function* and, moreover, *can always be integrated and differentiated termwise.*

To illustrate termwise integration, let us consider the power series

$$\frac{1}{1+x} = \sum_{n=0}^{\infty} (-1)^n x^n \qquad -1 < x < 1 \tag{1.40}$$

which is just the geometric series with x replaced by $-x$. It follows from above that it is uniformly convergent on every closed interval contained within $-1 < x < 1$. Termwise integration leads to

$$\int_0^x \frac{dt}{1+t} = \sum_{n=0}^{\infty} (-1)^n \int_0^x t^n \, dt \qquad -1 < x < 1$$

where we have introduced a dummy variable of integration. Completing the integration, we obtain the series

$$\ln(1+x) = \sum_{n=0}^{\infty} (-1)^n \frac{x^{n+1}}{n+1}$$

or, letting $n \to n - 1$, we obtain the more familiar form

$$\ln(1+x) = \sum_{n=1}^{\infty} (-1)^{n-1} \frac{x^n}{n} \qquad -1 < x < 1 \tag{1.41}$$

Notice that setting $x = 1$ in (1.41) yields

$$\ln 2 = \sum_{n=1}^{\infty} \frac{(-1)^{n-1}}{n}$$

where the right-hand side is the alternating harmonic series (see Sec. 1.2.3). It is interesting to observe that this result is valid even though $x = 1$ is outside the original interval of convergence [see Eq. (1.40)]. This example illustrates that the process of integration of an infinite series can sometimes produce a new series whose interval of convergence extends beyond that of the original series. By differentiating a series, just the opposite may be true. For example, the series

$$f(x) = \sum_{n=2}^{\infty} \frac{x^n}{n(n-1)} \qquad -1 \le x \le 1$$

converges uniformly for all $|x| \le 1$, but the differentiated series

$$f'(x) = \sum_{n=2}^{\infty} \frac{x^{n-1}}{n-1} = \sum_{n=1}^{\infty} \frac{x^n}{n} \qquad -1 \le x < 1$$

does not converge at $x = 1$. Moreover, the series

$$f''(x) = \sum_{n=1}^{\infty} x^{n-1} = \sum_{n=0}^{\infty} x^n \qquad -1 < x < 1$$

does not converge at either endpoint.

A power series about $x = a$ with a positive radius of convergence ρ converges to some function $f(x)$ in the interval $|x - a| < \rho$. In this case we say that f is **analytic** at $x = a$. Conversely, if f is analytic at $x = a$, then it has a power series representation of the form

$$f(x) = \sum_{n=0}^{\infty} \frac{f^{(n)}(a)}{n!} (x - a)^n \qquad |x - a| < \rho \qquad (1.42)$$

for which $c_n = f^{(n)}(a)/n!$, $n = 0, 1, 2, \ldots$. We refer to (1.42) as the **Taylor series** for the function f. The special case that occurs when $a = 0$ is known as a **Maclaurin series**, i.e.,

$$f(x) = \sum_{n=0}^{\infty} \frac{f^{(n)}(0)}{n!} x^n \qquad |x| < \rho \qquad (1.43)$$

Most of the elementary functions that arise in calculus can be represented by a Taylor series (or Maclaurin series). Familiar examples are given by the following:

$$\frac{1}{1-x} = \sum_{n=0}^{\infty} x^n \qquad -1 < x < 1 \qquad (1.44)$$

$$e^x = \sum_{n=0}^{\infty} \frac{x^n}{n!} \qquad -\infty < x < \infty \qquad (1.45)$$

$$\sin x = \sum_{n=0}^{\infty} (-1)^n \frac{x^{2n+1}}{(2n+1)!} \qquad -\infty < x < \infty \qquad (1.46)$$

$$\cos x = \sum_{n=0}^{\infty} (-1)^n \frac{x^{2n}}{(2n)!} \qquad -\infty < x < \infty \qquad (1.47)$$

$$\ln x = \sum_{n=1}^{\infty} \frac{(-1)^{n-1}}{n} (x - 1)^n \qquad 0 < x \leq 2 \qquad (1.48)$$

In addition, many of the special functions that we encounter in subsequent chapters can be represented by this form of power series.

Example 11 (Binomial series): Expand $f(x) = (1 + x)^\alpha$ in a Maclaurin series, where α is not restricted to integer values. Also find the radius of convergence of the resulting series.

Solution: Repeated differentiation of the function f reveals that

$$f'(x) = \alpha(1+x)^{\alpha-1}$$

$$f''(x) = \alpha(\alpha-1)(1+x)^{\alpha-2}$$

and, in general,

$$f^{(n)}(x) = \alpha(\alpha-1)\cdots(\alpha-n+1)(1+x)^{\alpha-n} \qquad n = 1, 2, 3, \ldots$$

Hence, we find that $f(0) = 1$, $f'(0) = \alpha$, $f''(0) = \alpha(\alpha-1)$, whereas in general

$$f^{(n)}(0) = \alpha(\alpha-1)\cdots(\alpha-n+1) \qquad n = 1, 2, 3, \ldots$$

Thus, the series we seek is

$$(1+x)^{\alpha} = 1 + \sum_{n=1}^{\infty} \frac{\alpha(\alpha-1)\cdots(\alpha-n+1)}{n!} x^n$$

or, by introducing the binomial coefficient, we can write this more compactly in the form

$$(1+x)^{\alpha} = \sum_{n=0}^{\infty} \binom{\alpha}{n} x^n$$

This series is called the **binomial series**. It is a generalization of the binomial formula (1.22) to nonintegral powers and is important in much of our work in subsequent chapters.

Finally, by application of the ratio test, we find that

$$\lim_{n\to\infty} \left| \frac{\alpha(\alpha-1)\cdots(\alpha-n)}{(n+1)!} x^{n+1} \cdot \frac{n!}{\alpha(\alpha-1)\cdots(\alpha-n+1)} x^{-n} \right|$$

$$= |x| \lim_{n\to\infty} \left| \frac{\alpha-n}{n+1} \right|$$

$$= |x|$$

and thus we conclude that the series converges for $|x| < 1$, that is, the radius of convergence is $\rho = 1$.

We previously used the notion that two polynomials are equal if and only if their corresponding terms are identical (recall Example 8). To extend this concept to infinite series, we have the following useful theorem.

Theorem 1.12. If $\sum c_n(x-a)^n$ and $\sum b_n(x-a)^n$ both have nonzero radii of convergence and $\sum c_n(x-a)^n = \sum b_n(x-a)^n$ wherever the two series converge, then $c_n = b_n$, $n = 0, 1, 2, \ldots$. Moreover, if $\sum c_n(x-a)^n = 0$, then $c_n = 0$, $n = 0, 1, 2, \ldots$.

1.3.3 Sums and products of power series

We have already found out that power series can always be integrated and differentiated termwise. In addition, we often need to combine two or more series through addition and multiplication. Suppose that

$$f(x) = \sum_{n=0}^{\infty} a_n x^n \qquad (1.49)$$

and

$$g(x) = \sum_{n=0}^{\infty} b_n x^n \qquad (1.50)$$

both have a nonzero radius of convergence. Then the series for their sum and product are defined, respectively, by

$$f(x) + g(x) = \sum_{n=0}^{\infty} (a_n + b_n) x^n \qquad (1.51)$$

and

$$f(x)g(x) = \sum_{n=0}^{\infty} \left(\sum_{k=0}^{n} a_k b_{n-k} \right) x^n = \sum_{n=0}^{\infty} \left(\sum_{k=0}^{n} a_{n-k} b_k \right) x^n \qquad (1.52)$$

and these series converge on the common interval of convergence. We recognize (1.52) as simply the Cauchy product (see Theorem 1.7) once again.

Example 12: Find the Maclaurin series for $e^x \sin x$.

Solution: By using Eqs. (1.45) and (1.46), we have

$$e^x = \sum_{n=0}^{\infty} \frac{x^n}{n!} \qquad \sin x = \sum_{n=0}^{\infty} (-1)^n \frac{x^{2n+1}}{(2n+1)!}$$

both of which converge for all x. However, since the series for $\sin x$ involves only odd powers of x while that for e^x involves both even and odd powers, we cannot directly apply (1.52). To remedy this situation, we wish to rewrite the above sine series in consecutive powers of

x. By recognizing the identity

$$\cos\frac{(n-1)\pi}{2} = \begin{cases} 0 & n \text{ even} \\ (-1)^{(n-1)/2} & n \text{ odd} \end{cases}$$

we can write

$$\sin x = \sum_{n=0}^{\infty} \left[\cos\frac{(n-1)\pi}{2} \right] \frac{x^n}{n!}$$

where all even terms are zero. The Cauchy product (1.52) now leads to

$$e^x \sin x = \sum_{n=0}^{\infty} c_n x^n$$

where $\qquad c_n = \sum_{k=0}^{n} \frac{\cos[(k-1)\pi/2]}{k!\,(n-k)!} \qquad n = 0, 1, 2, \ldots$

In many cases the expression for c_n cannot be simplified, and so we leave our answer in this form. However, in the present case we can actually sum the finite series for c_n. To do so, we use the Euler formula

$$\cos x = \tfrac{1}{2}(e^{ix} + e^{-ix})$$

and the finite binomial series. Replacing the cosine function in c_n with complex exponential functions and multiplying and dividing the series by $n!$, we obtain

$$c_n = \frac{1}{2n!} \sum_{k=0}^{n} \binom{n}{k} [e^{i(k-1)\pi/2} + e^{-i(k-1)\pi/2}]$$

$$= \frac{1}{2n!} \left[e^{-i\pi/2} \sum_{k=0}^{n} \binom{n}{k} (e^{i\pi/2})^k + e^{i\pi/2} \sum_{k=0}^{n} \binom{n}{k} (e^{-i\pi/2})^k \right]$$

$$= \frac{1}{2n!} [e^{-i\pi/2}(1 + e^{i\pi/2})^n + e^{i\pi/2}(1 + e^{-i\pi/2})^n]$$

Simplifying, we have

$$(1 + e^{i\pi/2})^n = e^{in\pi/4}(e^{i\pi/4} + e^{-i\pi/4})^n = 2^n e^{in\pi/4} \cos^n \frac{\pi}{4}$$

or $\qquad (1 + e^{i\pi/2})^n = 2^{n/2} e^{in\pi/4}$

Likewise, $\qquad (1 + e^{-i\pi/2})^n = 2^{n/2} e^{-in\pi/4}$

and we deduce that

$$c_n = \frac{1}{2n!} \left[e^{i(n-2)\pi/4} + e^{-i(n-2)\pi/4} \right] 2^{n/2}$$

$$= \frac{2^{n/2}}{n!} \cos \frac{(n-2)\pi}{4} \qquad n = 0, 1, 2, \ldots$$

Hence, our final result is

$$e^x \sin x = \sum_{n=0}^{\infty} \frac{2^{n/2}}{n!} \sin \left(\frac{n\pi}{4} \right) x^n \qquad -\infty < x < \infty$$

(Notice that the terms corresponding to $n = 0, 4, 8, \ldots$ are all zero.)

By generalizing the Cauchy product, we can obtain the **power formula**

$$\left(\sum_{n=0}^{\infty} a_n x^n \right)^p = \sum_{n=0}^{\infty} c_n x^n \qquad p = 2, 3, 4, \ldots \qquad (1.53)$$

where

$$c_0 = a_0^p \qquad c_n = \frac{1}{na_0} \sum_{k=1}^{n} (kp - n + k) a_k c_{n-k} \qquad n = 1, 2, 3, \ldots \quad (1.54)$$

We leave it to the reader to verify that this power formula is equivalent to the Cauchy product for $p = 2$. Then, by repeated application of the Cauchy product for $p = 3, 4, 5, \ldots$, the above result can be readily obtained.

Remark: The results presented here involving sums and products of Maclaurin series are readily extended to general Taylor series.

Exercises 1.3

In problems 1 through 4, use the ratio test to determine the radius of convergence.

1. $\displaystyle\sum_{n=1}^{\infty} \frac{x^n}{n}$

3. $\displaystyle\sum_{n=0}^{\infty} \frac{n!}{2^n} x^n$

2. $\displaystyle\sum_{n=1}^{\infty} \frac{1 \cdot 3 \cdot 5 \cdots (2n-1)}{2 \cdot 5 \cdot 8 \cdots (3n-1)} (x-1)^n$

4. $\displaystyle\sum_{n=1}^{\infty} n x^n$

In problems 5 through 8, use the M test to test the series for uniform convergence on the indicated interval.

5. $\displaystyle\sum_{n=0}^{\infty} \frac{\cos nx}{n^2}$, $-10 \le x \le 10$

7. $\displaystyle\sum_{n=0}^{\infty} \left(\frac{1}{nx+2} - \frac{1}{nx+x+2}\right)$, $0 \le x \le 1$

6. $\displaystyle\sum_{n=2}^{\infty} \frac{x^n}{n(\ln n)^2}$, $-1 \le x \le 1$

8. $\displaystyle\sum_{n=1}^{\infty} x(1+x)^{-n}$, $0 \le x \le 1$

9. Use termwise integration to show that

$$\int_0^1 \left(\sum_{n=0}^{\infty} \frac{x^n}{n!}\right) dx = e - 1$$

In problems 10 to 13, state whether the series can be differentiated termwise in the indicated interval.

10. $\displaystyle\sum_{n=1}^{\infty} \frac{x^n}{\sqrt{n}}$, $-1 \le x \le 10$

12. $\displaystyle\sum_{n=1}^{\infty} \left(\frac{x}{x-1}\right)^n$, $-4 \le x \le 3$

11. $\displaystyle\sum_{n=1}^{\infty} \frac{e^{-nx}}{n(n+1)^2}$, $0 \le x \le 10$

13. $\displaystyle\sum_{n=1}^{\infty} \left(\frac{x^n}{n} - \frac{x^{n+1}}{n+1}\right)$, $0 \le x \le 1$

14. Start with the geometric series

$$\frac{1}{1-x} = \sum_{n=0}^{\infty} x^n \qquad -1 < x < 1$$

   (a) Make a change of variable to derive the Maclaurin series for $1/(1+x^2)$.
   (b) Use the answer in (a) to determine the Maclaurin series for $\tan^{-1} x$. State the interval of convergence.

15. Start with the binomial series

$$(1+x)^\alpha = \sum_{n=0}^{\infty} \binom{\alpha}{n} x^n \qquad -1 < x < 1$$

   (a) Find the Maclaurin series for $(1-x^2)^{-1/2}$.
   (b) Use the answer in (a) to determine the Maclaurin series for $\sin^{-1} x$. Give the interval of convergence.

In problems 16 to 19, use the Cauchy product to find the Maclaurin series representation for the given function.

16. $f(x) = (1-x)^{-2}$

18. $f(x) = \sin^2 x$

17. $f(x) = \cos^2 x$

19. $f(x) = e^x \cos x$

20. Use Eqs. (1.53) and (1.54) to determine the first *four* terms of $f(x) = \cos^3 x$.

1.4 Fourier Trigonometric Series

The expansion of a function f in a power series requires (at least) that f be infinitely differentiable. However, many functions of practical interest do not satisfy such strong differentiability requirements, due to discontinuities, lack of smoothness, etc., and therefore cannot be represented in a power series. For such cases there are other types of series representations.

A particular type of series having a wide range of applications is the **Fourier trigonometric series** (or simply **Fourier series**)*

$$f(x) = \frac{1}{2}a_0 + \sum_{n=1}^{\infty} \left(a_n \cos \frac{n\pi x}{p} + b_n \sin \frac{n\pi x}{p} \right) \qquad (1.55)$$

where the constants a_0, a_n, and b_n are called the *Fourier coefficients* of the series. If the series representation is to be valid for all values of x, then clearly f must be a periodic function with period $2p$, since the right-hand side of (1.55) has this property. In other cases, the series (1.55) is useful for representing the function f only in the interval $-p \le x \le p$, so that the periodicity is of no concern.

Formally identifying the Fourier coefficients depends on the evaluation of the integrals

$$\int_{-p}^{p} \cos \frac{n\pi x}{p} \, dx = \int_{-p}^{p} \sin \frac{n\pi x}{p} \, dx = \int_{-p}^{p} \sin \frac{n\pi x}{p} \cos \frac{k\pi x}{p} \, dx = 0$$

$$(1.56a)$$

and

$$\int_{-p}^{p} \cos \frac{n\pi x}{p} \cos \frac{k\pi x}{p} \, dx = \int_{-p}^{p} \sin \frac{n\pi x}{p} \sin \frac{k\pi x}{p} \, dx = \begin{cases} 0 & k \ne n \\ p & k = n \end{cases}$$

$$(1.56b)$$

where n and k both assume positive integer values. The details of verifying these integral relations are left to the exercises.

Assuming that termwise integration of (1.55) is permitted, we find

$$\int_{-p}^{p} f(x) \, dx = \frac{1}{2}a_0 \int_{-p}^{p} dx + \sum_{n=1}^{\infty} \left(a_n \overset{0}{\overbrace{\int_{-p}^{p} \cos \frac{n\pi x}{p} \, dx}} + b_n \overset{0}{\overbrace{\int_{-p}^{p} \sin \frac{n\pi x}{p} \, dx}} \right)$$

---

* It is customary to write the constant term in (1.55) as $\frac{1}{2}a_0$.

from which we deduce

$$a_0 = \frac{1}{p} \int_{-p}^{p} f(x)\, dx \tag{1.57}$$

If we now multiply (1.55) by $\cos(k\pi x/p)$ and integrate once again, we have

$$\int_{-p}^{p} f(x) \cos \frac{k\pi x}{p}\, dx = \frac{1}{2} a_0 \overset{0}{\int_{-p}^{p} \cos \frac{k\pi x}{p}\, dx}$$

$$+ \sum_{n=1}^{\infty} \left(a_n \overset{0\,(n \neq k)}{\int_{-p}^{p} \cos \frac{n\pi x}{p} \cos \frac{k\pi x}{p}\, dx} + b_n \overset{0}{\int_{-p}^{p} \sin \frac{n\pi x}{p} \cos \frac{k\pi x}{p}\, dx} \right)$$

This time all terms on the right go to zero except for the coefficient of a_n corresponding to $n = k$, and here we find

$$\int_{-p}^{p} f(x) \cos \frac{k\pi x}{p}\, dx = a_k \int_{-p}^{p} \cos^2 \frac{k\pi x}{p}\, dx$$

$$= p a_k$$

or $$a_k = \frac{1}{p} \int_{-p}^{p} f(x) \cos \frac{k\pi x}{p}\, dx \qquad k = 1, 2, 3, \ldots \tag{1.58a}$$

By a similar process, the multiplication of (1.55) by $\sin(k\pi x/p)$ and subsequent integration provide the final formula

$$b_k = \frac{1}{p} \int_{-p}^{p} f(x) \sin \frac{k\pi x}{p}\, dx \qquad k = 1, 2, 3, \ldots \tag{1.58b}$$

In summary, we have formally shown that if f has the representation

$$f(x) = \frac{1}{2} a_0 + \sum_{n=1}^{\infty} \left(a_n \cos \frac{n\pi x}{p} + b_n \sin \frac{n\pi x}{p} \right) \tag{1.59}$$

then the Fourier coefficients are given by [changing the index back to

n and combining (1.57) and (1.58a)]

$$a_n = \frac{1}{p}\int_{-p}^{p} f(x)\cos\frac{n\pi x}{p}\,dx \qquad n = 0, 1, 2, \ldots \qquad (1.60)$$

and

$$b_n = \frac{1}{p}\int_{-p}^{p} f(x)\sin\frac{n\pi x}{p}\,dx \qquad n = 1, 2, 3, \ldots \qquad (1.61)$$

Example 13: Find the Fourier trigonometric series for the periodic function

$$f(x) = \begin{cases} 0 & -\pi < x < 0 \\ x & 0 < x < \pi \end{cases} \qquad f(x + 2\pi) = f(x)$$

Solution: The Fourier coefficients computed from (1.60) and (1.61) with $p = \pi$ lead to

$$a_0 = \frac{1}{\pi}\int_{-\pi}^{\pi} f(x)\,dx = \frac{1}{\pi}\int_0^{\pi} x\,dx = \frac{\pi}{2}$$

$$a_n = \frac{1}{\pi}\int_0^{\pi} x\cos nx\,dx = \begin{cases} 0 & n = 2, 4, 6, \ldots \\ -\dfrac{2}{\pi n^2} & n = 1, 3, 5, \ldots \end{cases}$$

and

$$b_n = \frac{1}{\pi}\int_0^{\pi} x\sin nx\,dx = \frac{(-1)^{n+1}}{n} \qquad n = 1, 2, 3, \ldots$$

Substituting these results into (1.59), we obtain

$$f(x) = \frac{\pi}{4} - \frac{2}{\pi}\left(\cos x + \frac{\cos 3x}{3^2} + \frac{\cos 5x}{5^2} + \cdots\right)$$

$$+ \left(\sin x - \frac{\sin 2x}{2} + \frac{\sin 3x}{3} - \cdots\right)$$

or more compactly,

$$f(x) = \frac{\pi}{4} - \sum_{n=1}^{\infty}\left[\frac{2}{\pi}\frac{\cos(2n-1)x}{(2n-1)^2} + \frac{(-1)^n}{n}\sin nx\right]$$

We might observe that the function f in Example 13 is not differentiable at $x = 0$ and multiples of π. Thus, while it surely does not have a power series expansion over any interval containing these points, its Fourier series converges for all x, even at the points of discontinuity (see Theorem 1.13).

Theorem 1.13 (Pointwise convergence). If $f(x+2p)=f(x)$ for some p and if f and f' are at least piecewise continuous in $-p \leq x \leq p$, then the Fourier series of f converges pointwise to $f(x)$ at all points of continuity of f. At points of discontinuity of f, the series converges to the average value $\frac{1}{2}[f(x^+)+f(x^-)]$.*

Remark: A function f is said to be *piecewise continuous* in an interval if it has only a finite number of discontinuities and, further, if all discontinuities are finite. This class of functions is discussed in more detail in Sec. 4.5.1. Also $f(x^+)$ and $f(x^-)$ denote the limits of f at x from the right and left, respectively.

Theorem 1.13 is also valid for nonperiodic functions which satisfy the other stated conditions in some interval $c \leq x \leq c + 2p$, where c is any real number. In such cases the convergence at the endpoints of the interval will lead to the value $\frac{1}{2}[f(c^+)+f(c+2p^-)]$. The Fourier coefficients are then computed by performing the integrations over the interval $c \leq x \leq c + 2p$. Finally, we remark that if we add to Theorem 1.13 the condition that f is also continuous, the Fourier series will then converge *uniformly*.

1.4.1   Cosine and sine series

If $f(-x)=f(x)$, we say that f is an *even function*, whereas if $f(-x)=-f(x)$, we say that f is an *odd function*. If the function f falls into one of these two classifications, certain simplifications in handling Fourier series occur. Such simplifications are primarily consequences of the following result (see problems 8 and 9 in Exercises 1.4):

$$\int_{-p}^{p} f(x)\, dx = \begin{cases} 2 \int_{0}^{p} f(x)\, dx & \text{if} \quad f(x) \text{ is even} \\ 0 & \text{if} \quad f(x) \text{ is odd} \end{cases} \quad (1.62)$$

If f is an even function, the product $f(x)\cos(n\pi x/p)$ is an even function while the product $f(x)\sin(n\pi x/p)$ is an odd function. (Why?) In this case, using (1.62), we see that the Fourier coefficients

---

*For a proof of Theorem 1.13, see L. C. Andrews, *Elementary Partial Differential Equations with Boundary Value Problems*, Academic, Orlando, 1986, pp. 159–163.

satisfy

$$a_n = \frac{1}{p} \int_{-p}^{p} f(x) \cos \frac{n\pi x}{p} \, dx$$

$$= \frac{2}{p} \int_{0}^{p} f(x) \cos \frac{n\pi x}{p} \, dx \qquad n = 0, 1, 2, \dots \qquad (1.63)$$

and $$b_n = \frac{1}{p} \int_{-p}^{p} f(x) \sin \frac{n\pi x}{p} \, dx = 0 \qquad n = 1, 2, 3, \dots \qquad (1.64)$$

Hence, for an even function f the Fourier series reduces to

$$f(x) = \frac{1}{2} a_0 + \sum_{n=1}^{\infty} a_n \cos \frac{n\pi x}{p} \qquad (1.65)$$

where a_n ($n = 0, 1, 2, \dots$) is defined by (1.63). We call such a series a *cosine series*.

By using a similar argument, when f is an odd function, the Fourier series reduces to the *sine series*

$$f(x) = \sum_{n=1}^{\infty} b_n \sin \frac{n\pi x}{p} \qquad (1.66)$$

where $a_n = 0$ ($n = 0, 1, 2, \dots$) and

$$b_n = \frac{2}{p} \int_{0}^{p} f(x) \sin \frac{n\pi x}{p} \, dx \qquad n = 1, 2, 3, \dots \qquad (1.67)$$

Exercises 1.4

In problems 1 through 6, determine the Fourier trigonometric series of each function.

1. $f(x) = \begin{cases} 1 & -p < x \leq 0 \\ \frac{1}{2} & 0 < x \leq p \end{cases}$

2. $f(x) = x, \ -\pi < x < \pi$

3. $f(x) = |x|, \ -\pi \leq x \leq \pi$

4. $f(x) = x^2, \ -1 < x \leq 1$

5. $f(x) = \begin{cases} x & -2 < x < 0 \\ 2 - x & 0 < x < 2 \end{cases}$

6. $f(x) = \begin{cases} x + \pi & -\pi < x < 0 \\ x - \pi & 0 < x < \pi \end{cases}$

7. Verify the integral relations (1.56a) and (1.56b).
   Hint: Use the trigonometric identities

$$\sin A \sin B = \tfrac{1}{2}[\cos (A - B) - \cos (A + B)]$$
$$\cos A \cos B = \tfrac{1}{2}[\cos (A - B) + \cos (A + B)]$$
$$\sin A \cos B = \tfrac{1}{2}[\sin (A - B) + \sin (A + B)]$$

8. Prove that if f is an even function,

$$\int_{-p}^{p} f(x)\, dx = 2 \int_{0}^{p} f(x)\, dx$$

9. Prove that if f is an odd function,

$$\int_{-p}^{p} f(x)\, dx = 0$$

10. Prove the following.
    (a)   The product of two odd functions is even.
    (b)   The product of two even functions is even.
    (c)   The product of an even and an odd function is odd.

11. To what numerical value will the Fourier series in problem 1 converge at
    (a) $x = 0$?
    (b) $x = -p$?
    (c) $x = p$?

12. A sinusoidal voltage $E \sin t$ is passed through a *half-wave rectifier*, which clips the negative portion of the wave. Find the Fourier series of the resulting waveform: $f(t) = 0$, $-\pi < t < 0$; $f(t) = E \sin t$, $0 < t < \pi$; and $f(t + 2\pi) = f(t)$.

13. Find the Fourier series of the periodic function resulting from passing the voltage $v(t) = E \cos 100\pi t$ through a half-wave rectifier (see problem 12).

14. A certain type of *full-wave rectifier* converts the input voltage $v(t)$ to its absolute value at the output, that is, $|v(t)|$. Assuming the input voltage is given by $v(t) = E \sin \omega t$, determine the Fourier series of the periodic output voltage.

15. From the Fourier series developed in Example 13, show that

(a) $1 + \dfrac{1}{3^2} + \dfrac{1}{5^2} + \dfrac{1}{7^2} + \cdots = \dfrac{\pi^2}{8}$

(b) $1 - \dfrac{1}{3} + \dfrac{1}{5} - \dfrac{1}{7} + \cdots = \dfrac{\pi}{4}$

16. Starting with the Fourier series representation

$$\frac{x}{2} = \sum_{n=1}^{\infty} \frac{(-1)^{n-1}}{n} \sin nx \qquad -\pi < x < \pi$$

obtain a Fourier series for x^2, $-\pi < x < \pi$, by integrating termwise.

17. If $f(x + 2p) = f(x)$, show that for any constant c

$$\int_{c-p}^{c+p} f(x)\, dx = \int_{-p}^{p} f(x)\, dx$$

Hint: Write

$$\int_{c-p}^{c+p} f(x)\, dx = \int_{c-p}^{-p} f(x)\, dx + \int_{-p}^{c+p} f(x)\, dx$$

and let $x = t - 2p$ in the first integral on the right-hand side.

1.5 Improper Integrals

If the Riemann integral $\int_a^b f(t)\, dt$ exists, it is said to be a **proper integral**. This generally means that the interval of integration is finite and that f is bounded on this interval. Integrals which have an infinite limit of integration or an unbounded integrand between the limits of integration are called **improper integrals**. If a certain amount of care is not exercised in the evaluation of such integrals, we may derive absurd results like

$$\int_{-1}^{1/2} \frac{dt}{t^2} = -\frac{1}{t}\Big|_{-1}^{1/2} = -3$$

Clearly, the integral of a positive quantity cannot be negative, so there is something wrong with this result.

The general theory of improper integrals closely parallels that of infinite series, and so our treatment of them will be somewhat brief.

1.5.1 Types of improper integrals

We say that a function f is **bounded** on the finite interval $a \le t \le b$ if there exists a constant B such that

$$|f(t)| < B \qquad a \le t \le b$$

Otherwise, it is *unbounded*. For example, the function $f(t) = e^{-t}$ is bounded on $0 \le t \le 1$ since $|e^{-t}| \le 1$ on this interval. On the other hand, $f(t) = 1/t$ is unbounded on any interval containing $t = 0$.

An **improper integral of the first kind** occurs when one or both limits of integration are infinite. For example, let us assume that f is bounded on the interval $a \le t < \infty$. If f is integrable on every finite interval $a \le t \le b$, then we write

$$\int_a^\infty f(t)\, dt = \lim_{b \to \infty} \int_a^b f(t)\, dt \tag{1.68}$$

If the limit on the right exists, we say the improper integral *converges,* and it *diverges* otherwise. Other improper integrals of this type are defined by

$$\int_{-\infty}^b f(t)\, dt = \lim_{a \to -\infty} \int_a^b f(t)\, dt \tag{1.69}$$

and

$$\int_{-\infty}^\infty f(t)\, dt = \lim_{a \to -\infty} \int_a^c f(t)\, dt + \lim_{b \to \infty} \int_c^b f(t)\, dt \tag{1.70}$$

Once again, we say the integrals converge when the limits on the right exist, and otherwise they diverge.

Example 14: Evaluate the integrals (if possible).

(a) $\displaystyle\int_1^\infty \frac{1}{t^2}\, dt$

(b) $\displaystyle\int_{-1}^\infty \frac{1}{\sqrt{t+5}}\, dt$

Solution

(a) Clearly, the function $f(t) = 1/t^2$ is bounded on the interval $1 \le t < \infty$. Therefore, we write

$$\int_1^\infty \frac{1}{t^2}\, dt = \lim_{b \to \infty} \int_1^b \frac{1}{t^2}\, dt = \lim_{b \to \infty} \left(1 - \frac{1}{b}\right)$$

or

$$\int_1^\infty \frac{1}{t^2}\, dt = 1$$

Thus, the integral converges to the value 1.

(b) The function $f(t) = 1/\sqrt{t+5}$ is bounded on the interval of integration, so that we write

$$\int_{-1}^\infty \frac{1}{\sqrt{t+5}}\, dt = \lim_{b \to \infty} \int_{-1}^b \frac{1}{\sqrt{t+5}}\, dt = \lim_{b \to \infty} 2(\sqrt{b+5} - 2)$$

This time the limit does not exist; thus, we conclude that the integral diverges.

If f is unbounded on the interval $a \le t \le b$, then the integral $\int_a^b f(t)\,dt$ is an **improper integral of the second kind**. Let us assume that f is unbounded at only one point on the interval, i.e., at $t = c$, $a \le c \le b$. In this case we write

$$\int_a^b f(t)\,dt = \lim_{\epsilon \to 0^+} \int_a^{c-\epsilon} f(t)\,dt + \lim_{\epsilon \to 0^+} \int_{v+\epsilon}^b f(t)\,dt \qquad (1.71)$$

If both limits on the right exist, we say the integral *converges* to the sum of the limits; otherwise, the integral *diverges*. (When c is either endpoint, then clearly only one limit is necessary.) If f is unbounded at several points on the interval, we apply the above procedure at all such points.

In some cases an integral may be classified as improper for more than one reason. For example, the integral

$$\int_0^\infty e^{-t} t^{-1/2}\,dt$$

is improper because of the infinite limit of integration, but also because the integrand becomes unbounded at $t = 0$. In such cases we write $(b > 0)$

$$\int_0^\infty e^{-t} t^{-1/2}\,dt = \int_0^b e^{-t} t^{-1/2}\,dt + \int_b^\infty e^{-t} t^{-1/2}\,dt$$

and evaluate (if possible) each improper integral on the right.

Because there are differences in the behavior of improper integrals for which the integrand f is always nonnegative and those for which f may be both positive and negative, we find it convenient to distinguish between the notions of *conditional convergence* and *absolute convergence* of improper integrals.

Definition 1.5. We say that $\int_a^\infty f(t)\,dt$ **converges absolutely** if $\int_a^\infty |f(t)|\,dt$ converges. On the other hand, if $\int_a^\infty f(t)\,dt$ converges but $\int_a^\infty |f(t)|\,dt$ diverges, then $\int_a^\infty f(t)\,dt$ **converges conditionally**.

If the integrand f is nonnegative on the interval $a \le t < \infty$ and $\int_a^\infty f(t)\,dt$ converges, it must necessarily converge absolutely. Hence, conditional convergence applies to only those integrals where the integrand can be both positive and negative on the interval of interest. Similar definitions and comments also apply to other types of improper integrals.

1.5.2 Convergence tests

Thus far our discussion of convergence and divergence of improper integrals has been based on direct evaluation of the related indefinite integral (using standard calculus techniques). In this regard we have chosen examples which could be integrated rather easily. This is not always possible in practice, however, since many integrals of interest are nonelementary. For instance, the improper integral

$$\int_0^\infty \frac{e^{-t}}{t^2+1}\, dt$$

cannot be evaluated by any direct method of integration from calculus (including integration by parts). Nonetheless, we may wish to arrive at a conclusion regarding its convergence beforehand if we are to use a numerical method. That is, it would be a waste of time (and money) to attempt to evaluate such an integral numerically on a computer if the integral in fact diverged. For this reason it is important to establish various *tests* of convergence of improper integrals that do not depend on direct evaluation of the integral. Many such tests are available, most of which are very similar to those used on infinite series. We refer the interested reader to any standard text on advanced calculus for a discussion of them. Here we simply state and illustrate the following two limit tests which are applicable to a particular type of improper integral. Similar theorems for other types of improper integrals have been developed.

Theorem 1.14. If f is continuous for all $t \geq a$ and if

$$\lim_{t\to\infty} t^p f(t) = A \qquad p > 1$$

where A is finite, then $\int_a^\infty f(t)\, dt$ converges absolutely.

Theorem 1.15. If f is continuous for all $t \geq a$ and if

$$\lim_{t\to\infty} t f(t) = A \neq 0$$

where A can be finite or infinite, then $\int_a^\infty f(t)\, dt$ diverges. If $A = 0$, the test fails.

Example 15: Test the following improper integrals for convergence:

(a) $\displaystyle\int_0^\infty e^{-t^2}\,dt$

(b) $\displaystyle\int_2^\infty \frac{1}{\ln t}\,dt$

Solution

(a) By taking $p = 2$ and applying the hypothesis of Theorem 1.14, we see that

$$\lim_{t\to\infty} t^2 e^{-t^2} = 0$$

and thus conclude that the integral converges absolutely.

(b) Here we find that

$$\lim_{t\to\infty} \frac{t}{\ln t} = \infty$$

and thus by Theorem 1.15 the integral diverges.

1.5.3 Pointwise and uniform convergence

Frequently the integral of interest is of the form

$$F(x) = \int_a^\infty f(x, t)\,dt \qquad (1.72)$$

where x is a parameter that can assume various values. Such integrals may converge for certain values of x and diverge for other values. If, for a fixed value of x, the integral in (1.72) sums to $F(x)$, we say the integral **converges pointwise** to $F(x)$. The collection of all such points constitutes the domain of the function F.

Remark: Integrals of the type (1.72) are similar to the series of functions discussed in Sec. 1.3.

To differentiate or integrate under the integral sign in (1.72), we need to establish a stronger type of convergence, called *uniform convergence*. The notion of uniform convergence of improper integrals can be introduced by analogy with that for infinite series. Corresponding to the concept of partial sums, therefore, we define the "partial integral"

$$S_R(x) = \int_a^R f(x, t)\,dt \qquad (1.73)$$

Definition 1.6. If, given some $\epsilon > 0$, there exists a number Q, independent of x in the interval $c \leq x \leq d$, such that

$$|F(x) - S_R(x)| < \epsilon \qquad c \leq x \leq d$$

whenever $R > Q$, then the integral $\int_a^\infty f(x, t)\, dt$ is said to **converge uniformly** in the interval $c \leq x \leq d$.

To test for uniform convergence, the following *Weierstrass M test* (analogous to Theorem 1.8) is frequently used.

Theorem 1.16 (Weierstrass M test). Let $M(t)$ be a nonnegative function such that $|f(x, t)| \leq M(t)$ for all $t > a$ and all x in the interval $c \leq x \leq d$. Then if the improper integral $\int_a^\infty M(t)\, dt$ converges, the integral $\int_a^\infty f(x, t)\, dt$ converges uniformly and absolutely in $c \leq x \leq d$.

Example 16: Show that $\int_1^\infty e^{-t} t^{x-1}\, dt$ converges uniformly in the interval $1 \leq x \leq 2$.

Solution: If we select $M(t) = t^2 e^{-t}$, then clearly

$$|e^{-t} t^{x-1}| \leq t^2 e^{-t} \qquad 1 \leq x \leq 2, \qquad t \geq 1$$

Also

$$\lim_{t \to \infty} t^2 M(t) = 0$$

so by virtue of Theorem 1.14 (with $p = 2$) the integral $\int_1^\infty M(t)\, dt$ converges. It now follows from the Weierstrass M test (Theorem 1.16) that the given integral $\int_1^\infty e^{-t} t^{x-1}\, dt$ converges uniformly in $1 \leq x \leq 2$.

The following theorems, which are important in much of our work in subsequent chapters, are simple consequences of having a uniformly convergent improper integral.

Theorem 1.17. If $f(x, t)$ is continuous in $c \leq x \leq d$, $t \geq a$, and $\int_a^\infty f(x, t)\, dt$ converges uniformly to $F(x)$ in $c \leq x \leq d$, then

(a) F is continuous in $c \leq x \leq d$.

(b) $\displaystyle\int_c^d F(x)\, dx = \int_a^\infty \int_c^d f(x, t)\, dx\, dt.$

Part (b) of Theorem 1.17 tells us that we can interchange the order of integration when the improper integral converges uniformly. This may not be valid under weaker conditions.

Theorem 1.19. If $f(x, t)$ and $(\partial f / \partial x)(x, t)$ are both continuous in $c \leq x \leq d$, $t \geq a$, the integral $\int_a^\infty f(x, t) \, dt$ converges to $F(x)$ in $c \leq x \leq d$, and if the integral $\int_a^\infty (\partial f / \partial x)(x, t) \, dt$ converges uniformly in the interval $c \leq x \leq d$, then

$$F'(x) = \int_a^\infty \frac{\partial f}{\partial x} (x, t) \, dt \qquad c \leq x \leq d$$

Notice that the conditions required to justify differentiation under the integral sign (Theorem 1.18) are much more stringent than those to justify integration under the integral sign. As with the infinite series, we see that the basic requirement for differentiation under the integral sign is uniform convergence of the integral of $\partial f / \partial x$.

Example 17: Use the concept of uniform convergence to derive the integral formula

$$\int_0^\infty \frac{e^{-pt} - e^{-qt}}{t} \, dt = \ln \frac{q}{p} \qquad 0 < p < q$$

Solution: Because the integral is not an elementary integral, we cannot derive the result directly. An indirect approach relies on the observation that the integrand above is obtained by integrating e^{-xt} between p and q. That is,

$$\int_p^q e^{-xt} \, dx = \frac{e^{-pt} - e^{-qt}}{t}$$

To use the above relation, we start with the integral

$$\frac{1}{x} = \int_0^\infty e^{-xt} \, dt$$

which can be shown to converge uniformly in $0 < p \leq x \leq q$. Hence by integrating each side from p to q and interchanging the order of integration on the right, we arrive at

$$\int_p^q \frac{1}{x} \, dx = \int_p^q \int_0^\infty e^{-xt} \, dt \, dx = \int_0^\infty \int_p^q e^{-xt} \, dx \, dt$$

from which we now deduce

$$\ln \frac{q}{p} = \int_0^\infty \frac{e^{-pt} - e^{-qt}}{t} \, dt$$

Exercises 1.5

In problems 1 through 6, determine if the integrals exist; if so, evaluate them by an appropriate method.

1. $\displaystyle\int_0^1 t^{-1/2}\,dt$

4. $\displaystyle\int_{-\infty}^{\infty} \frac{dt}{t^2+1}$

2. $\displaystyle\int_0^1 \frac{dt}{t}$

5. $\displaystyle\int_0^2 (4-t^2)^{-1/2}\,dt$

3. $\displaystyle\int_2^{\infty} t(t^2+1)^{-3}\,dt$

6. $\displaystyle\int_0^{\infty} \sin t\,dt$

In problems 7 to 10, use the limit tests (Theorems 1.14 and 1.15) to prove absolute convergence or divergence of the integrals.

7. $\displaystyle\int_0^{\infty} \frac{1}{\sqrt{1+t^3}}\,dt$

9. $\displaystyle\int_0^{\infty} e^{-t}t^{-1/2}\,dt$

8. $\displaystyle\int_0^{\infty} \frac{5e^{-t}-3}{(1+2t^2)^{1/3}}\,dt$

10. $\displaystyle\int_1^{\infty} t^{-2}(1+t)e^t\,dt$

In problems 11 to 15, use Theorem 1.16 to prove that the integral converges uniformly in the indicated interval.

11. $\displaystyle\int_1^{\infty} \frac{x\,dt}{3x^2+t^2}$, $1\le x\le 2$

14. $\displaystyle\int_0^{\infty} e^{-x^2t^2}\,dt$, $1\le x\le 10$

12. $\displaystyle\int_0^{\infty} \frac{\sin xt}{t}\,dt$, $1\le x\le 10$

15. $\displaystyle\int_0^{\infty} \frac{\cos xt}{t^2+4}\,dt$, $-10\le x\le 10$

13. $\displaystyle\int_0^1 e^{-t}t^{x-1}\,dt$, $0.1\le x\le 1$

16. Use the integral relation

$$\frac{1}{1+x^2}=\int_0^{\infty} e^{-t}\cos xt\,dt$$

to deduce the function $F(x)$ represented by

$$F(x)=\int_0^{\infty} e^{-t}\frac{\sin xt}{t}\,dt$$

17. Use the integral relation

$$\frac{x}{x^2+1}=\int_0^{\infty} e^{-xt}\cos t\,dt \qquad x>0$$

to deduce the value of the integral

$$I = \int_0^\infty te^{-2t}\cos t\, dt$$

18. Use the integral relation

$$(b^2 - x^2)^{-1/2} = \int_0^\infty (\cos xt)J_0(bt)\, dt \qquad b > x \geq 0$$

where $J_0(x)$ is a *Bessel function* (see Chap. 6), to deduce the relation

$$\sin^{-1}\frac{1}{b} = \int_0^\infty \frac{\sin t}{t} J_0(bt)\, dt \qquad b > 1$$

19. Given that

$$\frac{\pi}{2}x^{-1/2} = \int_0^\infty \frac{dt}{t^2 + x} \qquad x > 0$$

show that (for $n = 1, 2, 3, \dots$)

$$\int_0^\infty \frac{dt}{(t^2 + x)^{n+1}} - \frac{\pi(2n)!}{2^{2n+1}(n!)^2}x^{-n-1/2}$$

20. Given that (for $n = 0, 1, 2, \dots$)

$$P_n(x) = \frac{(-1)^n}{n!\sqrt{\pi}}\int_{-\infty}^\infty e^{-(1-x^2)t^2}\frac{d^n}{dx^n}(e^{-x^2t^2})\, dt$$

show that

$$xP_n'(x) - P_{n-1}'(x) = nP_n(x)$$

$[P_n(x)$ is the nth *Legendre polynomial*. See Chap. 4.]

1.6 Asymptotic Formulas

In computational analysis we often seek to represent a given function $f(x)$ by some simpler function, say $g(x)$, that accurately describes the numerical values of $f(x)$ in this vicinity of a particular point $x = a$. To accomplish this, we require that

$$\lim_{x \to a}\frac{f(x)}{g(x)} = 1$$

If this is true, we write

$$f(x) \sim g(x) \qquad x \to a$$

where the symbol $\sim$ means "behaves like" or "is asymptotic to." Although in theory we could choose any value of x, generally we confine our attention to either the case $x \to 0$ or $x \to \infty$, also denoted, respectively, by $|x| \ll 1$ and $x \gg 1$.

Example 18: Find asymptotic formulas for the given functions for $x \to 0$ and $x \to \infty$.

(a) $f(x) = \dfrac{bx}{1 + ax}$

(b) $f(x) = 2x + \sqrt{x}, \, x \geq 0$

Solution:

(a) We say that

$$f(x) \sim bx \qquad x \to 0$$

because

$$\lim_{x \to 0} \frac{f(x)}{bx} = \lim_{x \to 0} \frac{1}{1 + ax} = 1$$

Similarly, it follows that

$$f(x) \sim \frac{b}{a} \qquad x \to \infty$$

since

$$\lim_{x \to \infty} \frac{f(x)}{b/a} = \lim_{x \to \infty} \frac{ax}{1 + ax} = 1$$

(b) We see by inspection that

$$f(x) \sim \sqrt{x} \qquad x \to 0^{+}$$
$$f(x) \sim 2x \qquad x \to \infty$$

1.6.1 Small arguments

For $x \to 0$ and functions more complex than those illustrated in Example 18, we generally seek a representation of the form

$$f(x) \sim \sum_{n=0}^{\infty} c_n x^n \qquad x \to 0 \tag{1.74}$$

from which we can deduce *asymptotic formulas* such as

$$f(x) \sim c_0 \qquad\qquad x \to 0 \tag{1.75a}$$
$$f(x) \sim c_0 + c_1 x \qquad x \to 0 \tag{1.75b}$$

and so on. To illustrate, let us consider the sine series

$$\sin x = \sum_{n=0}^{\infty} \frac{(-1)^n x^{2n+1}}{(2n+1)!} \qquad -\infty < x < \infty \tag{1.76}$$

from which we obtain the asymptotic formulas

$$\sin x \sim x \qquad |x| \ll 1 \tag{1.77a}$$

$$\sin x \sim x - \frac{x^3}{3!} \qquad |x| \ll 1 \tag{1.77b}$$

The choice between (1.77a) and (1.77b) will usually depend on the particular application and on how close x is to zero, although (1.77b) will always give a more accurate result.

Like the sine series (1.76), we ordinarily obtain the representation (1.74) from the Maclaurin series expansion of $f(x)$. In such cases the series converges about $x = 0$ for all values of x in some interval $|x| < \rho$. That is, if

$$S_n(x) = \sum_{k=0}^{n} c_k x^k \tag{1.78}$$

and the series converges, then it follows that

$$\lim_{n \to \infty} |f(x) - S_n(x)| - 0 \tag{1.79}$$

for each fixed x in the interval $|x| < \rho$. By taking a sufficient number of terms in the asymptotic expression, our calculations for $f(x)$ can be made as precise as desired. However, the series representation (1.74) does *not* have to be a Maclaurin series, nor is there any requirement for the series to converge in order to be useful for computations! That is, we say that (1.74) is an **asymptotic series** for $f(x)$ as $x \to 0$ if and only if

$$\lim_{x \to 0} \frac{|f(x) - S_n(x)|}{|x|^n} = 0 \tag{1.80}$$

for each *fixed* n. By this condition we are requiring the difference between $f(x)$ and the sum of terms $S_n(x)$ to approach zero more rapidly than $|x|^n$, as x itself tends to zero. This means that if the series (1.74) diverges, the accuracy of computation for fixed x is closely tied to the number n in a most peculiar way. Namely, after a certain number of terms in this case, the accuracy of computation will actually get *worse* instead of better with increasing n—a sharp contrast to convergent series.

1.6.2   Large arguments

Given a function $f(x)$, we may be able to find useful approximations to it for large values of x by seeking a series of the form

$$f(x) \sim \sum_{n=0}^{\infty} \frac{a_n}{x^n} \qquad x \to \infty \qquad (1.81)$$

called an *asymptotic* or *semiconvergent series*. A precise definition of asymptotic series was first provided by J. H. Poincaré (1854–1912) in 1886. He stated that a series such as (1.81) is an **asymptotic series** if and only if, for every n,

$$\lim_{x \to \infty} x^n |f(x) - S_n(x)| = 0 \qquad (1.82)$$

where
$$S_n(x) = \sum_{k=0}^{n} \frac{a_k}{x^k} \qquad (1.83)$$

Asymptotic series like (1.81) are intriguing in that they usually *diverge* for all values of x. Nonetheless, they may still be useful for numerical computation if one does not take too many terms. As a rule of thumb, *one should stop just before the terms begin to increase.* The error incurred in most cases turns out to be less than the first term omitted in the approximation.

Not all functions have an asymptotic series of the form (1.81). For example, neither e^x nor $\sin x$ has such a series. If the function $f(x)$ itself has no asymptotic series of the form (1.81), there may exist a suitable function $h(x)$ such that an asymptotic series for the quotient $f(x)/h(x)$ does exist. In this case we write

$$\frac{f(x)}{h(x)} \sim \sum_{n=0}^{\infty} \frac{a_n}{x^n} \qquad x \to \infty$$

or
$$f(x) \sim h(x) \sum_{n=0}^{\infty} \frac{a_n}{x^n} \qquad x \to \infty \qquad (1.84)$$

If $f(x)$ has an asymptotic series, it may turn out that other functions have the same asymptotic series. That is, an asymptotic series does not uniquely determine the function from which it was generated.

There are several ways in which asymptotic series can be derived. For our first example, we consider the case where the function $f(x)$ is defined by an improper integral of the form

$$f(x) = \int_x^{\infty} g(t)\, dt$$

A simple and often effective way of developing the series in such instances consists of repeated *integration by parts*. Each new integration yields the next term in the expansion, and the error committed in stopping after n terms can be expressed by the remaining integral for which error bounds can usually be deduced.

Example 19: Find an asymptotic series for the function defined by

$$f(x) = \int_x^\infty \frac{e^{-t}}{t} dt \qquad x > 0$$

Solution: Using integration by parts with

$$u = \frac{1}{t} \qquad dv = e^{-t} dt$$

$$du = -\frac{dt}{t^2} \qquad v = -e^{-t}$$

we find

$$f(x) = -\frac{e^{-t}}{t}\Big|_x^\infty - \int_x^\infty \frac{e^{-t}}{t^2} dt$$

$$= \frac{e^{-x}}{x} - \int_x^\infty \frac{e^{-t}}{t^2} dt$$

Continued integration by parts leads to

$$f(x) = e^{-x}\left[\frac{1}{x} - \frac{1}{x^2} + \frac{1 \times 2}{x^3} - \cdots + (-1)^{n-1}\frac{1 \times 2 \times \cdots \times (n-1)}{x^n}\right]$$

$$+ (-1)^n (1 \times 2 \times \cdots \times n) \int_x^\infty \frac{e^{-t}}{t^{n+1}} dt$$

from which we deduce

$$f(x) \sim \frac{e^{-x}}{x} \sum_{n=0}^\infty \frac{(-1)^n n!}{x^n} \qquad x \to \infty$$

By use of the ratio test, we see that

$$\lim_{n\to\infty} \left|\frac{(n+1)!}{x^{n+1}} \cdot \frac{x^n}{n!}\right| = \frac{1}{|x|} \lim_{n\to\infty} (n+1) = \infty$$

and thus, the series diverges for all x. To show that this is indeed an asymptotic series, we first observe that

$$S_n(x) = e^{-x}\left[\frac{1}{x} - \frac{1}{x^2} + \frac{1 \times 2}{x^3} - \cdots + (-1)^{n-1}\frac{1 \times 2 \times \cdots \times (n-1)}{x^n}\right]$$

Next, we develop the inequality

$$|f(x) - S_n(x)| = 1 \times 2 \times \cdots \times n \int_x^\infty \frac{e^{-t}}{t^{n+1}} dt$$

$$= n! \int_x^\infty \frac{e^{-t}}{t^{n+1}} dt$$

$$\leq n! \int_x^\infty \frac{e^{-t}}{x^{n+1}} dt \qquad (t \geq x)$$

$$\leq \frac{n!}{x^{n+1}}$$

and thus deduce that

$$\lim_{x \to \infty} x^n |f(x) - S_n(x)| \leq \lim_{x \to \infty} \frac{n!}{x} = 0$$

Hence, this is indeed an asymptotic series.

Finally, we mention that, based on the above result, it is clear that the error $E_n(x)$ committed in approximating $f(x)$ by $S_n(x)$ is bounded by

$$|E_n(x)| \leq \frac{n!}{x^{n+1}} \qquad x \gg 1$$

When the function $f(x)$ is defined by an integral of the form

$$f(x) = \int_a^\infty g(x, t) \, dt$$

it may be advantageous to use another method besides integration by parts to develop the asymptotic series, as illustrated by the following example.

Example 20: Find an asymptotic series for the function defined by

$$f(x) = \int_0^\infty \frac{e^{-xt}}{1 + t^2} dt \qquad x > 0$$

Solution: Let us start by making the change of variable $s = xt$, which leads to the expression

$$f(x) = \frac{1}{x} \int_0^\infty e^{-s} \left(1 + \frac{s^2}{x^2}\right)^{-1} ds$$

For $x > s$, we have the binomial series (see Example 11)

$$\left(1 + \frac{s^2}{x^2}\right)^{-1} = \sum_{n=0}^{\infty} \binom{-1}{n} \frac{s^{2n}}{x^{2n}}$$

and using termwise integration, we obtain

$$f(x) \sim \sum_{n=0}^{\infty} \binom{-1}{n} \frac{1}{x^{2n}} \int_0^{\infty} e^{-s} s^{2n} \, ds \qquad x \to \infty$$

This last integral can be evaluated by repeated integration by parts to yield

$$\int_0^{\infty} e^{-s} s^{2n} \, ds = (2n)! \qquad n = 0, 1, 2, \dots .$$

and so we deduce that

$$f(x) \sim \sum_{n=0}^{\infty} (-1)^n \frac{(2n)!}{x^{2n+1}} \qquad x \to \infty$$

where we have made use of the identity

$$\binom{-1}{n} = (-1)^n \binom{n}{n} = (-1)^n$$

The technique used in Example 20 is not rigorous and is even somewhat incorrect in that the particular binomial series converges only for $s < x$ and we are allowing s to be arbitrarily large. Moreover, it has led once again to a series that diverges for all values of x. In spite of its apparent difficulties, we try to justify the use of this method by noting that for large x the major contribution to the integral exists for $s < x$.

Although they often diverge, asymptotic series behave very much like convergent power series. For example, the asymptotic series of two functions can be added to form the asymptotic series for the sum of the two functions. These same asymptotic series can be multiplied to form an asymptotic series of the product of the two functions. Also the asymptotic series of a function can be integrated termwise, and the result will be an asymptotic expansion of the integral of the original function. Under more stringent conditions, asymptotic series may even be differentiated termwise to produce series of the derivative of the original function.

Exercises 1.6

In problems 1 to 4, find asymptotic formulas for the given function for $x \to 0$ and $x \to \infty$

1. $f(x) = \dfrac{bx^2}{1 + ax}$

3. $f(x) = \sqrt{a + bx}$

2. $f(x) = \dfrac{bx}{1 + ax^2}$

4. $f(x) = (a + bx)^\alpha$

In problems 5 to 11, derive the given asymptotic series. Check convergence.

5. $\displaystyle\int_0^\infty e^{-xt} \cos t \, dt \sim \sum_{n=0}^\infty \frac{(-1)^n}{x^{2n+1}}$, $x \to \infty$

6. $\displaystyle\int_0^\infty e^{-xt} \sin t \, dt \sim \sum_{n=1}^\infty \frac{(-1)^{n-1}}{x^{2n}}$, $x \to \infty$

7. $\displaystyle\int_{-\infty}^x \frac{e^t}{t} \, dt \sim \frac{e^x}{x} \sum_{n=0}^\infty \frac{n!}{x^n}$, $x \to \infty$

8. $\displaystyle\int_0^x \frac{dt}{\ln t} \sim \frac{x}{\ln x} \sum_{n=0}^\infty \frac{n!}{(\ln x)^n}$, $x \to \infty$

   Hint: Let $u = \ln t$.

9. $\displaystyle\int_0^\infty \frac{e^{-xt}}{1+t} \, dt \sim \sum_{n=0}^\infty \frac{(-1)^n n!}{x^{n+1}}$, $x \to \infty$

10. $\displaystyle\int_x^\infty e^{-t} t^{a-1} \, dt \sim x^{a-1} e^{-x} \left[1 + \sum_{n=1}^\infty \frac{(a-1)(a-2)\cdots(a-n)}{x^n} \right]$,

    $x \to \infty$, $a > 0$

11. $\displaystyle\int_x^\infty e^{-t^2} \, dt \sim \frac{e^{-x^2}}{2x} \left[1 + \sum_{n=1}^\infty (-1)^n \frac{1 \times 3 \times \cdots \times (2n-1)}{(2x^2)^n} \right]$, $x \to \infty$

12. Given

$$f(x) = \int_0^\infty \frac{e^{-t}}{1 + xt} \, dt \qquad x \geq 0$$

    show that

$$f(x) \sim \sum_{n=0}^\infty n! \, (-1)^n x^n \qquad x \to 0^+$$

    Hint: Verify that Eq. (1.80) is satisfied by first establishing

$$S_n(x) = \sum_{k=0}^n k! \, (-1)^k x^k = \int_0^\infty e^{-t} \sum_{k=0}^n (-1)^k (xt)^k \, dt$$

1.7 Infinite Products

Given the infinite sequence of positive numbers $u_1, u_2, \ldots, u_n,$ we can express their product by the notation

$$u_1 \times u_2 \times u_3 \times \cdots \times u_n \times \cdots = \prod_{n=1}^{\infty} u_n \qquad (1.85)$$

By analogy with infinite series, we define the *partial product*

$$P_n = \prod_{k=1}^{n} u_k \qquad (1.86)$$

and investigate the limit

$$\lim_{n \to \infty} P_n = P \qquad (1.87)$$

If P is finite (but not zero), we say the infinite product (1.85) *converges* to P; otherwise, it *diverges*. The product (1.85) may diverge because the limit (1.87) fails to exist, but also because $P = 0$, in which case we say the infinite product *diverges to zero*. We do not discuss infinite products that diverge to zero.

Because the infinite product will become infinite if $\lim_{n \to \infty} u_n > 1$ or diverge to zero if $0 < \lim_{n \to \infty} u_n < 1$, we find it convenient to write $u_n = 1 + a_n$ and then discuss infinite products of the form $\prod_{n=1}^{\infty}(1 + a_n)$. Based on the above remarks, it is clear that a necessary (but not sufficient) condition for the infinite product $\prod_{n=1}^{\infty}(1 + a_n)$ to converge is that (see problem 1 in Exercises 1.7)

$$\lim_{n \to \infty} a_n = 0 \qquad (1.88)$$

Remark: Our original assumption was that the sequence $u_1, u_2, \ldots,$ $u_n, \ldots$ was composed of positive numbers. Hence it follows that $a_n > -1$ for all n. However, should m of the original numbers be negative, we can replace their product by $(-1)^m$ times the product of their absolute values.

Example 21: Find the value of the infinite product $\prod_{n=2}^{\infty} (1 - 1/n^2)$.

Solution: We first make the observation that

$$\lim_{n \to \infty} a_n = -\lim_{n \to \infty} \frac{1}{n^2} = 0$$

which is required for convergence. To find the value of the product, we try to obtain an expression for the partial product P_n and take its limit. The product of the first n terms leads to

$$P_n = \prod_{k=2}^{n} \left(1 - \frac{1}{k^2}\right)$$

$$= \prod_{k=2}^{n} \frac{(k-1)(k+1)}{k^2}$$

$$= \frac{1 \times 2 \times 3 \times \cdots \times (n-1) \times 3 \times 4 \times 5 \times \cdots \times (n+1)}{2 \times 2 \times 3 \times 3 \times 4 \times 4 \times \cdots \times n \times n}$$

$$= \frac{n+1}{2n}$$

where in the last step we have canceled all common factors. Thus, by taking the limit

$$\lim_{n \to \infty} P_n = \lim_{n \to \infty} \frac{n+1}{2n} = \frac{1}{2}$$

we conclude that

$$\prod_{n=2}^{\infty} \left(1 - \frac{1}{n^2}\right) = \frac{1}{2}$$

1.7.1 Associated infinite series

In many cases of interest we are unable to find an explicit expression for the partial product P_n and examine its limit as we did in Example 21. When this is the case, it is useful to have tests of convergence as we did in studying infinite series. Although we could devise convergence tests based directly on the product, there are related infinite series whose convergence or divergence will settle the question in regard to the infinite product.

For example, closely associated with all infinite products is the infinite series of logarithms derived from

$$\ln \prod_{n=1}^{\infty} (1 + a_n) = \sum_{n=1}^{\infty} \ln (1 + a_n) \tag{1.89}$$

where it is assumed that no $a_n = -1$. If we denote the partial product and partial sum, respectively, by

$$P_n = \prod_{k=1}^{n} (1 + a_k) \qquad S_n = \sum_{k=1}^{n} \ln (1 + a_k)$$

then clearly

$$\lim_{n\to\infty} F_n \quad \lim_{n\to\infty} \exp S_n = \exp\left(\lim_{n\to\infty} S_n\right)$$

through the properties of limits. Therefore we see that P_n approaches a limit value P $(P \neq 0)$ if and only if S_n has a limit value S.

A result more useful than considering the associated series of logarithms is contained in the following theorem.

Theorem 1.19 If $0 \leq a_n < 1$ for all $n > N$, the infinite products $\prod_{n=1}^{\infty}(1 + a_n)$ and $\prod_{n=1}^{\infty}(1 - a_n)$ converge or diverge according to whether the infinite series $\sum_{n=1}^{\infty} a_n$ converges or diverges.

1.7.2 Products of functions

When the general term of the product is a function of x, we are led to infinite products of the form

$$f(x) = \prod_{n=1}^{\infty} [1 + a_n(x)] \tag{1.90}$$

If, for a fixed value of x, the product (1.90) equals $f(x)$, we say the product *converges pointwise* to $f(x)$. The general theory of representing functions by infinite products of the form (1.90) goes beyond the intended scope of this text, and thus we treat only some special cases.

Remark: The notion of *uniform convergence* of infinite products plays an important role in the theory of infinite products, much as it does in the theory of infinite series and improper integrals. The usual way in which uniform convergence is established for infinite products is by (another) *Weierstrass M test*. The interested reader should consult E. D. Rainville, *Special Functions*, Chelsea, New York, 1960, p. 6.

Recall from algebra that an nth-degree polynomial $p_n(x)$ with n real roots (zeros) can be expressed in the product form*

$$p_n(x) = (x - x_1)(x - x_2) \cdots (x - x_n) = \prod_{k=1}^{n} (x - x_k) \tag{1.91}$$

---

*For simplicity of notation we are assuming $p_n(x) = x^n + \cdots$, where the leading coefficient is unity.

We might well wonder if functions with an infinite number of zeros have similar product representations. This is sometimes indeed the case. For example, the zeros of $\sin \pi x$ occur at $x = \pm n$ ($n = 0, 1, 2, \ldots$), and it can be shown that (see problem 8 in Exercises 1.7)

$$\sin \pi x = \pi x \prod_{n=1}^{\infty} \left(1 - \frac{x}{n}\right)\left(1 + \frac{x}{n}\right)$$

or

$$\sin \pi x = \pi x \prod_{n=1}^{\infty} \left(1 - \frac{x^2}{n^2}\right) \tag{1.92}$$

whereas for the cosine function

$$\cos \pi x = \prod_{n=1}^{\infty} \left[1 - \frac{4x^2}{(2n-1)^2}\right] \tag{1.93}$$

An interesting result can be derived from (1.92) by setting $x = \frac{1}{2}$. That is,

$$1 = \frac{\pi}{2} \prod_{n=1}^{\infty} \left[1 - \frac{1}{(2n)^2}\right] = \frac{\pi}{2} \prod_{n=1}^{\infty} \frac{(2n-1)(2n+1)}{(2n)^2}$$

and by solving for $\pi/2$, we see that

$$\frac{\pi}{2} = \frac{2 \times 2}{1 \times 3} \cdot \frac{4 \times 4}{3 \times 5} \cdot \frac{6 \times 6}{5 \times 7} \cdots \tag{1.94}$$

which is Wallis' famous formula for $\pi/2$. In Sec. 2.2.4 we again use (1.92) to derive another interesting relation between the sine function and the gamma function.

Exercises 1.7

1. If $\prod_{n=1}^{\infty} (1 + a_n)$ converges to the value $P \neq 0$, show that

$$\lim_{n \to \infty} a_n = 0$$

Hint: Consider the ratio

$$\lim_{n \to \infty} \frac{\prod_{k=1}^{n} (1 + a_k)}{\prod_{k=1}^{n-1} (1 + a_k)}$$

In problems 2 to 7, show that the infinite product converges by finding its value.

2. $\displaystyle\prod_{n=1}^{\infty}\left[1-\frac{2}{(n+1)(n+2)}\right]$

5. $\displaystyle\prod_{n=2}^{\infty}\left[1+\frac{1}{n(n+2)}\right]$

3. $\displaystyle\prod_{n=1}^{\infty}\left[1+\frac{6}{(n+1)(2n+9)}\right]$

6. $\displaystyle\prod_{n=1}^{\infty}\left[1+\frac{1}{(4n-1)(4n-3)}\right]$

4. $\displaystyle\prod_{n=1}^{\infty}\left(1-\frac{1}{4n^2}\right)$

7. $\displaystyle\prod_{n=1}^{\infty}\left[1+\frac{(-1)^{n+1}}{n}\right]$

8. Use (1.92) and (1.93) to verify the identity
$$2\sin x\cos x = \sin 2x$$

9. You are given the function $f(x)=\cos kx$, $-\pi\leq x\leq\pi$, where k is not an integer.
   (a) Find its Fourier trigonometric series.
   (b) Letting $k=z$ in (a), and substituting $x=0$ and $x=\pi$, obtain the series expansions
$$\csc \pi z = \frac{1}{\pi z}+\frac{2z}{\pi}\sum_{n=1}^{\infty}\frac{(-1)^n}{z^2-n^2}$$
$$\cot \pi z = \frac{1}{\pi z}+\frac{2z}{\pi}\sum_{n=1}^{\infty}\frac{1}{z^2-n^2}$$
   (c) Assume $0<z<1$ and integrate the series in (b) for $\cot \pi z$ from 0 to x, $0<x<1$, and show that
$$\ln\frac{\sin \pi x}{\pi x}=\sum_{n=1}^{\infty}\ln\left(1-\frac{x^2}{n^2}\right)$$
   so that
$$\sin \pi x = \pi x\prod_{n=1}^{\infty}\left(1-\frac{x^2}{n^2}\right)$$
   (d) From (c), deduce that
$$\sin x = x\prod_{n=1}^{\infty}\left(1-\frac{x^2}{n^2\pi^2}\right)$$

10. By using the results of problem 9, show that
$$\int_0^\infty\frac{x^{z-1}}{1+x}\,dx = \frac{\pi}{\sin \pi z}\qquad 0<z<1$$

Hint: Express the integral as a sum of two integrals, the first having $(0,1)$ as the interval of integration and the second $(1,\infty)$. Then let $x=1/t$ in the second integral, and use the geometric series for $(1+x)^{-1}$.

2

The Gamma Function and Related Functions

2.1 Introduction

In the eighteenth century, L. Euler (1707–1783) concerned himself with the problem of interpolating between the numbers

$$n! = \int_0^\infty e^{-t} t^n \, dt \qquad n = 0, 1, 2, \ldots$$

with nonintegral values of n. This problem led Euler in 1729 to the now famous *gamma function*, a generalization of the factorial function that gives meaning to $x!$ when x is any positive number. His result can be extended to certain negative numbers and even to complex numbers. The notation $\Gamma(x)$ that is now widely accepted for the gamma function is not due to Euler, however, but was introduced in 1809 by A. Legendre (1752–1833), who was also responsible for the *duplication formula* for the gamma function. Nearly 150 years after Euler's discovery of it, the theory concerning the gamma function was greatly expanded by means of the theory of entire functions developed by K. Weierstrass (1815–1897).

Because it is a generalization of $n!$, the gamma function has been examined over the years as a means of generalizing certain functions, operations, etc., that are commonly defined in terms of factorials. In addition, the gamma function is useful in the evaluation of many nonelementary integrals and in the definition of other special functions.

Another function useful in various applications is the related *beta function*, often called the *eulerian integral of the first kind*. In 1771, some 43 years after discovering the gamma function, Euler found that the beta function was actually a particular combination of gamma functions. The logarithmic derivative of the gamma function leads to the *digamma function,* while further differentiations produce the family of *polygamma functions,* all of which are related to the *zeta function* of G. Riemann (1826–1866).

2.2 Gamma Function

One of the simplest but very important special functions is the **gamma function**, denoted by $\Gamma(x)$. It appears occasionally by itself in physical applications (mostly in the form of some integral), but much of its importance stems from its usefulness in developing other functions such as *Bessel functions* (Chaps. 6 to 8) and *hypergeometric functions* (Chaps. 9 to 12) which have more direct physical application.

The gamma function has several equivalent definitions, most of which are due to Euler. To begin, however, we use the definition

$$\Gamma(x) = \lim_{n\to\infty} \frac{n!n^x}{x(x+1)(x+2)\cdots(x+n)} \qquad (2.1)$$

which was actually due to C. Gauss (1777–1855).* If x is not zero or a negative integer, it can be shown that the limit (2.1) exists.† It is apparent, however, that $\Gamma(x)$ cannot be defined at $x = 0, -1, -2, \ldots$, since the limit becomes infinite for any of these values. As a consequence we have the following theorem.

Theorem 2.1. If $x = -n$ $(n = 0, 1, 2, \ldots)$, then $|\Gamma(x)| = \infty$, or equivalently,

$$\frac{1}{\Gamma(-n)} = 0 \qquad n = 0, 1, 2, \ldots$$

By setting $x = 1$ in Eq. (2.1), we see that

$$\Gamma(1) = \lim_{n\to\infty} \frac{n!n}{1\times 2\times 3\times\cdots\times n(n+1)} = \lim_{n\to\infty}\frac{n}{n+1}$$

---

*A variation of (2.1), called *Euler's infinite product* (see problem 43 in Exercises 2.2), was actually the starting point of Euler's work on the interpolation problem for $n!$.

† See E. D. Rainville, *Special Functions,* Chelsea, New York, 1960, p. 5.

from which we deduce the special value

$$\Gamma(1) = 1 \tag{2.2}$$

Other values of $\Gamma(x)$ are not obtained so easily, but the substitution of $x + 1$ for x in (2.1) leads to

$$\Gamma(x + 1) = \lim_{n\to\infty} \frac{n!n^{x+1}}{(x+1)(x+2)\cdots(x+n)(x+n+1)}$$

$$= \lim_{n\to\infty} \frac{nx}{x+n+1} \cdot \lim_{n\to\infty} \frac{n!n^x}{x(x+1)\cdots(x+n)}$$

from which we deduce the **recurrence formula**

$$\Gamma(x + 1) = x\Gamma(x) \tag{2.3}$$

Equation (2.3) is the basic functional relation for the gamma function; it is in the form of a *difference equation*. While many of the special functions satisfy some linear *differential equation*, it has been shown that the gamma function does not satisfy any linear differential equation with rational coefficients.*

A direct connection between the gamma function and factorials can be obtained from (2.2) and (2.3). That is, if we combine these relations, we have

$$\Gamma(2) = 1 \times \Gamma(1) = 1$$

$$\Gamma(3) = 2 \times \Gamma(2) = 2 \times 1 = 2!$$

$$\Gamma(4) = 3 \times \Gamma(3) = 3 \times 2! = 3!$$

$$\vdots$$

and through mathematical induction it can be shown that

$$\Gamma(n + 1) = n! \qquad n = 0, 1, 2, \ldots \tag{2.4}$$

Thus the gamma function is a generalization of the factorial function from the domain of positive integers to the domain of all real numbers (except as noted in Theorem 2.1).

It is sometimes considered a nuisance that $n!$ is not $\Gamma(n)$, but $\Gamma(n + 1)$. Because of this, some authors adopt the notation $x!$ for the gamma function, whether or not x is an integer. Gauss introduced the notation $\Pi(x)$, where $\Pi(x) = x!$, but this notation is seldom utilized.

---

*See R. Campbell, *Les intégrals Eulériennes et leurs applications*, Dunod, Paris, 1966, pp. 152–159.

The symbol $\Gamma(x)$, due to Legendre, is the most widely used today. We will not use the notation of Gauss, nor will we use the factorial notation except when dealing with nonnegative integer values.

2.2.1 Integral representations

Our reason for using the limit definition (2.1) of the gamma function is that it defines the gamma function for negative values of x as well as positive values. The gamma function rarely appears in the form (2.1) in applications. Instead, it most often arises in the evaluation of certain integrals; for example, Euler was able to show that*

$$\Gamma(x) = \int_0^\infty e^{-t} t^{x-1}\, dt \qquad x > 0 \tag{2.5}$$

This **integral representation** of $\Gamma(x)$ is the most common way in which the gamma function is now defined. Since integrals are fairly easy to manipulate, (2.5) is often preferred to (2.1) for developing properties of this function. Equation (2.5) is less general than (2.1), however, since the variable x is restricted in (2.5) to positive values. Last, we note that (2.5) is an improper integral, due to the infinite limit of integration and because the factor t^{x-1} becomes infinite at $t = 0$ for values of x in the interval $0 < x < 1$. Nonetheless, the integral (2.5) is *uniformly convergent* for all $a \le x \le b$, where $0 < a \le b < \infty$ (e.g., see problem 16 in Exercises 2.2).

Let us first establish the equivalence of (2.1) and (2.5) for positive values of x. To do so, we set

$$F(x) = \int_0^\infty e^{-t} t^{x-1}\, dt$$

$$= \lim_{n \to \infty} \int_0^n \left(1 - \frac{t}{n}\right)^n t^{x-1}\, dt \qquad x > 0 \tag{2.6}$$

where we are making the observation

$$e^{-t} = \lim_{n \to \infty} \left(1 - \frac{t}{n}\right)^n \tag{2.7}$$

---

*Legendre termed the right-hand side of (2.5) the *eulerian integral of the second kind*.

Using successive integration by parts, after making the change of variable $z = t/n$, we find

$$F(x) = \lim_{n \to \infty} n^x \int_0^1 (1-z)^n z^{x-1} \, dz$$

$$= \lim_{n \to \infty} n^x \left[(1-z)^n \frac{z^x}{x} \Big|_0^1 + \frac{n}{x} \int_0^1 (1-z)^{n-1} z^x \, dz \right]$$

$$\vdots \tag{2.8}$$

$$= \lim_{n \to \infty} n^x \left[\frac{n(n-1) \cdots 2 \times 1}{x(x+1) \cdots (x+n-1)} \int_0^1 z^{x+n-1} \, dz \right]$$

$$= \lim_{n \to \infty} \frac{n! \, n^x}{x(x+1)(x+2) \cdots (x+n)}$$

and thus we have shown that

$$F(x) = \int_0^\infty e^{-t} t^{x-1} \, dt = \Gamma(x) \qquad x > 0 \tag{2.9}$$

It follows from the uniform convergence of the integral (2.5) that $\Gamma(x)$ is a continuous function for all $x > 0$ (see Theorem 1.17). To investigate the behavior of $\Gamma(x)$ as x approaches the value zero from the right, we use the recurrence formula (2.3) written in the form

$$\Gamma(x) = \frac{\Gamma(x+1)}{x}$$

Thus, we see that

$$\lim_{x \to 0^+} \Gamma(x) = \lim_{x \to 0^+} \frac{\Gamma(x+1)}{x} = +\infty \tag{2.10}$$

Another consequence of the uniform convergence of the defining integral for $\Gamma(x)$ is that we may differentiate the function under the integral sign to obtain*

$$\Gamma'(x) = \int_0^\infty e^{-t} t^{x-1} \ln t \, dt \qquad x > 0 \tag{2.11}$$

---

*Actually, to completely justify the derivative relations (2.11) and (2.12) requires that we first establish the uniform convergence of the integrals in them. See Theorem 1.18 in Sec. 1.5.3.

and

$$\Gamma''(x) = \int_0^\infty e^{-t} t^{x-1} (\ln t)^2 \, dt \qquad x > 0 \qquad (2.12)$$

The integrand in (2.12) is positive over the entire interval of integration, and thus it follows that $\Gamma''(x) > 0$. This implies that the graph of $y = \Gamma(x)$ is *concave upward* for all $x > 0$. While maxima and minima are ordinarily found by setting the derivative of the function to zero, here we make the observation that since $\Gamma(1) = \Gamma(2) = 1$ and $\Gamma(x)$ is always concave upward, the gamma function has *only a minimum* on the interval $x > 0$. Moreover, the minimum occurs on the interval $1 < x < 2$. The exact position of the minimum was first computed by Gauss and found to be $x_0 = 1.4616\ldots$, which leads to the minimum value $\Gamma(x_0) = 0.8856\ldots$. Last, from the continuity of $\Gamma(x)$ and its concavity, we deduce that

$$\lim_{x \to +\infty} \Gamma(x) = +\infty \qquad (2.13)$$

With this last result, we have determined the fundamental characteristics of the graph of $\Gamma(x)$ for $x > 0$ (see Fig. 2.1). Values of $\Gamma(x)$ are commonly tabulated for the interval $1 \le x \le 2$ (e.g., see Table

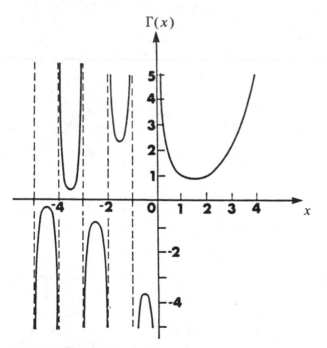

Figure 2.1 The gamma function.

TABLE 2.1 Values of $\Gamma(x)$ for $1 \le x \le 2$

x	$\Gamma(x)$	x	$\Gamma(x)$
1.00	1.0000	1.50	0.8862
1.05	0.9735	1.55	0.8889
1.10	0.9514	1.60	0.8935
1.15	0.9330	1.65	0.9001
1.20	0.9182	1.70	0.9086
1.25	0.9064	1.75	0.9191
1.30	0.8975	1.80	0.9314
1.35	0.8912	1.85	0.9456
1.40	0.8873	1.90	0.9618
1.45	0.8857	1.95	0.9799
1.50	0.8862	2.00	1.0000

2.1), and then other values of $\Gamma(x)$ can be generated through use of the recurrence formula. For example, $\Gamma(2.75) = \Gamma(1.75 + 1) = 1.75\Gamma(1.75) = 1.608$.

The gamma function is defined for both positive and negative values of x by Eq. (2.1), although generally it is more convenient to use the recurrence formula (2.3) when dealing with negative values. For example, if x is in the range $-1 < x < 0$, we rewrite (2.3) as

$$\Gamma(x) = \frac{\Gamma(x+1)}{x} \qquad x > -1, \qquad x \ne 0 \tag{2.14}$$

and we use the right-hand side for computational purposes. Also using (2.14), we obtain the left- and right-hand limits

$$\lim_{x \to 0^-} \Gamma(x) = \lim_{x \to 0^-} \frac{\Gamma(x+1)}{x} = -\infty \tag{2.15}$$

and

$$\lim_{x \to -1^+} \Gamma(x) = \lim_{x \to -1^+} \frac{\Gamma(x+1)}{x} = -\infty \tag{2.16}$$

If $x + 1$ is still a negative number, we can replace x with $x + 1$ in (2.14) to get

$$\Gamma(x+1) = \frac{\Gamma(x+2)}{x+1}$$

which, combined with (2.14), leads to

$$\Gamma(x) = \frac{\Gamma(x+2)}{x(x+1)} \qquad x > -2, \qquad x \ne 0, -1$$

Using this last expression, we find the limiting values

$$\lim_{x \to -1^-} \Gamma(x) = \lim_{x \to -1^-} \frac{\Gamma(x+2)}{x(x+1)} = +\infty \qquad (2.17)$$

and

$$\lim_{x \to -2^+} \Gamma(x) = \lim_{x \to -2^+} \frac{\Gamma(x+2)}{x(x+1)} = +\infty \qquad (2.18)$$

Continuing this process, we finally derive the formula

$$\Gamma(x) = \frac{\Gamma(x+k)}{x(x+1)(x+2)\cdots(x+k-1)} \qquad k = 1, 2, 3, \ldots \quad (2.19)$$

which defines the gamma function over the interval $x > -k$ (except $x \neq 0, -1, -2, \ldots, -k+1$) in terms of a gamma function with positive argument. From (2.19) we see that the above pattern of alternating infinite limits at the negative integers continues indefinitely (see Fig. 2.1).

Example 1: Given that $\Gamma(\frac{1}{2}) = \sqrt{\pi}$, evaluate $\Gamma(-\frac{3}{2})$.

Solution: Repeated use of (2.14) yields

$$\Gamma\left(-\frac{3}{2}\right) = \frac{\Gamma(-\frac{1}{2})}{-\frac{3}{2}} = \frac{\Gamma(\frac{1}{2})}{(-\frac{3}{2})(-\frac{1}{2})}$$

which simplifies to

$$\Gamma(-\tfrac{3}{2}) = \tfrac{4}{3}\sqrt{\pi}$$

In addition to

$$\Gamma(x) = \int_0^\infty e^{-t}t^{x-1}\,dt \qquad x > 0$$

there are a variety of other integral representations of $\Gamma(x)$, most of which can be derived from that one by simple changes of variable. For example, if we set $t = u^2$ in the above integral, we get

$$\Gamma(x) = 2\int_0^\infty e^{-u^2}u^{2x-1}\,du \qquad x > 0 \qquad (2.20)$$

whereas the substitution $t = \ln(1/u)$ yields

$$\Gamma(x) = \int_0^1 \left(\ln\frac{1}{u}\right)^{x-1}\,du \qquad x > 0 \qquad (2.21)$$

A slightly more complicated relation can be derived by using representation (2.20) and forming the product

$$\Gamma(x)\Gamma(y) = 2\int_0^\infty e^{-u^2}u^{2x-1}\,du \cdot 2\int_0^\infty e^{-v^2}v^{2y-1}\,dv$$

$$= 4\int_0^\infty \int_0^\infty e^{-(u^2+v^2)}u^{2x-1}v^{2y-1}\,du\,dv$$

The presence of the term $u^2 + v^2$ in the integrand suggests the change of variables

$$u = r\cos\theta \qquad v = r\sin\theta$$

which leads to

$$\Gamma(x)\Gamma(y) = 4\int_0^{\pi/2}\int_0^\infty e^{-r^2}r^{2x-1}\cos^{2x-1}\theta\, r^{2y-1}\sin^{2y-1}\theta\, r\,dr\,d\theta$$

$$= 4\int_0^\infty e^{-r^2}r^{2(x+y)-1}\,dr \cdot \int_0^{\pi/2}\cos^{2x-1}\theta\sin^{2y-1}\theta\,d\theta$$

$$= 2\Gamma(x+y)\int_0^{\pi/2}\cos^{2x-1}\theta\sin^{2y-1}\theta\,d\theta$$

Finally, solving for the integral, we get the interesting relation

$$\int_0^{\pi/2}\cos^{2x-1}\theta\sin^{2y-1}\theta\,d\theta = \frac{\Gamma(x)\Gamma(y)}{2\Gamma(x+y)} \qquad x>0, \qquad y>0 \ (2.22)$$

By setting $x = y = 1/2$ in (2.22), we have

$$\int_0^{\pi/2} d\theta = \frac{\Gamma(1/2)\Gamma(1/2)}{2\Gamma(1)}$$

from which we deduce the special value

$$\Gamma(1/2) = \sqrt{\pi} \tag{2.23}$$

Example 2: Evaluate $\int_0^\infty x^4 e^{-x^3}\,dx$.

Solution: Let $t = x^3$, and then

$$\int_0^\infty x^4 e^{-x^3}\,dx = 1/3\int_0^\infty e^{-t}t^{2/3}\,dt = 1/3\Gamma(5/3)$$

Example 3: Derive the asymptotic formula

$$\Gamma(x+1) \sim \sqrt{2\pi x}\,x^x e^{-x} \qquad x\to\infty$$

Solution: By making the substitution $t = x + s$ in the integral representation

$$\Gamma(x + 1) = \int_0^\infty t^x e^{-t}\, dt$$

we obtain

$$\Gamma(x + 1) = x^x e^{-x} \int_{-x}^\infty e^{x \ln(1 + s/x)} e^{-s}\, ds$$

For large x, we use the approximation [see (1.41)]

$$\ln\left(1 + \frac{s}{x}\right) \sim \frac{s}{x} - \frac{s^2}{2x^2} \qquad x \to \infty$$

which leads to

$$\Gamma(x + 1) \sim x^x e^{-x} \int_{-\infty}^\infty e^{-s^2/2x}\, ds$$

$$\sim \sqrt{2x}\, x^x e^{-x} \int_{-\infty}^\infty e^{-u^2}\, du$$

the last step following the substitution $u = s/\sqrt{2x}$. Using properties of even functions, and recalling Eqs. (2.20) and (2.23), this last integral yields

$$\int_{-\infty}^\infty e^{-u^2}\, du = 2 \int_0^\infty e^{-u^2}\, du = \Gamma(\tfrac{1}{2}) = \sqrt{\pi}$$

Combining results, we deduce that

$$\Gamma(x + 1) \sim \sqrt{2\pi x}\, x^x e^{-x} \qquad x \to \infty$$

known as *Stirling's formula.**

2.2.2 Legendre duplication formula

A formula involving gamma functions that is somewhat comparable to the double-angle formulas for trigonometric functions is the **Legendre duplication formula**

$$2^{2x-1} \Gamma(x) \Gamma(x + \tfrac{1}{2}) = \sqrt{\pi}\, \Gamma(2x) \tag{2.24}$$

---

* See also Sec. 2.6.2.

To derive this relation, first we set $y = x$ in (2.22) to get

$$\frac{\Gamma(x)\Gamma(x)}{2\Gamma(2x)} = \int_0^{\pi/2} \cos^{2x-1}\theta \sin^{2x-1}\theta \, d\theta$$

$$= 2^{1-2x} \int_0^{\pi/2} \sin^{2x-1} 2\theta \, d\theta$$

where we have used the double-angle formula for the sine function. Next we make the variable change $\phi = 2\theta$, which yields

$$\frac{\Gamma(x)\Gamma(x)}{2\Gamma(2x)} = 2^{-2x} \int_0^{\pi} \sin^{2x-1}\phi \, d\phi$$

$$= 2^{1-2x} \int_0^{\pi/2} \sin^{2x-1}\phi \, d\phi$$

$$= \frac{2^{1-2x}\Gamma(\tfrac{1}{2})\Gamma(x)}{2\Gamma(x + \tfrac{1}{2})}$$

where the last step follows from (2.22). Simplification of this identity leads to (2.24).

An important special case of (2.24) occurs when $x = n$ ($n = 0, 1, 2, \ldots$), i.e.,

$$\Gamma\left(n + \frac{1}{2}\right) = \frac{(2n)!}{2^{2n}n!}\sqrt{\pi} \qquad n = 0, 1, 2, \ldots \qquad (2.25)$$

the verification of which is left to the exercises (see problem 39 in Exercises 2.2).

Example 4: Compute $\Gamma(\tfrac{3}{2})$.

Solution: The substitution of $n = 1$ in (2.25) yields

$$\Gamma\left(\frac{3}{2}\right) = \Gamma\left(1 + \frac{1}{2}\right) = \frac{2!\sqrt{\pi}}{2^2 \times 1!} = \frac{1}{2}\sqrt{\pi}$$

2.2.3 Weierstrass' infinite product

Although it was originally found by Schlömilch in 1844, thirty-two years before Weierstrass' famous work on entire functions, Weierstrass is usually credited with the infinite product definition of the gamma function

$$\frac{1}{\Gamma(x)} = xe^{\gamma x} \prod_{n=1}^{\infty} \left(1 + \frac{x}{n}\right)e^{-x/n} \qquad (2.26)$$

where γ is the *Euler–Mascheroni constant* defined by*

$$\gamma = \lim_{n \to \infty} \left(\sum_{k=1}^{n} \frac{1}{k} - \ln n \right) = 0.577215 \ldots \qquad (2.27)$$

We can derive this representation of $\Gamma(x)$ directly from (2.1) by first observing that

$$\frac{1}{\Gamma(x)} = \lim_{n \to \infty} \frac{x(x+1)(x+2) \cdots (x+n)}{n! n^x}$$

$$= x \lim_{n \to \infty} n^{-x} \left(\frac{x+1}{1} \cdot \frac{x+2}{2} \cdots \frac{x+n}{n} \right)$$

$$= x \lim_{n \to \infty} \exp[-(\ln n)x] \prod_{k=1}^{n} \left(1 + \frac{x}{k} \right) \qquad (2.28)$$

where we have written $n^{-x} = \exp[-(\ln n)x]$. Next, relying on repeated use of $e^{a+b} = e^a e^b$, we recognize the identity

$$\exp\left[\left(\sum_{k=1}^{n} \frac{1}{k} \right) x \right] = \prod_{k=1}^{n} e^{x/k}$$

Thus, if we multiply (2.28) by the left-hand side of this expression and divide by the right-hand side, we arrive at

$$\frac{1}{\Gamma(x)} = x \lim_{n \to \infty} \exp\left[\left(\sum_{k=1}^{n} \frac{1}{k} - \ln n \right) x \right] \cdot \lim_{n \to \infty} \prod_{k=1}^{n} \left(1 + \frac{x}{k} \right) e^{-x/k}$$

which reduces to (2.26).

An important identity involving the gamma function and sine function can now be derived by using (2.26). We begin with the product of gamma functions

$$\frac{1}{\Gamma(x)\Gamma(-x)} = x e^{\gamma x} \prod_{n=1}^{\infty} \left(1 + \frac{x}{n} \right) e^{-x/n} \cdot (-x) e^{-\gamma x} \prod_{n=1}^{\infty} \left(1 - \frac{x}{n} \right) e^{x/n}$$

or

$$\frac{1}{\Gamma(x)\Gamma(-x)} = -x^2 \prod_{n=1}^{\infty} \left(1 - \frac{x^2}{n^2} \right) \qquad (2.29)$$

where we assume that x is nonintegral. Recalling Eq. (1.92) in Sec. 1.7.2, which gives the infinite product definition of the sine function,

---

* The constant γ is commonly called (simply) *Euler's constant*.

we have

$$\prod_{n=1}^{\infty}\left(1-\frac{x^2}{n^2}\right)=\frac{\sin \pi x}{\pi x} \tag{2.30}$$

Comparison of (2.29) and (2.30) reveals that

$$\Gamma(x)\Gamma(-x)=-\frac{\pi}{x \sin \pi x} \qquad (x \text{ nonintegral}) \tag{2.31}$$

Also by writing the recurrence formula (2.3) in the form

$$-x\Gamma(-x)=\Gamma(1-x)$$

we deduce the identity

$$\Gamma(x)\Gamma(1-x)=\frac{\pi}{\sin \pi x} \qquad (x \text{ nonintegral}) \tag{2.32}$$

Example 5: Evaluate the integral $\int_0^{\pi/2} \tan^{1/2}\theta\, d\theta$.

Solution: Making use of (2.22) and (2.32), we get

$$\int_0^{\pi/2} \tan^{1/2}\theta\, d\theta = \int_0^{\pi/2} \sin^{1/2}\theta \cos^{-1/2}\theta\, d\theta$$

$$= \frac{\Gamma(^3/_4)\Gamma(^1/_4)}{2\Gamma(1)}$$

$$= \frac{1}{2}\frac{\pi}{\sin (\pi/4)}$$

$$= \frac{\pi}{\sqrt{2}}$$

Remark: An *entire function* is one that is analytic for all finite values of its argument. Weierstrass was the first to show that any entire function (under appropriate restrictions) with an infinite number of zeros, such as $\sin x$ and $\cos x$, is essentially determined by its zeros. This result led to the infinite product representations of such functions and, in particular, to the infinite product representation of the gamma function.

A list of some of the most important properties of the gamma function can be found in the Appendix for easy reference.

Exercises 2.2

1. Use Eq. (2.1) directly to evaluate

    (*a*) $\Gamma(2)$ (*b*) $\Gamma(3)$

In problems 2 to 7, give numerical values for the expressions.

2. $\dfrac{\Gamma(6)}{\Gamma(3)}$ 5. $\Gamma(-\tfrac{1}{2})$

3. $\dfrac{\Gamma(7)}{\Gamma(4)\Gamma(3)}$ 6. $\dfrac{\Gamma(-\tfrac{5}{2})}{\Gamma(\tfrac{1}{2})}$

4. $\Gamma(\tfrac{7}{2})$ 7. $\dfrac{\Gamma(\tfrac{8}{3})}{\Gamma(\tfrac{2}{3})}$

In problems 8 to 14, verify the given identity.

8. $\Gamma(a+n)=a(a+1)(a+2)\cdots(a+n-1)\Gamma(a),\ n=1,2,3,\ldots$

9. $\dfrac{\Gamma(n-a)}{\Gamma(-a)}=(-1)^n a(a-1)(a-2)\cdots(a-n+1),\ n=1,2,3,\ldots$

10. $\dfrac{\Gamma(a)}{\Gamma(a-n)}=(a-1)(a-2)\cdots(a-n),\ n=1,2,3,\ldots$

11. $\dfrac{\Gamma(k-n)}{\Gamma(-n)}=\begin{cases}\dfrac{(-1)^k n!}{(n-k)!} & 0\le k\le n\ (k,n\ \text{nonnegative integers})\\[2mm] 0 & k>n\end{cases}$

    Hint: See problem 9.

12. $\dbinom{a}{n}=\dfrac{\Gamma(a+1)}{n!\,\Gamma(a-n+1)},\ n=0,1,2,\ldots$

    Hint: See problem 10.

13. $\dbinom{-\tfrac{1}{2}}{n}=\dfrac{(-1)^n(2n)!}{2^{2n}(n!)^2},\ n=0,1,2,\ldots$

14. $\dbinom{-2k-1}{m}=(-1)^m\dfrac{(m+2k)!}{(2k)!\,m!},\ k,m=0,1,2,\ldots$

15. In problems in electromagnetic theory it is quite common to come across products like

$$2\times4\times6\times\cdots\times 2n\equiv(2n)!!$$

    and $1\times3\times5\times\cdots\times(2n+1)\equiv(2n+1)!!$

Use these definitions of the !! notation to show that

(a) $(2n)!! = 2^n n!$

(c) $(-2n-1)!! = \dfrac{(-1)^n 2^n n!}{(2n)!}$

(b) $(2n+1)!! = \dfrac{(2n+1)!}{2^n n!}$

(d) $(-1)!! = 1$

Hint: See problem 10 for (c) and (d).

16. Prove that $\displaystyle\int_0^\infty e^{-t} t^{x-1}\, dt$ converges uniformly in $1 \le x \le 2$.

In problems 17 to 20, verify the given integral representation.

17. $\Gamma(x) = s^x \displaystyle\int_0^\infty e^{-st} t^{x-1}\, dt,\ x,\ s > 0$

18. $\Gamma(x) = \displaystyle\int_{-\infty}^\infty \exp(xt - e^t)\, dt,\ x > 0$
   Hint: Let $u = e^t$.

19. $\Gamma(x) = \displaystyle\int_1^\infty e^{-t} t^{x-1}\, dt + \sum_{n=0}^\infty \dfrac{(-1)^n}{n!(x+n)},\ x > 0$

20. $\Gamma(x) = (\ln b)^x \displaystyle\int_0^\infty t^{x-1} b^{-t}\, dt,\ x > 0,\ b > 1$
   Hint: Let $u = t \ln b$.

In problems 21 to 29, use properties of the gamma function to obtain the result.

21. $\displaystyle\int_a^\infty e^{2ax - x^2}\, dx = \tfrac{1}{2}\sqrt{\pi}\, e^{a^2}$
   Hint: $2ax - x^2 = -(x-a)^2 + a^2$.

22. $\displaystyle\int_0^\infty e^{-2x} x^6\, dx = {}^{45}\!/_8$

23. $\displaystyle\int_0^\infty \sqrt{x}\, e^{-x^3}\, dx = \dfrac{\sqrt{\pi}}{3}$

24. $\displaystyle\int_0^1 \dfrac{du}{\sqrt{-\ln u}} = \sqrt{\pi}$

25. $\displaystyle\int_0^1 x^k (\ln x)^n\, dx = \dfrac{(-1)^n n!}{(k+1)^{n+1}},\ k > -1,\ n = 0, 1, 2, \ldots$

26. $\displaystyle\int_0^{\pi/2} \cos^6 \theta\, d\theta = \dfrac{5\pi}{32}$

27. $\displaystyle\int_0^{\pi/2} \sin^3 \theta \cos^2 \theta \, d\theta = {}^2\!/_{15}$

28. $\displaystyle\int_0^{\pi} \cos^4 x \, dx = \frac{3\pi}{8}$

29. $\displaystyle\int_0^{\pi/2} \sin^{2n+1} \theta \, d\theta = \int_0^{\pi/2} \cos^{2n+1} \theta \, d\theta = \frac{2^{2n}(n!)^2}{(2n+1)!}, \; n = 0, 1, 2, \ldots$

In problems 30 to 35, evaluate the integral in terms of the gamma function and simplify when possible.

30. $\displaystyle\int_0^{\infty} \frac{e^{-st}}{\sqrt{t}} \, dt, s > 0$

33. $\displaystyle\int_0^1 t^{x-1}\left(\ln\frac{1}{t}\right)^{y-1} dt, x, y > 0$

31. $\displaystyle\int_0^{\infty} \frac{dx}{1 + x^4}$

34. $\displaystyle\int_0^{\pi/2} \cot^{1/2} \theta \, d\theta$

   Hint: Let $x^2 = \tan \theta$.

32. $\displaystyle\int_0^{\pi/2} \sqrt{\sin 2x} \, dx$

35. $\displaystyle\int_0^{\infty} e^{-st^p} t^{x-1} \, dt, p, s, x > 0$

36. Using the recurrence formula (2.3), deduce

   (a) $\Gamma(x) = \Gamma'(x+1) - x\Gamma'(x)$

   (b) $\Gamma(x) = \displaystyle\int_0^{\infty} e^{-t}(t - x)t^{x-1} \ln t \, dt, x > 0$

In problems 37 and 38, use the Euler formulas

$$\cos x = \frac{e^{ix} + e^{-ix}}{2} \qquad \sin x = \frac{e^{ix} - e^{-ix}}{2i}$$

and properties of the gamma function to derive the result. Assume that $b, x > 0$ and $-{}^1\!/_2\pi < a < {}^1\!/_2\pi$.

37. $\Gamma(x) \cos ax = b^x \displaystyle\int_0^{\infty} t^{x-1} e^{-bt \cos a} \cos (bt \sin a) \, dt$

38. $\Gamma(x) \sin ax = b^x \displaystyle\int_0^{\infty} t^{x-1} e^{-bt \cos a} \sin (bt \sin a) \, dt$

39. Based on the Legendre duplication formula, show that (for $n = 0, 1, 2, \ldots$)

   (a) $\Gamma\left(n + \dfrac{1}{2}\right) = \dfrac{(2n)!\sqrt{\pi}}{2^{2n}n!}$

   (b) $\Gamma\left(\dfrac{1}{2} - n\right) = \dfrac{(-1)^n 2^{2n-1}(n-1)!\sqrt{\pi}}{(2n-1)!}$

   (c) $\Gamma({}^1\!/_2 + n)\Gamma({}^1\!/_2 - n) = (-1)^n \pi$

40. Show that

$$\Gamma(3x) = \frac{1}{2\pi} 3^{2x-1/2} \Gamma(x)\Gamma\left(x + \frac{1}{3}\right)\Gamma\left(x + \frac{2}{3}\right)$$

41. Show that

$$|\Gamma'(x)|^2 \le \Gamma(x)\Gamma''(x) \qquad x > 0$$

42. Show that
   (a) $\Gamma(1+x)\Gamma(1-x) = \pi x \csc \pi x$ (x nonintegral)
   (b) $\Gamma(\frac{1}{2}+x)\Gamma(\frac{1}{2}-x) = \pi \sec \pi x$, $x \ne n + \frac{1}{2}$, $n = 0, 1, 2, \ldots$

43. Derive Euler's infinite product representation

$$\frac{1}{\Gamma(x)} = x \prod_{n=1}^{\infty} \frac{1 + x/n}{(1 + 1/n)^x}$$

44. Derive the recurrence relation $\Gamma(x + 1) = x\Gamma(x)$ by use of the (a) integral definition (2.5) and (b) Weierstrass' infinite product (2.26).

45. Use $\Gamma(x + 1) = x\Gamma(x)$ and the result of Example 3.
   (a) Deduce that, for fixed a,

$$\frac{\Gamma(x + a)}{x^a \Gamma(x)} \sim \left(1 + \frac{a}{x}\right)^x \left(1 + \frac{a}{x}\right)^{a-1/2} e^{-a}, x \to \infty$$

   (b) Using the relation (from calculus)

$$\left(1 + \frac{a}{x}\right)^x \sim e^a \qquad x \to \infty$$

   show that

$$\frac{\Gamma(x + a)}{x^a \Gamma(x)} \sim \sum_{n=0}^{\infty} \binom{a - \frac{1}{2}}{n}\left(\frac{a}{x}\right)^n \qquad x \to \infty$$

2.3 Applications

The gamma function arises naturally in a variety of applications involving certain kinds of integrals that can be transformed to one of the integral representations for $\Gamma(x)$ by appropriate changes of variable, as well as in more novel applications like those involving fractional derivatives.

2.3.1 Miscellaneous problems

Many standard problems in calculus, such as finding areas, arc-length, volume, and so on, can lead to nonelementary integrals that are simple variations of the gamma function. Consider the following example.

Example 6: For $x > 0$, find the area between $y = 4x^{3/2}e^{-x^2/2}$ and its asymptote (see Fig. 2.2).

Solution: From the figure it is clear that the curve lies entirely above the positive x axis with the x axis acting as an asymptote as $x \to \infty$. Therefore, the area under the curve is given by the standard formula

$$A = \int_0^\infty y \, dx = 4 \int_0^\infty x^{3/2} e^{-x^2/2} \, dx$$

By making the change of variable $t = x^2/2$, we get

$$A = 2^{9/4} \int_0^\infty t^{1/4} e^{-t} \, dt = 2^{9/4} \Gamma(5/4)$$

Integrals that lead to the gamma function are also prominent in computing statistical quantities. For example, the *moments* of a random variable **x** are defined by the integral

$$E[\mathbf{x}^n] \equiv \langle \mathbf{x}^n \rangle = \int_{-\infty}^\infty x^n p_{\mathbf{x}}(x) \, dx \qquad n = 1, 2, 3, \ldots \qquad (2.33)$$

where $p_{\mathbf{x}}(x)$ is the probability density function for **x**. The special case

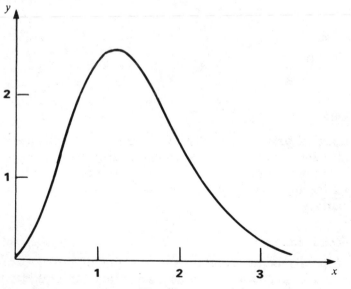

Figure 2.2 Graph of $y = 4x^{3/2}e^{-x^2/2}$.

corresponding to $n = 1$ is called the *mean value* while the particular combination of first and second moments

$$\sigma_{\mathbf{x}}^2 = \langle \mathbf{x}^2 \rangle - \langle \mathbf{x} \rangle^2 \qquad (2.34)$$

is called the *variance*.

A density function prominent in many applications is the **gamma density function**

$$p_{\mathbf{x}}(x) = \begin{cases} \dfrac{1}{\beta \Gamma(\alpha)} \left(\dfrac{x}{\beta} \right)^{\alpha - 1} e^{-x/\beta} & x > 0 \\[2ex] 0 & x < 0 \end{cases} \qquad (2.35)$$

where α and β are positive parameters of the distribution. Other important probability density functions are introduced in Exercises 2.3.

Example 7: Calculate the moments of the gamma distribution defined by (2.35).

Solution: From the definition, we find that

$$\langle \mathbf{x}^n \rangle = \frac{1}{\beta \Gamma(\alpha)} \int_0^\infty x^n \left(\frac{x}{\beta} \right)^{\alpha - 1} e^{-x/\beta} \, dx$$

$$= \frac{1}{\beta^\alpha \Gamma(\alpha)} \int_0^\infty x^{n + \alpha - 1} e^{-x/\beta} \, dx$$

By making the change of variable $t = x/\beta$, this integral becomes

$$\langle \mathbf{x}^n \rangle = \frac{\beta^n}{\Gamma(\alpha)} \int_0^\infty t^{n + \alpha - 1} e^{-t} \, dt = \frac{\beta^n \Gamma(n + \alpha)}{\Gamma(\alpha)} \qquad n = 1, 2, 3, \ldots$$

Observe that the first and second moments are, respectively, $\langle \mathbf{x} \rangle = \alpha\beta$ and $\langle \mathbf{x}^2 \rangle = \alpha(\alpha + 1)\beta^2$, leading to the variance

$$\sigma_{\mathbf{x}}^2 = \alpha(\alpha + 1)\beta^2 - (\alpha\beta)^2 = \alpha\beta^2$$

2.3.2 Fractional-order derivatives

Besides generalizing the notion of factorials, the gamma function can be used in a variety of situations to generalize discrete processes into the continuum. Such generalizations are not new, however; mathematicians over the years have concerned themselves with this concept. In particular, the question concerning derivatives of nonintegral order was first raised by Leibniz in 1695, many years before Euler introduced the gamma function.

The general procedure for developing fractional derivatives is too involved for our purposes.* However, we can illustrate the concept by first recalling the familiar derivative formula from calculus

$$D^n x^a = a(a-1)\cdots(a-n+1)x^{a-n} \quad a \geq 0 \quad (2.36)$$

where $D^n = d^n/dx^n$, $n = 1, 2, 3, \ldots$. In terms of the gamma function, we can rewrite (2.36) as (see problem 10 in Exercises 2.2)

$$D^n x^a = \frac{\Gamma(a+1)}{\Gamma(a-n+1)}x^{a-n}$$

The right-hand side of this expression is meaningful for any real number n for which $\Gamma(a-n+1)$ is defined. Hence, we assume that the same is true of the left-hand side and write

$$D^v x^a = \frac{\Gamma(a+1)}{\Gamma(a-v+1)}x^{a-v} \quad a \geq 0 \quad (2.37)$$

where v is not restricted to integer values. Equation (2.37) provides a simple method of computing *fractional-order derivatives* of polynomials.

Example 8. Compute $D^{1/2}x^2$.

Solution: Directly from (2.37), we obtain

$$D^{1/2}x^2 = \frac{\Gamma(3)}{\Gamma(5/2)}x^{3/2}$$

the simplification of which yields

$$D^{1/2}x^2 = \frac{8}{3\sqrt{\pi}}x^{3/2}$$

Generalization of the differentiation formula for $D^n x^{-a}$, which covers the case of negative exponents, is left to the exercises (see problem 12 in Exercises 2.3).

Exercises 2.3

1. Find the area enclosed by the curve $x^4 + y^4 = 1$.
2. Find the total arclength of the lemniscate $r^2 = a^2 \cos 2\theta$.

---

*For a deeper discussion of fractional derivatives, see L. Debnath, "Generalized Calculus and Its Applications," *Int. J. Math. Educ. Sci. Technol.*, **9**(4): 399–416, 1978.

3. Find the area inside the curve $x^{2/3} + y^{2/3} = 1$.

4. Find the volume in the first octant below the surface

$$x^{1/2} + y^{1/2} + z^{1/2} = 1$$

5. A particle of mass m starts from rest at $r = 1$ and moves along a radial line toward the origin $r = 0$ under the reciprocal force law $f = -k/r$, where k is a positive constant. The energy equation of the particle is given by

$$\frac{1}{2} m \left(\frac{dr}{dt}\right)^2 + k \ln r = 0$$

(a) Show that the time required for the particle to reach the origin is $\sqrt{m\pi/(2k)}$.

(b) If the particle starts from rest at $r = a$ $(a > 0)$, the energy equation becomes

$$\frac{1}{2} m \left(\frac{dr}{dt}\right)^2 + k \ln r = k \ln a$$

Again find the time required for the particle to reach the origin.

6. The *Rayleigh distribution* is defined by $p_x(x) = 0$, $x < 0$, and

$$p_x(x) = \frac{x}{b^2} e^{-x^2/(2b^2)} \qquad x > 0$$

(a) Calculate the moments $\langle x^n \rangle$, $n = 1, 2, 3, \ldots$.

(b) Find the variance σ_x^2.

7. The *normal distribution* with zero mean and variance σ^2 is defined by

$$p_x(x) = \frac{1}{\sigma\sqrt{2\pi}} e^{-x^2/(2\sigma^2)} \qquad -\infty < x < \infty$$

Show that (a) all odd-order moments are zero and (b) the even-order moments are given by

$$\langle x^{2n} \rangle = \frac{(2n)!}{2^n n!} \sigma^{2n} \qquad n = 1, 2, 3, \ldots$$

8. The output of a *half-wave rectifier* is given by the random variable $y = 0$, $x < 0$, $y = x$, $x > 0$, where x is a normal random variable whose density function is that in problem 7. Show that the variance of y is given by

$$\sigma_y^2 = \frac{1}{2}\left(1 - \frac{1}{\pi}\right)\sigma^2$$

9. Calculate the variance for the output of a *square-law device* described by $y = x^2$, where x is a normal random variable whose density function is that in problem 7.

10. Compute the fractional-order derivatives.

   (a) $D^{1/2}c$, where c is constant

   (b) $D^{1/2}(3x^2 - 7x + 4)$

   (c) $D^{3/2}x^2$

   (d) $D^{\nu}x^{\nu}$, where ν is not a positive integer

11. Show that

   (a) $D^{1/2}(D^{1/2}x^2) = Dx^2$

   (b) $D^{-1/2}(D^{1/2}x^2) = x^2$

   (c) $D^{\nu}(D^{\mu}x^a) = D^{\nu+\mu}x^a$

12. By generalizing the formula for $D^n x^{-a}$, $n = 1, 2, 3, \ldots$, show that

$$D^{\nu}x^{-a} = (-1)^{\nu}\frac{\Gamma(\nu + a)}{\Gamma(a)}x^{-(a+\nu)} \qquad a > 0$$

2.4 Beta Function

A useful function of two variables is the **beta function***

$$B(x, y) = \int_0^1 t^{x-1}(1-t)^{y-1}\, dt \qquad x > 0, \qquad y > 0 \qquad (2.38)$$

The utility of the beta function is often overshadowed by that of the gamma function, partly perhaps because it can be evaluated in terms of the gamma function. However, since it occurs so frequently in practice, a special designation for it is widely accepted.

If we make the change of variable $u = 1 - t$ in (2.38), we find

$$B(x, y) = \int_0^1 (1-u)^{x-1}u^{y-1}\, du$$

from which we deduce the *symmetry property*

$$B(x, y) = B(y, x) \qquad (2.39)$$

Another representation of the beta function results if we make the

---

* This is called the *eulerian integral of the first kind*.

variable change $t = u/(1 + u)$, leading to

$$B(x, y) = \int_0^\infty \frac{u^{x-1}}{(1+u)^{x+y}} du \qquad x > 0, \qquad y > 0 \qquad (2.40)$$

Finally, to show how the beta function is related to the gamma function, we set $t = \cos^2 \theta$ in (2.38) to find

$$B(x, y) = 2\int_0^{\pi/2} \cos^{2x-1} \theta \sin^{2y-1} \theta \, d\theta$$

and hence from (2.22) we obtain the relation

$$B(x, y) = \frac{\Gamma(x)\Gamma(y)}{\Gamma(x+y)} \qquad x > 0, \qquad y > 0 \qquad (2.41)$$

Example 9: Evaluate the integral $I = \int_0^\infty x^{-1/2}(1+x)^{-2} \, dx$.

Solution: By comparison with (2.40), we recognize

$$I = B(\tfrac{1}{2}, \tfrac{3}{2})$$

$$= \frac{\Gamma(\tfrac{1}{2})\Gamma(\tfrac{3}{2})}{\Gamma(2)}$$

Hence, we deduce that

$$\int_0^\infty x^{-1/2}(1+x)^{-2} \, dx = \frac{\pi}{2}$$

Example 10: Show that

$$\int_0^\infty \frac{\cos x}{x^p} \, dx = \frac{\pi}{2\Gamma(p)\cos(p\pi/2)} \qquad 0 < p < 1$$

Solution: Making the observation (problem 17 in Exercises 2.2)

$$\frac{1}{x^p} = \frac{1}{\Gamma(p)}\int_0^\infty e^{-xt}t^{p-1} \, dt$$

it follows that

$$\int_0^\infty \frac{\cos x}{x^p} \, dx = \frac{1}{\Gamma(p)}\int_0^\infty \cos x \int_0^\infty e^{-xt}t^{p-1} \, dt \, dx$$

$$= \frac{1}{\Gamma(p)}\int_0^\infty t^{p-1}\int_0^\infty e^{-xt}\cos x \, dx \, dt$$

$$= \frac{1}{\Gamma(p)}\int_0^\infty \frac{t^p}{1+t^2} \, dt$$

where we have reversed the order of integration. If we now let $u = t^2$, then

$$\int_0^\infty \frac{\cos x}{x^p}\,dx = \frac{1}{2\Gamma(p)}\int_0^\infty \frac{u^{(p-1)/2}}{1+u}\,du$$

$$= \frac{1}{2\Gamma(p)} B\left(\frac{1+p}{2}, \frac{1-p}{2}\right)$$

However (see problem 10 in Exercises 2.4),

$$B\left(\frac{1+p}{2}, \frac{1-p}{2}\right) = \pi \sec\frac{p\pi}{2}$$

and thus we have our result.

Example 10 illustrates one of the basic approaches we use in the evaluation of nonelementary integrals. That is, we replace part of (or all) the integrand by its series representation or integral representation and then interchange the order in which the operations are carried out.

Exercises 2.4

In problems 1 to 4, evaluate the beta function.

1. $B(^2/_3, \,^1/_3)$ 3. $B(^1/_2, 1)$

2. $B(^3/_4, \,^1/_4)$ 4. $B(x, 1-x), 0 < x < 1$

In problems 5 to 10, verify the identity.

5. $B(x+1, y) + B(x, y+1) = B(x, y),\ x, y > 0$

6. $B(x, y+1) = \frac{y}{x}B(x+1, y) = \frac{y}{x+y}B(x, y),\ x, y > 0$

7. $B(x, x) = 2^{1-2x}B(x, \,^1/_2),\ x > 0$

8. $B(x, y)B(x+y, z)B(x+y+z, w) = \frac{\Gamma(x)\Gamma(y)\Gamma(z)\Gamma(w)}{\Gamma(x+y+z+w)}$,

$x, y, z, w > 0$

9. $B(n, n)B(n + \,^1/_2, n + \,^1/_2) = \pi 2^{1-4n}n^{-1},\ n = 1, 2, 3, \ldots$

10. $B\left(\frac{1+p}{2}, \frac{1-p}{2}\right) = \pi \sec\frac{p\pi}{2},\ 0 < p < 1$

In problems 11 to 18, use properties of the beta and gamma functions to evaluate the integral.

11. $\int_0^1 \sqrt{x(1-x)}\, dx$

12. $\int_0^1 x^4(1-x^2)^{-1/2}\, dx$

13. $\int_0^\infty \dfrac{x}{(1+x^3)^2}\, dx$
    Hint: Set $t = x^3/(1+x^3)$.

14. $\int_{-1}^1 \left(\dfrac{1+x}{1-x}\right)^{1/2} dx$
    Hint: Set $x = 2t - 1$.

15. $\int_a^b (b-x)^{m-1}(x-a)^{n-1}\, dx$, where m, n are positive integers

16. $\int_0^2 x^2(2-x)^{-1/2}\, dx$

17. $\int_0^a x^4\sqrt{a^2-x^2}\, dx$

18. $\int_0^2 x\sqrt[3]{8-x^3}\, dx$

In problems 19 to 30, verify the integral formula.

19. $\int_0^\infty \dfrac{x^{p-1}}{1+x}\, dx = \pi\csc p\pi,\ 0<p<1$

20. $\int_0^\infty \dfrac{\sin x}{x^p}\, dx = \dfrac{\pi}{2\Gamma(p)\sin(p\pi/2)},\ 0<p<1$

21. $\int_0^\infty \sin x^2\, dx = \dfrac{1}{2}\sqrt{\dfrac{\pi}{2}}$
    Hint: Use problem 20.

22. $\int_0^\infty \cos x^2\, dx = \dfrac{1}{2}\sqrt{\dfrac{\pi}{2}}$

23. $\int_0^{\pi/2} \tan^p x\, dx = \int_0^{\pi/2} \cot^p x\, dx = \dfrac{\pi}{2\cos(p\pi/2)},\ 0<p<1$

24. $\displaystyle\int_0^\infty \frac{x^{p-1}\ln x}{1+x}\,dx = -\pi^2 \csc p\pi \cot p\pi,\; 0<p<1$

25. $\displaystyle\int_0^\infty \frac{x^{p-1}}{1+x^a}\,dx = \frac{\pi}{a\sin(p\pi/a)},\; 0<p<a$

26. $\displaystyle\int_0^\infty e^{-st}(1-e^{-t})^n\,dt = \frac{n!\,\Gamma(s)}{\Gamma(s+n+1)},$ where $s>0,\; n=0,1,2,\ldots$

27. $\displaystyle\int_{-\infty}^\infty \frac{e^{2x}}{ae^{3x}+b}\,dx = \frac{2\pi}{3\sqrt{3}}a^{-2/3}b^{-1/3},$ where $a,b>0$

28. $\displaystyle\int_{-\infty}^\infty \frac{e^{2x}}{(e^{3x}+1)^2}\,dx = \frac{2\pi}{9\sqrt{3}}$

    Hint: Differentiate with respect to b in problem 27.

29. $\displaystyle\int_0^1 \frac{t^{x-1}+t^{y-1}}{(t+1)^{x+y}}\,dt = B(x,y),$ where $x,y>0$

30. $\displaystyle\int_0^1 \frac{t^{x-1}(1-t)^{y-1}}{(t+p)^{x+y}}\,dt = \frac{B(x,y)}{p^x(1+p)^{x+y}},$ where $x,y,p>0$

31. Using the notation of problem 15 in Exercises 2.2, show that

    (a) $\displaystyle\int_{-1}^1 (1-x^2)^{1/2}x^{2n}\,dx = \begin{cases} \dfrac{\pi}{2} & n=0 \\[2mm] \pi\dfrac{(2n-1)!!}{(2n+2)!!} & n=1,2,3,\ldots \end{cases}$

    (b) $\displaystyle\int_{-1}^1 (1-x^2)^{-1/2}x^{2n}\,dx = \begin{cases} \pi & n=0 \\[2mm] \pi\dfrac{(2n-1)!!}{(2n)!!} & n=1,2,3,\ldots \end{cases}$

32. Show that
$$\int_{-1}^1 (1-x^2)^n\,dx = 2^{2n+1}\frac{(n!)^2}{(2n+1)!} \qquad n=0,1,2,\ldots$$

33. The *incomplete beta function* is defined by
$$B_x(p,q) = \int_0^x t^{p-1}(1-t)^{q-1}\,dt \qquad 0\le x\le 1, \qquad p,q>0$$

    (a) Show that
$$B_x(p,q) = x^p\Gamma(q)\sum_{n=0}^\infty \frac{(-1)^n x^n}{\Gamma(q-n)(p+n)n!} \qquad 0\le x\le 1$$

    (b) From (a), deduce that
$$\sum_{n=0}^\infty \frac{(-1)^n}{\Gamma(q-n)(p+n)n!} = \frac{\Gamma(p)}{\Gamma(p+q)}$$

2.5 Incomplete Gamma Function

Generalizing the Euler integral (2.5), we introduce the related function

$$\gamma(a, x) = \int_0^x e^{-t} t^{a-1}\, dt \qquad a > 0 \tag{2.42}$$

called the **incomplete gamma function**. This function most commonly arises in probability theory, particularly those applications involving the chi-square distribution. It is customary to also introduce the companion function

$$\Gamma(a, x) = \int_x^\infty e^{-t} t^{a-1}\, dt \qquad a > 0 \tag{2.43}$$

which is known as the **complementary incomplete gamma function.** Thus, it follows that

$$\gamma(a, x) + \Gamma(a, x) = \Gamma(a) \tag{2.44}$$

Because of the close relationship between these two functions, the choice of using $\gamma(a, x)$ or $\Gamma(a, x)$ in practice is simply a matter of convenience.

By substituting the series respresentation for e^{-t} in (2.42), we get

$$\gamma(a, x) = \int_0^x \left[\sum_{n=0}^\infty \frac{(-1)^n}{n!} t^{n+a-1} \right] dt$$

and then, performing termwise integration, we are led to the series representation

$$\gamma(a, x) = x^a \sum_{n=0}^\infty \frac{(-1)^n x^n}{n!(n+a)} \qquad a > 0 \tag{2.45}$$

It immediately follows from (2.44) that

$$\Gamma(a, x) = \Gamma(a) - x^a \sum_{n=0}^\infty \frac{(-1)^n x^n}{n!(n+a)} \qquad a > 0 \tag{2.46}$$

2.5.1 Asymptotic series

The integration of Eq. (2.43) by parts gives us

$$\Gamma(a, x) = \int_x^\infty e^{-t} t^{a-1} \, dt$$

$$= -e^{-t} t^{a-1} \Big|_x^\infty + (a-1) \int_x^\infty e^{-t} t^{a-2} \, dt$$

$$= e^{-x} x^{a-1} + (a-1) \int_x^\infty e^{-t} t^{a-2} \, dt \qquad (2.47)$$

while continued integration by parts yields

$$\Gamma(a, x) = e^{-x} x^{a-1} + (a-1) e^{-x} x^{a-2} + (a-1)(a-2) \int_x^\infty e^{-t} t^{a-3} \, dt$$

and so on. Thus we generate the *asymptotic series**

$$\Gamma(a, x) \sim e^{-x} x^{a-1} \left[1 + \frac{a-1}{x} + \frac{(a-1)(a-2)}{x^2} + \cdots \right] \qquad x \to \infty$$
$$(2.48)$$

which can be expressed as

$$\Gamma(a, x) \sim \Gamma(a) x^{a-1} e^{-x} \sum_{k=0}^\infty \frac{1}{\Gamma(a-k) x^k} \qquad a > 0, \qquad x \to \infty \quad (2.49)$$

If we set $a = n + 1$ $(n = 0, 1, 2, \ldots)$ in (2.49), we find that

$$\Gamma(n + 1, x) = n! x^n e^{-x} \sum_{k=0}^n \frac{x^{-k}}{(n-k)!} \qquad (2.50)$$

where the series truncates because $1/\Gamma(n + 1 - k) = 0$ for $k > n$ (Theorem 2.1). The change of variable $j = n - k$ further simplifies (2.50) to

$$\Gamma(n + 1, x) = n! e^{-x} \sum_{j=0}^n \frac{x^j}{j!} \qquad (2.51)$$

or
$$\Gamma(n + 1, x) = n! e^{-x} e_n(x) \qquad n = 0, 1, 2, \ldots \qquad (2.52)$$

---

*The asymptotic series (2.48) or (2.49) diverges for all finite x.

where $e_n(x)$ denotes the first $n+1$ terms of the Maclaurin series for e^x.[*] By a similar analysis, it can be shown that

$$\gamma(n+1,x) = n![1 - e^{-x}e_n(x)] \qquad n = 0,1,2,\ldots \qquad (2.53)$$

Remark: It is interesting to note that both (2.52) and (2.53) are valid representations for all $x > 0$, while the asymptotic series (2.49) [from which (2.52) and (2.53) were derived] diverges for all x.

Exercises 2.5

1 Show that
   (a) $\gamma(a+1,x) = a\gamma(a,x) - x^a e^{-x}$
   (b) $\Gamma(a+1,x) = a\Gamma(a,x) + x^a e^{-x}$

2. Show that

   (a) $\dfrac{d}{dx}[x^{-a}\Gamma(a,x)] = -x^{-a-1}\Gamma(a+1,x)$

   (b) $\dfrac{d^m}{dx^m}[x^{-a}\Gamma(a,x)] = (-1)^m x^{-a-m}\Gamma(a+m,x),\ m = 1,2,3,\ldots$

3. Show that
   $$\Gamma(a)\Gamma(a+n,x) - \Gamma(a+n)\Gamma(a,x) = \Gamma(a+n)\gamma(a,x)$$
   $$- \Gamma(a)\gamma(a+n,x)$$

4. Verify the integral formula
   $$\Gamma(a,xy) = y^a e^{-xy}\int_0^\infty e^{-yt}(t+x)^{a-1}\,dt \qquad x,y>0, \qquad a>1$$

5. Verify the integral representation
   $$\gamma(a,x) = x^{a/2}\int_0^\infty e^{-t}t^{a/2-1}J_a(2\sqrt{xt})\,dt \qquad a>0$$

   where $J_a(z)$ is the *Bessel function* defined by (see Chap. 6)
   $$J_a(z) = \sum_{n=0}^\infty \frac{(-1)^n(z/2)^{2n+a}}{n!\Gamma(n+a+1)}$$

6. Formally derive the asymptotic series (2.49) by setting $y=1$ in the result of problem 4 and using the binomial series
   $$\left(1+\frac{t}{x}\right)^{a-1} = \sum_{k=0}^\infty \binom{a-1}{k}\left(\frac{t}{x}\right)^k \qquad x>t$$

---

[*] For additional properties of the function $e_n(x)$, see problems 11 and 12 in Exercises 4.2.

2.6 Digamma and Polygamma Functions

Closely associated with the derivative of the gamma function is the **logarithmic derivative function**, or **digamma function**, defined by*

$$\psi(x) \equiv \frac{d}{dx} \ln \Gamma(x) = \frac{\Gamma'(x)}{\Gamma(x)} \qquad x \neq 0, -1, -2, \ldots \qquad (2.54)$$

To find an infinite series representation of $\psi(x)$, we first take the natural logarithm of both sides of the Weierstrass infinite product

$$\frac{1}{\Gamma(x)} = xe^{\gamma x} \prod_{n=1}^{\infty} \left(1 + \frac{x}{n}\right) e^{-x/n}$$

which yields

$$-\ln \Gamma(x) = \ln x + \gamma x + \sum_{n=1}^{\infty} \left[\ln\left(1 + \frac{x}{n}\right) - \frac{x}{n} \right] \qquad x > 0 \qquad (2.55)$$

Then, negating both sides of (2.55) and differentiating the result with respect to x, we find

$$\psi(x) \equiv \frac{d}{dx} \ln \Gamma(x) = -\frac{1}{x} - \gamma + \sum_{n=1}^{\infty} \left(\frac{1}{n} - \frac{1}{x+n}\right)$$

which we choose to write as

$$\psi(x) = -\gamma + \sum_{n=0}^{\infty} \left(\frac{1}{n+1} - \frac{1}{n+x}\right) \qquad x > 0 \qquad (2.56)$$

The restriction $x > 0$ follows from Eq. (2.52).†
  Noteworthy here is the special value

$$\psi(1) = \frac{\Gamma'(1)}{\Gamma(1)} = -\gamma \qquad (2.57)$$

---

* The function $\psi(x)$ is also commonly called the *psi function*.

  † Actually, (2.56) is valid for all x except $x = 0, -1, -2, \ldots$, although we will not prove it.

and by recalling Eq. (2.11), we see that

$$\Gamma'(1) = -\gamma = \int_0^\infty e^{-t} \ln t\, dt \qquad (2.58)$$

Based upon Eq. (2.54), it is clear that the digamma function has the same domain of definition as the gamma function. It has characteristics quite distinct from those of the gamma function, however, since it is related to the derivative of $\Gamma(x)$. For example, unlike the gamma function, the function $\psi(x)$ crosses the x axis. In fact, it has infinitely many zeros, corresponding to the extrema of $\Gamma(x)$, that is, points where $\Gamma'(x) = 0$. For positive x the only extremum of the gamma function occurs at $x_0 = 1.4616\ldots$. Because x_0 corresponds to a minimum of $\Gamma(x)$, it follows that $\Gamma'(x)$ and $\psi(x)$ are both negative on the interval $0 < x < x_0$ and both positive for $x > x_0$. For large values of x, the digamma function is approximately equal to $\ln x$ [see Eq. (2.77)]. The general characteristics of $\psi(x)$ for both positive and negative values of x are illustrated in Fig. 2.3.

The function $\psi(x)$ satisfies relations somewhat analogous to those for the gamma function, which can be derived by taking logarithmic

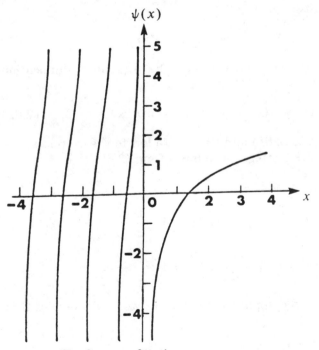

Figure 2.3 The digamma function.

derivatives of the latter. As an example, let us consider the recurrence formula

$$\Gamma(x+1) = x\Gamma(x) \tag{2.59}$$

By taking the natural logarithm we have

$$\ln \Gamma(x+1) = \ln x + \ln \Gamma(x)$$

which upon differentiation yields

$$\frac{d}{dx}\ln\Gamma(x+1) = \frac{1}{x} + \frac{d}{dx}\ln\Gamma(x)$$

Thus,

$$\psi(x+1) = \psi(x) + \frac{1}{x} \tag{2.60}$$

Also the logarithmic derivative of

$$\Gamma(x)\Gamma(1-x) = \pi \csc \pi x$$

results in the identity

$$\psi(1-x) - \psi(x) = \pi \cot \pi x \tag{2.61}$$

Finally, the logarithmic derivative of the Legendre duplication formula (2.24) leads to

$$\psi(x) + \psi(x + \tfrac{1}{2}) + 2\ln 2 = 2\psi(2x) \tag{2.62}$$

The details of deriving (2.61) and (2.62) are left to the exercises.
  If n denotes a positive integer, it follows from (2.60) that

$$\psi(2) = \psi(1) + 1$$

$$\psi(3) = \psi(2) + \frac{1}{2} = \psi(1) + 1 + \frac{1}{2}$$

$$\psi(4) = \psi(3) + \frac{1}{3} = \psi(1) + 1 + \frac{1}{2} + \frac{1}{3}$$

and so forth. By repeated application of (2.60), we finally deduce that

$$\psi(n+1) = \psi(1) + 1 + \frac{1}{2} + \frac{1}{3} + \cdots + \frac{1}{n}$$

Since $\psi(1) = -\gamma$, we can write this as

$$\psi(n+1) = -\gamma + \sum_{k=1}^{n} \frac{1}{k} \qquad n = 1, 2, 3, \ldots \qquad (2.63)$$

Example 11: Use properties of the digamma function to sum the series

$$\sum_{n=2}^{\infty} \frac{1}{n^2 - 1}$$

Solution: By use of partial fractions,

$$\frac{1}{n^2 - 1} = \frac{1}{2}\left(\frac{1}{n-1} - \frac{1}{n+1}\right)$$

and therefore

$$\sum_{n=2}^{\infty} \frac{1}{n^2 - 1} = \frac{1}{2} \sum_{n=2}^{\infty} \left(\frac{1}{n-1} - \frac{1}{n+1}\right)$$

$$= \frac{1}{2} \sum_{k=0}^{\infty} \left(\frac{1}{k+1} - \frac{1}{k+3}\right)$$

where we have introduced the change of index $n - 2 = k$. Now, from Eqs. (2.56) and (2.63), it follows that

$$\sum_{n=2}^{\infty} \frac{1}{n^2 - 1} = \frac{1}{2}[\psi(3) + \gamma]$$

$$= \frac{1}{2}(-\gamma + 1 + \frac{1}{2} + \gamma)$$

or

$$\sum_{n=2}^{\infty} \frac{1}{n^2 - 1} = \frac{3}{4}$$

2.6.1 Integral representations

Like the gamma function, the digamma function also has various integral representations. Let us start with the known relation

$$\Gamma'(x) = \int_0^{\infty} e^{-t} t^{x-1} \ln t \, dt \qquad x > 0 \qquad (2.64)$$

and replace $\ln t$ with the *Frullani integral representation* (see Example 17 in Sec. 1.5.3)

$$\ln t = \int_0^\infty \frac{e^{-u} - e^{-ut}}{u} \, du \qquad t > 0 \tag{2.65}$$

Hence,
$$\Gamma'(x) = \int_0^\infty e^{-t} t^{x-1} \left(\int_0^\infty \frac{e^{-u} - e^{-ut}}{u} \, du \right) dt$$

$$= \int_0^\infty \int_0^\infty e^{-t} t^{x-1} \left(\frac{e^{-u} - e^{-ut}}{u} \right) dt \, du$$

where we have reversed the order of integration. Next, splitting the inside integral into a sum of integrals, and recalling the integral relation (see problem 17 in Exercises 2.2)

$$\int_0^\infty e^{-t(u+1)} t^{x-1} \, dt = \frac{\Gamma(x)}{(u+1)^x} \qquad x > 0 \tag{2.66}$$

we see that

$$\Gamma'(x) = \int_0^\infty \frac{1}{u} \left(e^{-u} \int_0^\infty e^{-t} t^{x-1} \, dt - \int_0^\infty e^{-t(u+1)} t^{x-1} \, dt \right) du$$

$$= \int_0^\infty \frac{1}{u} \left[e^{-u} \Gamma(x) - \frac{\Gamma(x)}{(u+1)^x} \right] du$$

Finally, division of this last result by $\Gamma(x)$ leads to the desired integral relation

$$\psi(x) = \int_0^\infty \frac{1}{u} [e^{-u} - (u+1)^{-x}] \, du \qquad x > 0 \tag{2.67}$$

Another integral representation can be derived by first writing (2.67) as

$$\psi(x) = \int_0^\infty \frac{e^{-u}}{u} \, du - \int_0^\infty \frac{(u+1)^{-x}}{u} \, du$$

and then making the substitution $u + 1 = e^t$ in the second integral to get

$$\psi(x) = \int_0^\infty \frac{e^{-u}}{u} \, du - \int_0^\infty \frac{e^{-t(x-1)}}{e^t - 1} \, dt$$

Combining the last two integrals once again as a single integral yields

$$\psi(x) = \int_0^\infty \left(\frac{e^{-t}}{t} - \frac{e^{-t(x-1)}}{e^t - 1} \right) dt \qquad x > 0 \qquad (2.68)$$

Remark: Although (2.67) is a convergent integral, it is not technically correct to write it as the difference of two integrals, since each integral by itself is divergent. We are simply using a mathematical gimmick here in order to formally derive (2.68), which happens also to be a convergent integral.

2.6.2 Asymptotic series

Our next task is to derive asymptotic series for both the digamma and gamma functions. We begin with the integral representation

$$\psi(x + 1) = \int_0^\infty \left(\frac{e^{-t}}{t} - \frac{e^{-xt}}{e^t - 1} \right) dt \qquad (2.69)$$

which comes from (2.68) with x replaced by $x + 1$. We then rewrite (2.69) in the form

$$\psi(x + 1) = \int_0^\infty \left(\frac{e^{-t}}{t} - \frac{e^{-xt}}{t} + \frac{e^{-xt}}{t} - \frac{e^{-xt}}{e^t - 1} \right) dt$$

$$= \int_0^\infty \frac{e^{-t} - e^{-xt}}{t} dt + \int_0^\infty \left(\frac{1}{t} - \frac{1}{e^t - 1} \right) e^{-xt} dt$$

$$= \ln x + I \qquad (2.70)$$

where we recognize the Frullani integral (2.65) and define

$$I = \int_0^\infty \left(\frac{1}{t} - \frac{1}{e^t - 1} \right) e^{-xt} dt \qquad x > 0 \qquad (2.71)$$

To perform the integration in (2.71), we need to represent the function $(e^t - 1)^{-1}$ in a series and to integrate termwise. Since this function is not defined at $t = 0$, it does not have a Maclaurin series about this point. However, the related function $t(e^t - 1)^{-1}$ and all its derivatives are well defined at $t = 0$, so we write

$$\frac{t}{e^t - 1} = \sum_{n=0}^\infty B_n \frac{t^n}{n!} \qquad |t| < \infty \qquad (2.72)$$

where

$$B_n = \frac{d^n}{dt^n}\left[t(e^t-1)^{-1}\right]\big|_{t=0} \qquad n = 0, 1, 2, \ldots \qquad (2.73)$$

The constants B_n are called the *Bernoulli numbers*.* The first few are found to be

$$B_0 = 1$$
$$B_1 = -\tfrac{1}{2}$$
$$B_2 = \tfrac{1}{6} \qquad\qquad (2.74)$$
$$B_3 = 0$$
$$B_4 = -\tfrac{1}{30}$$
$$\vdots$$

All Bernoulli numbers with odd index, except B_1, are zero. To show this, we simply replace t by $-t$ in (2.72) and then subtract the result from (2.72) itself, finding

$$\frac{t}{e^t-1} - \frac{-t}{e^{-t}-1} = -t = \sum_{n=0}^{\infty} [1-(-1)^n]B_n \frac{t^n}{n!}$$

and by equating coefficients of like powers of t, we see that $B_1 = -\tfrac{1}{2}$ and $B_3 = B_5 = B_7 = \cdots = 0$.

If we divide both sides of (2.72) by t, we get

$$\frac{1}{e^t-1} = \sum_{n=0}^{\infty} B_n \frac{t^{n-1}}{n!}$$

$$= \frac{1}{t} - \frac{1}{2} + \sum_{n=2}^{\infty} B_n \frac{t^{n-1}}{n!}$$

and since all odd B_n values are zero for n greater than 1, we replace n by $2n$ in the sum to obtain the result

$$\frac{1}{e^t-1} = \frac{1}{t} - \frac{1}{2} + \sum_{n=1}^{\infty} B_{2n} \frac{t^{2n-1}}{(2n)!} \qquad (2.75)$$

---

*The Bernoulli numbers are named after Jacob Bernoulli (1654–1705), who first introduced them.

Hence, the substitution of (2.75) into (2.71) gives us

$$I = \int_0^\infty \left[\frac{1}{t} - \frac{1}{t} + \frac{1}{2} - \sum_{n=1}^\infty B_{2n} \frac{t^{2n-1}}{(2n)!} \right] e^{-xt} \, dt$$

$$= \frac{1}{2} \int_0^\infty e^{-xt} \, dt - \sum_{n=1}^\infty \frac{B_{2n}}{(2n)!} \int_0^\infty e^{-xt} t^{2n-1} \, dt$$

Evaluating the above integrals in terms of gamma functions, the expression for I becomes

$$I = \frac{1}{2x} - \frac{1}{2} \sum_{n=1}^\infty \frac{B_{2n}}{n} \frac{1}{x^{2n}} \tag{2.76}$$

and this in turn, substituted into (2.70), leads to the asymptotic series

$$\psi(x+1) \sim \ln x + \frac{1}{2x} - \frac{1}{2} \sum_{n=1}^\infty \frac{B_{2n}}{n} \frac{1}{x^{2n}} \qquad x \to \infty \tag{2.77}$$

Unlike many of the asymptotic series that we derive, (2.77) converges for all $x > 0$.

In statistical mechanics, probability theory, and so forth, often we are dealing with large factorials or gamma functions with large arguments. To facilitate the computations involving such expressions, it is helpful to have an accurate asymptotic formula from which to approximate $\Gamma(x)$. Our approach to finding such a formula will be to first find a suitable asymptotic relation for $\ln \Gamma(x+1)$ and then exponentiate this result.

Since, by definition,

$$\psi(x+1) = \frac{d}{dx} \ln \Gamma(x+1)$$

it follows that the indefinite integral of (2.77) leads to the asymptotic series

$$\ln \Gamma(x+1) \sim C + (x + \tfrac{1}{2}) \ln x - x + \frac{1}{2} \sum_{n=1}^\infty \frac{B_{2n}}{n(2n-1)} \frac{1}{x^{2n-1}} \tag{2.78}$$

where C is a constant of integration. To evaluate C, we would normally need to know the exact behavior of series (2.78) for some value of x. However, by allowing $x \to \infty$, we can eliminate the series in

(2.78), and thus we see that

$$C = \lim_{x \to \infty} [\ln \Gamma(x+1) - (x + \tfrac{1}{2}) \ln x + x]$$

$$= \lim_{x \to \infty} \left\{ \ln \left[\frac{\Gamma(x+1)}{x^{x+1/2}} \right] + x \right\} \tag{2.79}$$

Now, by defining

$$K = e^C = \lim_{x \to \infty} \frac{\Gamma(x+1)}{x^{x+1/2}} e^x \tag{2.80}$$

we have the limit relation

$$\lim_{x \to \infty} \Gamma(x+1) = K \lim_{x \to \infty} (e^{-x} x^{x+1/2}) \tag{2.81}$$

The constant K can be determined by substituting (2.81) into the Legendre duplication formula written as

$$\sqrt{\pi} = \lim_{x \to \infty} \frac{2^{2x-1}\Gamma(x)\Gamma(x + \tfrac{1}{2})}{\Gamma(2x)} \tag{2.82}$$

The result is $K = \sqrt{2\pi}$ (see problem 21 in Exercises 2.6), and therefore (2.81) leads to the asymptotic formula (recall Example 3)

$$\Gamma(x+1) \sim \sqrt{2\pi x}\, x^x e^{-x} \qquad x \to \infty \tag{2.83}$$

In particular, if we set $x = n$, where n is a large positive integer, we get the well-known expression

$$n! \sim \sqrt{2\pi n}\, n^n e^{-n} \qquad n \gg 1 \tag{2.84}$$

called **Stirling's formula.***

It is interesting to note that Stirling's formula is remarkably accurate even for small values of n. For example, when $n = 6$, we find $6! \approx 710.08$, an error of only 1.4 percent from the exact value of 720. Of course, for larger values of n the formula is even more accurate.

Our original intent was to find an asymptotic *series* for the gamma function, and to do this, we substitute $K = \sqrt{2\pi}$ into (2.80), which identifies

$$C = \ln K = \tfrac{1}{2} \ln 2\pi \tag{2.85}$$

---

*Equation (2.84), which is a special case of the asymptotic series for the gamma function, was published in 1730 by James Stirling (1692–1770).

Then, returning to the series (2.78), we have

$$\ln \Gamma(x+1) \sim \tfrac{1}{2} \ln 2\pi + (x + \tfrac{1}{2}) \ln x - x$$

$$+ \frac{1}{2} \sum_{n=1}^{\infty} \frac{B_{2n}}{n(2n-1)} \frac{1}{x^{2n-1}} \qquad x \to \infty \quad (2.86)$$

This last expression is called **Stirling's series**. It represents a convergent series for $\ln \Gamma(x+1)$ for all positive values of x. Moreover, the absolute value of the error incurred in using this series to evaluate $\ln \Gamma(x+1)$ is less than the absolute value of the first term neglected in the series.

Although Stirling's series is valid for all positive x, it is used primarily for evaluating the gamma function for large arguments. We can eliminate the logarithm terms by exponentiating both sides, to get (retaining only the first few terms of the series)

$$\Gamma(x+1) \sim \sqrt{2\pi}\, x^{x+1/2} e^{-x} \exp\left(\frac{1}{12x} - \frac{1}{360x^3} + \cdots\right)$$

or

$$\Gamma(x+1) \sim \sqrt{2\pi x}\, x^x e^{-x}\left(1 + \frac{1}{12x} + \frac{1}{288x^2} + \cdots\right) \qquad x \to \infty \quad (2.87)$$

In this final step, we have replaced the last exponential function by the first few terms of its Maclaurin series.

Finally, if we set $x = n$, where n is a large positive integer, and retain only the first two terms of (2.87), we get a more accurate version of Stirling's formula [Eq. (2.84)]:

$$n! \sim \sqrt{2\pi n}\, n^n e^{-n}\left(1 + \frac{1}{12n}\right) \qquad n \gg 1 \qquad (2.88)$$

Here we find for $n = 6$ that $6! \approx 719.94$, which has an error of only 8.3×10^{-3} percent. Perhaps even more remarkable is that if we let $n = 1, 2, 3, \ldots$, we calculate from (2.88) the values

$$1! \approx 0.99898$$

$$2! \approx 1.99896$$

$$3! \approx 5.99833$$

$$\vdots$$

and thus conclude that (2.88) is accurate enough for many applications for *all* positive integers.

2.6.3 Polygamma functions

By repeated differentiation of the digamma function

$$\psi(x) = \frac{d}{dx}\ln\Gamma(x) \tag{2.89}$$

we form the family of **polygamma functions**

$$\psi^{(m)}(x) = \frac{d^{m+1}}{dx^{m+1}}\ln\Gamma(x) \qquad m = 1, 2, 3, \ldots \tag{2.90}$$

Recalling Eq. (2.56),

$$\psi(x) = -\gamma + \sum_{n=0}^{\infty}\left(\frac{1}{n+1} - \frac{1}{n+x}\right) \tag{2.91}$$

we readily determine the representation

$$\psi^{(m)}(x) = (-1)^{m+1}m!\sum_{n=0}^{\infty}\frac{1}{(n+x)^{m+1}} \qquad m = 1, 2, 3, \ldots \tag{2.92}$$

Of special interest is the evaluation of (2.92) when $x = 1$, that is,

$$\psi^{(m)}(1) = (-1)^{m+1}m!\sum_{n=0}^{\infty}\frac{1}{(n+1)^{m+1}}$$

$$= (-1)^{m+1}m!\sum_{n=1}^{\infty}\frac{1}{n^{m+1}}$$

or $\qquad \psi^{(m)}(1) = (-1)^{m+1}m!\zeta(m+1) \qquad m = 1, 2, 3, \ldots \tag{2.93}$

where $\qquad \zeta(p) = \sum_{n=1}^{\infty}\frac{1}{n^p} \qquad p > 1 \tag{2.94}$

is the *Riemann zeta function* (see Sec. 2.6.4). The evaluation of $\psi^{(m)}(x)$ for other values of x also leads to the zeta function (see problems 29 and 30 in Exercises 2.6).

   Although (2.91) and (2.92) are valid representations of $\psi(x)$ and $\psi^{(m)}(x)$, respectively, for all values of x except $x = 0, -1, -2, \ldots$, they are not the most convenient series to use for computational purposes, particularly in the neighborhood of $x = 1$. Instead, it may be preferable to have power series expansions for such calculations.

   To begin, we seek a power series of the form

$$\ln\Gamma(x+1) = \sum_{n=0}^{\infty}c_n x^n \tag{2.95}$$

where we choose $\ln \Gamma(x+1)$ instead of $\ln \Gamma(x)$ so that we can expand about $x = 0$. The constants in this Maclaurin expansion are defined by

$$c_0 = \ln \Gamma(x+1)\big|_{x=0} = \ln \Gamma(1) = 0$$

$$c_1 = \frac{d}{dx} \ln \Gamma(x+1)\bigg|_{x=0} = \psi(1) = -\gamma \tag{2.96}$$

and for $n > 2$,

$$c_n = \frac{1}{n!} \frac{d^n}{dx^n} \ln \Gamma(x+1)\bigg|_{x=0} = \frac{1}{n!} \psi^{(n-1)}(1)$$

which in view of (2.93), becomes

$$c_n = \frac{(-1)^n}{n} \zeta(n) \qquad n = 2, 3, 4, \ldots \tag{2.97}$$

Hence, the substitution of (2.96) and (2.97) into (2.95) yields

$$\ln \Gamma(x+1) = -\gamma x + \sum_{n=2}^{\infty} \frac{(-1)^n \zeta(n)}{n} x^n \qquad -1 < x \le 1 \tag{2.98}$$

where the interval of convergence is shown.

Termwise differentiation of (2.98) is permitted and leads to

$$\psi(x+1) = \frac{d}{dx} \ln \Gamma(x+1)$$

$$= -\gamma + \sum_{n=2}^{\infty} (-1)^n \zeta(n) x^{n-1}$$

or, by making a change of index,

$$\psi(x+1) = -\gamma + \sum_{n=1}^{\infty} (-1)^{n+1} \zeta(n+1) x^n \qquad -1 < x < 1 \tag{2.99}$$

This last series no longer converges at the endpoint $x = 1$, as was the case in (2.98). Continued differentiation of (2.99) finally leads to the following relation for $m = 1, 2, 3, \ldots$ (see problem 23 in Exercises 2.6):

$$\psi^{(m)}(x+1) = (-1)^{m+1} \sum_{n=0}^{\infty} (-1)^n \frac{(m+n)!}{n!} \zeta(m+n+1) x^n$$

$$-1 < x < 1 \tag{2.100}$$

which also converges for $-1 < x < 1$.

Both the digamma and polygamma functions are used at times for summing series, particularly those series involving rational functions with the power of the denominator at least 2 greater than that in the numerator. In such cases, the infinite series can be expressed as a finite sum of digamma or polygamma functions by the use of partial fraction expansions (see Example 11). Of course, the values of the digamma and polygamma functions must usually be obtained from tables.*

2.6.4   Riemann zeta function

The **Riemann zeta function**

$$\zeta(x) = \sum_{n=1}^{\infty} \frac{1}{n^x} \qquad x > 1 \tag{2.101}$$

first arose in Sec. 1.2.2 as a series that is useful in proving convergence or divergence of other series by means of a comparison test. We also found that the zeta function is closely related to the logarithm of the gamma function and to the polygamma functions. Although the zeta function was known to Euler, it was Riemann in 1859 who established most of its properties, which now are very important in the field of number theory, among others. Thus it bears his name.

An interesting relation for the zeta function can be derived by first making the observation

$$\zeta(x)(1 - 2^{-x}) = 1 + \frac{1}{2^x} + \frac{1}{3^x} + \frac{1}{4^x} + \cdots - \left(\frac{1}{2^x} + \frac{1}{4^x} + \frac{1}{6^x} + \cdots \right)$$

where all terms are eliminated from (2.101) in which n is a multiple of 2. Therefore we deduce that

$$\zeta(x)(1 - 2^{-x}) = \sum_{n=1}^{\infty} \frac{1}{(2n-1)^x} \tag{2.102}$$

One of the advantages of (2.102) is that, using it, $\zeta(x)$ can be computed to the same accuracy as given by (2.101), but with only half as many terms. Similarly, the product

$$\zeta(x)(1 - 2^{-x})(1 - 3^{-x}) = 1 + \frac{1}{3^x} + \frac{1}{5^x} + \frac{1}{7^x} + \cdots - \left(\frac{1}{3^x} + \frac{1}{9^x} + \frac{1}{15^x} + \cdots \right) \tag{2.103}$$

---

*See M. Abramowitz and I. A. Stegun (eds.), *Handbook of Mathematical Tables*, Dover, New York, 1965, chap. 6.

eliminates all terms from (2.102) in which n is a multiple of 3. Continuing in this fashion, it can eventually be shown that the infinite product over all prime numbers greater than 1 leads to

$$\zeta(x)(1-2^{-x})(1-3^{-x})\cdots(1-P^{-x})\cdots=1 \qquad (2.104)$$

where P denotes a prime number. Hence, we have *Euler's infinite product representation*

$$\zeta(x) = \prod_{P=2}^{\infty} (1-P^{-x})^{-1} \qquad P \text{ prime} \qquad (2.105)$$

It can readily be shown that the zeta function has the integral representation (see problem 17 in Exercises 2.6)

$$\zeta(x) = \frac{1}{\Gamma(x)} \int_0^\infty \frac{t^{x-1}}{e^t - 1} dt \qquad x>1 \qquad (2.106)$$

Also by using complex variable methods, it can be shown that[*]

$$\zeta(1-x) = 2^{1-x}\pi^{-x} \cos \tfrac{1}{2}\pi x \, \Gamma(x)\zeta(x) \qquad (2.107)$$

which is the famous formula of Riemann. Other relations involving this function, as well as some special values, are taken up in the exercises.

The graph of $\zeta(x)-1$ is shown in Fig. 2.4 for $x>1$. For comparison, the dashed line is the graph of 2^{-x}.

Exercises 2.6

1. Show that

$$\psi(x) - \psi(y) = \sum_{n=0}^{\infty} \left(\frac{1}{y+n} - \frac{1}{x+n} \right)$$

2. Take the logarithmic derivative of $\Gamma(x)\Gamma(1-x) = \pi \csc \pi x$ to deduce the identity

$$\psi(1-x) - \psi(x) = \pi \cot \pi x$$

3. (a) By taking the logarithmic derivative of the Legendre duplication formula

$$2^{2x-1}\Gamma(x)\Gamma(x + \tfrac{1}{2}) = \sqrt{\pi}\,\Gamma(2x)$$

---

[*] See E. T. Whittaker and G. N. Watson, *A Course of Modern Analysis*, Cambridge University Press, London, 1965, p. 269.

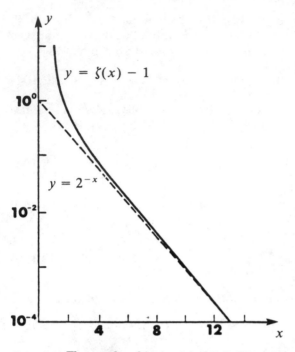

Figure 2.4 The graphs of $\zeta(x) - 1$ and 2^{-x}. (Note the logarithmic scale on the vertical axis.)

deduce that

$$\psi(x) + \psi(x + \tfrac{1}{2}) + 2\ln 2 = 2\psi(2x)$$

(*b*)   From (*a*), deduce that $\psi(\tfrac{1}{2}) = -\gamma - 2\ln 2$.

(*c*)   For $n = 1, 2, 3, \ldots$, show that

$$\psi(n + \tfrac{1}{2}) = -\gamma - 2\ln 2 + 2\sum_{k=1}^{n}(2k - 1)^{-1}$$

4. Derive the formula

$$3\psi(3x) = \psi(x) + \psi(x + \tfrac{1}{3}) + \psi(x + \tfrac{2}{3}) + 3\ln 3$$

Hint: Recall problem 40 in Exercises 2.2.

In problems 5 to 8, verify the given relation.

5. $\psi(n + 1) = -\gamma + \sum_{k=1}^{\infty} \dfrac{n}{k(k + n)}$, $n = 0, 1, 2, \ldots$

6. $\lim_{n \to \infty} [\psi(x + n) - \ln n] = 0$

7. $\psi(\tfrac{1}{2} + p) = \psi(\tfrac{1}{2} - p) + \pi \tan \pi p$

8. $\exp\left[\psi(x)\right] = x \prod_{n=0}^{\infty} \left(1 + \frac{1}{x+n}\right) e^{-1/(x+n)}$

9. Show that

(a) $\gamma = -\int_0^1 \ln\left(\ln\frac{1}{t}\right) dt$

(b) $\gamma = \frac{1}{2} - 2\int_0^{\infty} \frac{t \, dt}{(1+t^2)(e^{2\pi t} - 1)}$

10. Starting with Eq. (2.58), use integration by parts followed by a change of variable to show that

$$\gamma = \int_0^1 \frac{1 - e^{-t} - e^{-1/t}}{t} dt$$

11. Derive the Maclaurin series expansion

$$t \coth t = \sum_{m=0}^{\infty} B_{2m} \frac{(2t)^{2m}}{(2m)!}$$

Hint: First show that

$$\coth t = \frac{e^t + e^{-t}}{e^t - e^{-t}} = \frac{1}{e^{2t} - 1} + \frac{1}{1 - e^{-2t}}$$

12. (a) Starting with the infinite product representation

$$\sin x = x \prod_{n=1}^{\infty} \left(1 - \frac{x^2}{n^2 \pi^2}\right)$$

show that the logarithmic derivative leads to

$$x \cot x = 1 - 2\sum_{n=1}^{\infty} \frac{[x/(n\pi)]^2}{1 - x^2/(n^2 \pi^2)} \qquad -\pi < x < \pi$$

(b) From (a), deduce that

$$x \cot x = 1 - 2\sum_{m=1}^{\infty} \zeta(2m)\left(\frac{x}{\pi}\right)^{2m}$$

13. (a) By using the identity $\coth ix = -i \cot x$ $(i^2 = -1)$ and the result of problem 11, deduce that

$$x \cot x = 1 + \sum_{m=1}^{\infty} B_{2m} \frac{(-1)^m}{(2m)!} (2x)^{2m}$$

(b) Comparing the result of (a) with that of problem 12b, deduce the relation

$$\zeta(2m) = \frac{(2\pi)^{2m}(-1)^{m-1}}{2(2m)!} B_{2m} \qquad m = 1, 2, 3, \ldots$$

14. Show that
(a) $\zeta(2) = \pi^2/6$ (b) $\zeta(4) = \pi^4/90$ (c) $\zeta(6) = \pi^6/945$
Hint: Use problem 13b.

15. Show that

$$\gamma = \sum_{n=2}^{\infty} \frac{(-1)^n}{n} \zeta(n)$$

16. The total energy radiated by a blackbody (*Stefan-Boltzmann law*) is proportional to the integral

$$I = \int_0^{\infty} \frac{x^3}{e^x - 1} dx$$

Show that $I = \pi^4/15$.

Hint: Observe that $(1 - e^{-x})^{-1} = \sum_{n=0}^{\infty} e^{-nx}$ and use problem 14.

17. Starting with the observation

$$\frac{1}{n^x} = \frac{1}{\Gamma(x)} \int_0^{\infty} e^{-nt} t^{x-1} dt \qquad x > 1 \qquad n = 1, 2, 3, \ldots$$

sum over all values of n to deduce that

$$\zeta(x) = \frac{1}{\Gamma(x)} \int_0^{\infty} \frac{t^{x-1}}{e^t - 1} dt \qquad x > 1$$

18. Show that $(p > 1)$

$$\int_0^{\infty} t^{p-1} \left(\frac{1}{e^t - 1} - \frac{1}{e^t + 1} \right) dt = 2^{1-p} \Gamma(p) \zeta(p)$$

19. Using the results of problems 17 and 18, deduce that $(p > 1)$

$$\frac{1}{\Gamma(p)} \int_0^{\infty} \frac{t^{p-1}}{e^t + 1} dt = \sum_{n=1}^{\infty} \frac{(-1)^{n-1}}{n^p} = (1 - 2^{1-p}) \zeta(p)$$

20. By expressing $\ln(1 + x)$ in its Maclaurin series, show that

$$\int_0^1 \frac{\ln(1+x)}{x} dx = \frac{\pi^2}{12}$$

Hint: See problems 14 and 19.

21. By substituting the limit expression

$$\lim_{x \to \infty} \Gamma(x + 1) = K \lim_{x \to \infty} (e^{-x} x^{x+1/2})$$

into the Legendre duplication formula written in the form

$$\sqrt{\pi} = \lim_{x \to \infty} \frac{2^{2x-1} \Gamma(x) \Gamma(x + 1/2)}{\Gamma(2x)}$$

deduce that $K = \sqrt{2\pi}$.

22. Use problem 21 to establish that

$$\lim_{x \to \infty} x^{b-a} \frac{\Gamma(x + a + 1)}{\Gamma(x + b + 1)} = 1$$

23. Show that the mth derivative of Eq. (2.99) leads to

$$\psi^{(m)}(x+1) = (-1)^{m+1} \sum_{n=0}^{\infty} (-1)^n \frac{(m+n)!}{n!} \zeta(m+n+1)x^n$$

24. Show that

(a) $\psi'(1) = \dfrac{\pi^2}{6}$

(c) $\psi'\left(\dfrac{1}{2}\right) = \dfrac{\pi^2}{2}$

(b) $\psi'(2) = \dfrac{\pi^2}{6} - 1$

(d) $\psi'''(2) = \dfrac{\pi^4}{15} - 6$

25. Write the sum of the series in terms of the digamma and polygamma functions and evaluate.

(a) $\displaystyle\sum_{n=1}^{\infty} \frac{1}{n(n+1)}$

(c) $\displaystyle\sum_{n=1}^{\infty} \frac{1}{n(n+1)^2}$

(b) $\displaystyle\sum_{n=0}^{\infty} \frac{1}{(n+2)(n+4)}$

(d) $\displaystyle\sum_{n=1}^{\infty} \frac{1}{n(4n^2-1)}$

26. Show that $(m = 1, 2, 3, \ldots)$

$$\psi^{(m)}(x) = (-1)^{m+1} \int_0^{\infty} \frac{t^m e^{-xt}}{1-e^{-t}} dt$$

27. Derive the asymptotic series

$$\psi'(x+1) \sim \frac{1}{x} - \frac{1}{2x^2} + \sum_{n=1}^{\infty} B_{2n} x^{-(2n+1)} \qquad x \to \infty$$

Note: This series diverges for all x.

28. Use the first *four* terms of the series in problem 27 (including the terms outside the summation) to approximate $\psi'(4)$, and compare with the exact value $\psi'(4) = \pi^2/6 - {}^{49}/_{36}$.

29. For $k = 2, 3, 4, \ldots$, show that

$$\psi^{(m)}(k) = (-1)^{m+1} m! \left[\zeta(m+1) - \sum_{n=1}^{k-1} \frac{1}{n^{m+1}} \right]$$

30. For $k = 2, 3, 4, \ldots$, show that

$$\psi^{(m)}\left(k - \frac{1}{2}\right)$$
$$= (-1)^{m+1} m! \left[(2^{m+1} - 1)\zeta(m+1) - 2^{m+1} \sum_{n=1}^{k-1} \frac{1}{(n-1)^{m+1}} \right]$$

3

Other Functions Defined by Integrals

3.1 Introduction

In addition to the gamma function, there are numerous other special functions whose primary definition involves an integral. Some of these functions were introduced in Chap. 2 along with the gamma function, and in this chapter we consider several others.

The *error function* derives its name from its importance in the theory of errors, but it also occurs in probability theory and in certain heat conduction problems on infinite domains. The closely related *Fresnel integrals,* which are fundamental in the theory of optics, can be derived directly from the error function. A special case of the incomplete gamma function (Sec. 2.5) leads to the *exponential integral* and related functions—the *logarithmic integral,* which is important in analysis and number theory, and the *sine* and *cosine integrals,* which arise in Fourier transform theory.

Elliptic integrals first arose in the problems associated with computing the arclength of an ellipse and a lemniscate (a curve in the shape of a figure eight). Some early results concerning elliptic integrals were discovered by L. Euler and J. Landen, but virtually the whole theory of these integrals was developed by Legendre over a period spanning 40 years. The inverses of the elliptic integrals, called *elliptic functions,* were independently introduced in 1827 by C. G. J. Jacobi (1802–1859) and N. H. Abel (1802–1829). Many of the properties of elliptic functions, however, had already been developed as early as 1809 by Gauss. Elliptic functions have the distinction of being doubly periodic, with one real period and one imaginary period. Among other areas of application, the elliptic functions are important in solving the pendulum problem (Sec. 3.5.2).

3.2   The Error Function and Related Functions

The **error function** is defined by the integral

$$\text{erf}\, x = \frac{2}{\sqrt{\pi}} \int_0^x e^{-t^2}\, dt \qquad -\infty < x < \infty \tag{3.1}$$

This function is encountered in probability theory, the theory of errors, the theory of heat conduction, and various branches of mathematical physics. By representing the exponential function in (3.1) in terms of its power series expansion, we have

$$\text{erf}\, x = \frac{2}{\sqrt{\pi}} \int_0^x \sum_{n=0}^{\infty} \frac{(-1)^n}{n!} t^{2n}\, dt$$

from which we deduce (termwise integration of power series is permitted)

$$\text{erf}\, x = \frac{2}{\sqrt{\pi}} \sum_{n=0}^{\infty} \frac{(-1)^n x^{2n+1}}{n!(2n+1)} \qquad |x| < \infty \tag{3.2}$$

Examination of series (3.2) reveals that the error function is an *odd function,* i.e.,

$$\text{erf}(-x) = -\text{erf}\, x \tag{3.3}$$

Also we see that

$$\text{erf}\, 0 = \frac{2}{\sqrt{\pi}} \int_0^0 e^{-t^2}\, dt = 0 \tag{3.4}$$

and by using properties of the gamma function, we find that (in the limit)

$$\text{erf}\, \infty = \frac{2}{\sqrt{\pi}} \int_0^{\infty} e^{-t^2}\, dt = \frac{\Gamma(\frac{1}{2})}{\sqrt{\pi}} = 1 \tag{3.5}$$

The graph of $\text{erf}\, x$ is shown in Fig. 3.1, and a list of values for $0 \le x \le 2$ is provided in Table 3.1.

In some applications it is useful to introduce the **complementary error function**

$$\text{erfc}\, x = \frac{2}{\sqrt{\pi}} \int_x^{\infty} e^{-t^2}\, dt \tag{3.6}$$

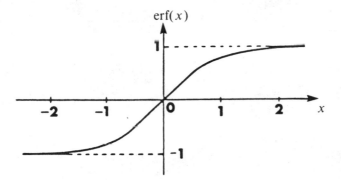

Figure 3.1 The error function

Clearly it follows that

$$\operatorname{erfc} x = \frac{2}{\sqrt{\pi}} \int_0^\infty e^{-t^2} \, dt - \frac{2}{\sqrt{\pi}} \int_0^x e^{-t^2} \, dt$$

from which we deduce

$$\operatorname{erfc} x = 1 - \operatorname{erf} x \qquad (3.7)$$

Hence, all properties of $\operatorname{erfc} x$ can be derived from those of $\operatorname{erf} x$.

Example 1: Find the Laplace transform of $f(t) = \operatorname{erfc}(1/\sqrt{t})$.

Solution: The Laplace transform is defined by

$$\mathscr{L}\left\{\operatorname{erfc}\left(\frac{1}{\sqrt{t}}\right); s\right\} = \int_0^\infty e^{-st} \operatorname{erfc}\left(\frac{1}{\sqrt{t}}\right) dt$$

$$= \int_0^\infty e^{-st} \frac{2}{\sqrt{\pi}} \int_{1/\sqrt{t}}^\infty e^{-u^2} \, du \, dt$$

TABLE 3.1 Values of erf x for $0 \le x \le 2$

x	$\operatorname{erf} x$	x	$\operatorname{erf} x$
0.00	0.0000	1.00	0.8427
0.10	0.1125	1.10	0.8802
0.20	0.2227	1.20	0.9103
0.30	0.3286	1.30	0.9340
0.40	0.4284	1.40	0.9523
0.50	0.5205	1.50	0.9661
0.60	0.6039	1.60	0.9763
0.70	0.6778	1.70	0.9838
0.80	0.7421	1.80	0.9891
0.90	0.7969	1.90	0.9928
1.00	0.8427	2.00	0.9953

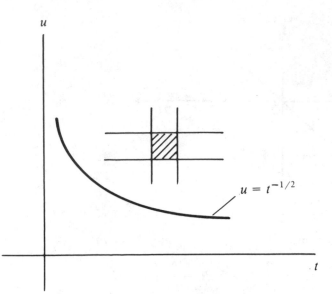

Figure 3.2

If we interpret this last expression as an iterated integral, we can interchange the order of integration (see Fig. 3.2). Hence,

$$\mathscr{L}\left\{\operatorname{erfc}\left(\frac{1}{\sqrt{t}}\right); s\right\} = \frac{2}{\sqrt{\pi}} \int_0^\infty e^{-u^2} \int_{u^{-2}}^\infty e^{-st}\,dt\,du$$

$$= \frac{2}{s\sqrt{\pi}} \int_0^\infty e^{-u^2 - su^{-2}}\,du$$

and by calling upon the integral formula (see problem 6 in Exercises 3.2)

$$\int_0^\infty e^{-a^2 x^2 - b^2 x^{-2}}\,dx = \frac{\sqrt{\pi}}{2a} e^{-2ab} \qquad a > 0, \qquad b \geq 0$$

we deduce that

$$\mathscr{L}\left\{\operatorname{erfc}\left(\frac{1}{\sqrt{t}}\right); s\right\} = \frac{1}{s} e^{-2\sqrt{s}} \qquad s > 0$$

3.2.1 Asymptotic series

An asymptotic series for the complementary error function can be obtained through repeated integration by parts. To obtain this series, we first observe that integration by parts leads to

$$\int_x^\infty e^{-t^2} dt = \frac{e^{-x^2}}{2x} - \frac{1}{2} \int_x^\infty \frac{e^{-t^2}}{t^2} dt$$

and by integrating by parts again, we get

$$\int_x^\infty e^{-t^2} dt = \frac{e^{-x^2}}{2x} - \frac{e^{-x^2}}{2^2 x^3} + \frac{1 \times 3}{2^2} \int_x^\infty \frac{e^{-t^2}}{t^4} dt$$

Continuing this process indefinitely, we finally derive the asymptotic series

$$\text{erfc}\, x \sim \frac{e^{-x^2}}{\sqrt{\pi}x}\left[1 + \sum_{n=1}^\infty (-1)^n \frac{1 \times 3 \times \cdots \times (2n-1)}{(2x^2)^n}\right] \qquad x \to \infty \quad (3.8)$$

3.2.2 Fresnel integrals

Closely associated with the error function are the **Fresnel integrals**

$$C(x) = \int_0^x \cos \tfrac{1}{2}\pi t^2 \, dt \qquad\qquad (3.9)$$

and

$$S(x) = \int_0^x \sin \tfrac{1}{2}\pi t^2 \, dt \qquad\qquad (3.10)$$

These integrals come up in various branches of physics and engineering, such as in diffraction theory and the theory of vibrations.

From definition, we have the immediate results

$$C(0) = S(0) = 0 \qquad\qquad (3.11)$$

The derivatives of these functions are

$$C'(x) = \cos \tfrac{1}{2}\pi x^2 \qquad S'(x) = \sin \tfrac{1}{2}\pi x^2 \qquad (3.12)$$

and thus we deduce that both $C(x)$ and $S(x)$ are oscillatory. Namely, $C(x)$ has extrema at the points where $x^2 = 2n + 1$ $(n = 0, 1, 2, \ldots)$, and $S(x)$ has extrema where $x^2 = 2n$ $(n = 1, 2, 3, \ldots)$. The largest maxima occur first and are found to be $C(1) = 0.77989\ldots$ and $S(\sqrt{2}) = 0.71397\ldots$. For $x \to \infty$, we can use the integral formulas (see problem 27 in Exercises 3.2)

$$\int_0^\infty \cos t^2 \, dt = \int_0^\infty \sin t^2 \, dt = \frac{1}{2}\sqrt{\frac{\pi}{2}} \qquad (3.13)$$

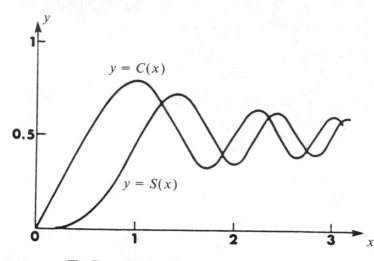

Figure 3.3 The Fresnel integrals.

to obtain the results

$$C(\infty) = S(\infty) = {}^1\!/_2 \qquad (3.14)$$

The graphs of $C(x)$ and $S(x)$ for positive x are shown in Fig. 3.3.

To derive the relation between the Fresnel integrals and the error function, we start with

$$\operatorname{erf} z = \frac{2}{\sqrt{\pi}} \int_0^z e^{-u^2} \, du \qquad (3.15)$$

where z may be real or complex.* Substituting $z = x\sqrt{i\pi/2}$ and $u = t\sqrt{i\pi/2}$ into (3.15) leads to

$$\operatorname{erf}\left(x\sqrt{\frac{i\pi}{2}}\right) = \sqrt{2i} \int_0^x e^{-i\pi t^2/2} \, dt$$

$$= \sqrt{2i} \left(\int_0^x \cos {}^1\!/_2 \pi t^2 \, dt - i \int_0^x \sin {}^1\!/_2 \pi t^2 \, dt \right)$$

from which it follows that

$$\frac{1}{\sqrt{2i}} \operatorname{erf}\left(x\sqrt{\frac{i\pi}{2}}\right) = C(x) - iS(x) \qquad (3.16)$$

Other properties of these functions are taken up in the exercises.

---

* For a discussion of the error function with complex argument, see N. N. Lebedev, *Special Functions and Their Applications,* Dover, New York, 1972, chap. 2.

Exercises 3.2

In problems 1 to 10, use properties of the error function to develop the given relation.

1. $\dfrac{d}{dx}\operatorname{erf} x = \dfrac{2}{\sqrt{\pi}}e^{-x^2}$

2. $\displaystyle\int_{-a}^{a} e^{-t^2}\,dt = \sqrt{\pi}\operatorname{erf} a$

3. $\displaystyle\int_{a}^{b} e^{-t^2}\,dt = \tfrac{1}{2}\sqrt{\pi}(\operatorname{erf} b - \operatorname{erf} a) = \tfrac{1}{2}\sqrt{\pi}(\operatorname{erfc} a - \operatorname{erfc} b)$

4. $\displaystyle\int_{0}^{\infty} e^{-\alpha^2 t^2}\,dt = \dfrac{\sqrt{\pi}}{2\alpha},\qquad \alpha > 0$

5. $\displaystyle\int_{0}^{\infty} e^{-(at^2+2bt+c)}\,dt = \dfrac{1}{2}\sqrt{\dfrac{\pi}{a}}\,e^{(b^2-ac)/c}\operatorname{erfc}\dfrac{b}{\sqrt{a}},\ a > 0$

6. $\displaystyle\int_{0}^{\infty} \dfrac{e^{-st}}{\sqrt{t+x}}\,dt = \sqrt{\dfrac{\pi}{s}}\,e^{sx}\operatorname{erfc}\sqrt{sx},\ x > 0,\ s > 0$

7. $\displaystyle\int_{x}^{\infty} \dfrac{e^{-t^2}}{t^2}\,dt = \dfrac{1}{x}e^{-x^2} - \sqrt{\pi}\operatorname{erfc} x$

8. $\displaystyle\int_{0}^{\infty} e^{-st-t^2/4}\,dt = \sqrt{\pi}\,e^{s^2}\operatorname{erfc} s,\ s > 0$

   Hint: Write $t^2/4 + st = (t/2+s)^2 - s^2$ and make the change of variable $u = t/2 + s$.

9. $\displaystyle\int_{0}^{\infty} e^{-st}\operatorname{erf} t\,dt = \dfrac{1}{s}e^{s^2/4}\operatorname{erfc}\dfrac{s}{2},\ s > 0$

   Hint: Reverse the order of integration.

10. $\displaystyle\int_{-\infty}^{\infty} e^{-ax^2}\operatorname{erf}\left(\dfrac{x+b}{\sqrt{2}}\right)dx = \sqrt{\dfrac{\pi}{a}}\operatorname{erf}\left(b\sqrt{\dfrac{a}{2a+1}}\right),\ a > 0$

11. Evaluate.
    (a) $\operatorname{erfc} 0$
    (b) $\operatorname{erfc}\infty$

12. Establish the relations (see Sec. 2.5)

    (a) $\operatorname{erf} x = \dfrac{1}{\sqrt{\pi}}\gamma\left(\dfrac{1}{2}, x^2\right)$

    (b) $\operatorname{erfc} x = \dfrac{1}{\sqrt{\pi}}\Gamma\left(\dfrac{1}{2}, x^2\right)$

In problems 13 to 16, establish the indefinite integral relations, where C is an arbitrary constant.

13. $\displaystyle\int e^{-(ax^2+2bx+x)}\,dx = \frac{1}{2}\sqrt{\frac{\pi}{a}}\,e^{(b^2-ac)/c}\,\mathrm{erf}\left(\sqrt{ax}+\frac{b}{\sqrt{a}}\right)+C$

14. $\displaystyle\int \mathrm{erf}\,x\,dx = x\,\mathrm{erf}\,x + \frac{1}{\sqrt{\pi}}e^{-x^2}+C$

    Hint: Use integration by parts.

15. $\displaystyle\int e^{-x^2}\mathrm{erf}\,x\,dx = \frac{\sqrt{\pi}}{4}\mathrm{erf}^2 x + C$

16. $\displaystyle\int e^{ax}\,\mathrm{erf}\,bx\,dx = \frac{1}{a}\left[e^{ax}\,\mathrm{erf}\,bx - e^{a^2/(4b^2)}\,\mathrm{erf}\left(bx-\frac{a}{2b}\right)\right]+C,$

    $a \neq 0$

17. Both $F(x)$ and $F'(x)$ are uniformly convergent integrals for all x, where

$$F(x) = \int_0^\infty e^{-t^2}\cos xt\,dt$$

   (a)   Use integration by parts to show that

$$F'(x) = -\frac{x}{2}F(x)$$

   (b)   Verify directly that $F(0) = \sqrt{\pi}/2$.

   (c)   Solve the differential equation (DE) in (a) subject to the initial condition in (b) to deduce that

$$\int_0^\infty e^{-t^2}\cos xt\,dt = \frac{\sqrt{\pi}}{2}e^{-x^2/4}$$

18. Consider the integral

$$I(b) = \int_0^\infty e^{-a^2x^2 - b^2x^{-2}}\,dx \qquad a>0, \qquad b \geq 0$$

as a function of the parameter b.

   (a)   Show that I satisfies the first-order linear DE

$$\frac{dI}{db} + 2aI = 0$$

   (b)   Evaluate $I(0)$ directly from the integral

   (c)   Solve the DE in (a) subject to the initial condition in (b) to deduce the result

$$I(b) = \frac{\sqrt{\pi}}{2a}e^{-2ab}$$

19. Consider the integral

$$I(a) = \int_{-\infty}^{\infty} \frac{e^{-a^2 x^2}}{x^2 + b^2} \, dx \qquad a \geq 0, \qquad b > 0$$

as a function of the parameter a.
  (a) Show that I satisfies the first-order linear DE

$$\frac{dI}{da} - 2ab^2 I = -2\sqrt{\pi}$$

  (b) Evaluate $I(0)$ directly from the integral.
  (c) Solve the DE in (a) subject to the initial condition in (b) to
       deduce that

$$I(a) = \frac{\pi}{b} e^{a^2 b^2} \operatorname{erfc} ab$$

20. Use the result of problem 19 to show that

$$\int_0^{\pi/2} e^{-a^2 \tan^2 x} \, dx = \int_0^{\pi/2} e^{-a^2 \cot^2 x} \, dx = \frac{\pi}{2} e^{a^2} \operatorname{erfc} a \qquad a \geq 0$$

In problems 21 to 23, derive the integral representation.

21. $\operatorname{erf} x = \dfrac{2}{\pi} \displaystyle\int_0^\infty e^{-t^2} \dfrac{\sin 2xt}{t} \, dt$

  Hint: Write $\sin 2xt$ in a power series.

22. $(\operatorname{erf} x)^2 = 1 - \dfrac{4}{\pi} \displaystyle\int_0^1 \dfrac{e^{-x^2(1+t^2)}}{1+t^2} \, dt$

  Hint: Write the expansion on the left as a double integral and
  transform to polar coordinates.

23. $(\operatorname{erfc} x)^2 = \dfrac{4}{\sqrt{\pi}} e^{-2x^2} \displaystyle\int_0^\infty e^{-t^2 - 2\sqrt{2}xt} \operatorname{erf} t \, dt, \qquad x > 0$

  Hint: Use the result of problem 8.

24. Show that the Fresnel integrals satisfy
  (a) $C(-x) = -C(x)$
  (b) $S(-x) = -S(x)$

25. Obtain the series representations.

  (a) $C(x) = \displaystyle\sum_{n=0}^\infty \frac{(-1)^n (\pi/2)^{2n}}{(2n)!(4n+1)} x^{4n+1}$

  (b) $S(x) = \displaystyle\sum_{n=0}^\infty \frac{(-1)^n (\pi/2)^{2n+1}}{(2n+1)!(4n+3)} x^{4n+3}$

26. The Fresnel integrals are sometimes defined by

$$C_1(x) = \sqrt{\frac{2}{\pi}} \int_0^x \cos t^2 \, dt \qquad C_2(x) = \frac{1}{\sqrt{2\pi}} \int_0^x \frac{\cos t}{\sqrt{t}} \, dt$$

$$S_1(x) = \sqrt{\frac{2}{\pi}} \int_0^x \sin t^2 \, dt \qquad S_2(x) = \frac{1}{\sqrt{2\pi}} \int_0^x \frac{\sin t}{\sqrt{t}} \, dt$$

Show that
(a) $C(x) = C_1(x\sqrt{\pi/2}) = C_2(\pi x^2/2)$
(b) $S(x) = S_1(x\sqrt{\pi/2}) = S_2(\pi x^2/2)$

27. Set $\alpha = (1 - i)/\sqrt{2}$ in the result of problem 4, and separate into real and imaginary parts.
    (a)   Deduce that

$$\int_0^\infty \cos t^2 \, dt = \int_0^\infty \sin t^2 \, dt = \frac{1}{2} \sqrt{\frac{\pi}{2}}$$

   (b)   Use (a) to evaluate

$$\int_0^\infty \int_0^\infty \sin (x^2 + y^2) \, dx \, dy$$

28. Establish the integral formula

$$\frac{2}{\sqrt{\pi}} \int_0^x e^{-a^2 t^2} \, dt = \frac{1}{a} \operatorname{erf} ax$$

Then, following the suggestion in problem 27 and using the asymptotic series (3.8), derive the asymptotic formulas

$$C(x) \sim \frac{1}{2} - \frac{1}{\pi x} \left[B(x) \cos \frac{1}{2} \pi x^2 - A(x) \sin \frac{1}{2} \pi x^2 \right] \qquad x \to \infty$$

$$S(x) \sim \frac{1}{2} - \frac{1}{\pi x} \left[A(x) \cos \frac{1}{2} \pi x^2 + B(x) \sin \frac{1}{2} \pi x^2 \right] \qquad x \to \infty$$

where $A(x)$ and $B(x)$ are each asymptotic series related to (3.8).

3.3   Applications

In this section we briefly discuss some classical examples involving *probability theory, heat conduction,* and *vibrating beams* which lead to solutions in terms of error functions or Fresnel integrals.

3.3.1   Probability and statistics

The error function is prominent in problems involving the normal distribution in probability theory. We say that **x** is a **normal** (also called **gaussian**) **random variable** if it has the probability density function

$$p_{\mathbf{x}}(x) = \frac{1}{\sigma\sqrt{2\pi}} e^{-(x-m)^2/(2\sigma^2)} \tag{3.17}$$

where m is the mean value of $\mathbf{x}$ and σ^2 is the variance. Among other applications, this density function appears in the analysis of *random noise,* such as that found at the input to radar and sonar detection devices.

In general, the probability that a random variable $\mathbf{x} \leq x$ is defined by the integral

$$\Pr(\mathbf{x} \leq x) = \int_{-\infty}^{x} p_{\mathbf{x}}(u)\, du \tag{3.18}$$

This is also called the *cumulative distribution function.* For a normal distribution, this probability integral leads to

$$\Pr(\mathbf{x} \leq x) = \frac{1}{\sigma\sqrt{2\pi}} \int_{-\infty}^{x} e^{-(u-m)^2/(2\sigma^2)}\, du$$

$$= \frac{1}{\sigma\sqrt{2\pi}} \int_{-\infty}^{\infty} e^{-(u-m)^2/(2\sigma^2)}\, du$$

$$- \frac{1}{\sigma\sqrt{2\pi}} \int_{x}^{\infty} e^{-(u-m)^2/(2\sigma^2)}\, du$$

$$= 1 - \frac{1}{2} \operatorname{erfc} \frac{x-m}{\sigma\sqrt{2}} \tag{3.19}$$

where the last step is obtained after making the change of variable $t = (u-m)/(\sigma\sqrt{2})$. Using the identity $\operatorname{erfc} x = 1 - \operatorname{erf} x$, we can rewrite (3.19) in the alternative form

$$\Pr(\mathbf{x} \leq x) = \frac{1}{2}\left(1 + \operatorname{erf} \frac{x-m}{\sigma\sqrt{2}}\right) \tag{3.20}$$

Observe that $\Pr(\mathbf{x} < -\infty) = 0$ and $\Pr(\mathbf{x} < \infty) = 1$, which follow from the properties of the error funlction.

3.3.2 Heat conduction in solids

Another application involving error functions concerns the problem of heat flow in a long homogeneous rod, one end of which is exposed to a time-varying heat reservoir. If we assume the initial temperature distribution is 0°C along the rod and that the lateral surface of the rod is insulated, then the problem we wish to solve can be mathematically described by*

$$\frac{\partial^2 u}{\partial x^2} = a^{-2}\frac{\partial u}{\partial t} \qquad 0 < x < \infty, \qquad t > 0$$

$$u(0, t) = f(t) \qquad u(x, t) \to 0 \qquad \text{as} \qquad x \to \infty \qquad (3.21)$$

$$u(x, 0) = 0 \qquad 0 < x < \infty$$

Here $u(x, t)$ denotes the temperature in the rod at time t and position x, $f(t)$ is the prescribed time-varying end temperature at the finite boundary, and a^2 is a physical constant called the *diffusivity*. The governing equation is a *partial differential equation* called the **one-dimensional heat equation**.

We assume the reader is familiar with the *Laplace transform* method for solving (ordinary) differential equations. Because the functions involved in solving a partial differential equation depend on more than one independent variable, we define the Laplace transform of $u(x, t)$, for example, by the notation

$$\mathscr{L}\{u(x, t); t \to s\} = \int_0^\infty e^{-st} u(x, t)\, dt = U(x, s) \qquad (3.22)$$

Based on properties of the Laplace transform, it follows that

$$\mathscr{L}\left\{\frac{\partial^2 u}{\partial x^2}(x, t); t \to s\right\} = \frac{\partial^2}{\partial x^2}\int_0^\infty e^{-st}u(x, t)\, dt = \frac{\partial^2 U}{\partial x^2} \qquad (3.23)$$

and

$$\mathscr{L}\left\{\frac{\partial u}{\partial t}(x, t); t \to s\right\} = sU(x, s) - u(x, 0) \qquad (3.24)$$

If we also define $F(s) = \mathscr{L}\{f(t); t \to s\}$, then the problem described by (3.21) is transformed to

$$\frac{\partial^2 U}{\partial x^2} - \left(\frac{s}{a^2}\right)U = 0 \qquad 0 < x < \infty$$

$$\qquad (3.25)$$

$$U(0, s) = F(s) \qquad U(x, s) \to 0 \qquad \text{as} \qquad x \to \infty$$

---

* See L. C. Andrews, *Elementary Partial Differential Equations with Boundary Value Problems*, Academic, Orlando, 1986, chap. 6.

We recognize (3.25) as a second-order linear differential equation whose general solution is

$$U(x, s) = A(s)e^{x\sqrt{s/a}} + B(s)e^{-x\sqrt{s/a}}$$

where $A(s)$ and $B(s)$ are arbitrary functions of s. However, to satisfy the condition $U(x, s) \to 0$ as $x \to \infty$, we must choose $A(s) = 0$. The remaining boundary condition in (3.25) then demands that $B(s) = F(s)$, leading to the particular solution

$$U(x, s) = F(s)e^{-x\sqrt{s/a}} \tag{3.26}$$

To invert (3.26), we observe that (see problem 8 in Exercises 3.3)

$$\mathscr{L}^{-1}\{e^{-x\sqrt{s/a}}; s \to t\} = \frac{x}{2a\sqrt{\pi}t^{3/2}}e^{-x^2/(4a^2t)} \tag{3.27}$$

and through use of the *convolution theorem*, we obtain the formal solution

$$u(x, t) = \frac{x}{2a\sqrt{\pi}}\int_0^t \frac{f(\tau)}{(t - \tau)^{3/2}}\exp\left[-\frac{x^2}{4a^2(t - \tau)}\right]d\tau \tag{3.28}$$

Example 2: Solve the problem described by (3.21) when $f(t) = T_0$ (constant).

Solution: The formal solution is that given by (3.28). Setting $f(t) = T_0$ and making the change of variable

$$z = \frac{x}{2a\sqrt{t - \tau}}$$

we find that (3.28) becomes

$$u(x, t) = \frac{2T_0}{\sqrt{\pi}}\int_{x/(2a\sqrt{t})}^{\infty} e^{-z^2}\, dz$$

Now recalling the definition of the complementary error function, we get

$$u(x, t) = T_0\, \text{erfc}\, \frac{x}{2a\sqrt{t}}$$

The physical interpretation of the solution suggests that for any fixed value of x, the temperature in the rod at that point will eventually approach T_0 if we wait long enough $(t \to \infty)$. We also recognize that the temperature remains constant along any member of the family of parabolas in the xt plane defined by

$$\frac{x}{2a\sqrt{t}} = \text{constant}$$

3.3.3 Vibrating beams

Let us consider a uniform semi-infinite beam that is initially at rest along the x axis and then at time $t = 0$ is given a transverse displacement b at the end $x = 0$. We assume the beam is fixed at the end $x \to \infty$. If $u(x, t)$ denotes the subsequent displacement at time t and position x, then $u(x, t)$ is a solution of the boundary-value problem

$$\frac{\partial^4 u}{\partial x^4} + \left(\frac{1}{a^2}\right)\frac{\partial^2 u}{\partial t^2} = 0 \qquad 0 < x < \infty \qquad t > 0$$

$$u(0, t) = b \qquad \frac{\partial^2 u}{\partial x^2}(0, t) = 0^* \qquad u(x, t) \to 0 \qquad \text{as} \qquad x \to \infty \quad (3.29)$$

$$u(x, 0) = 0 \qquad \frac{\partial u}{\partial t}(x, 0) = 0 \qquad 0 < x < \infty$$

Once again we solve the problem by use of the Laplace transform. Following the procedure used in Sec. 3.3.2, the transformed problem becomes

$$\frac{\partial^4 U}{\partial x^4} + \left(\frac{s}{a}\right)^2 U = 0 \qquad 0 < x < \infty$$

$$U(0, s) = \frac{b}{s} \qquad \frac{\partial^2 U}{\partial x^2}(0, s) = 0 \qquad (3.30)$$

$$U(x, s) \to 0 \qquad \text{as} \qquad x \to \infty$$

where $U(x, s) = \mathcal{L}\{u(x, t); t \to s\}$. The general solution of this fourth-

---

*The end condition $\partial^2 u / \partial x^2(0, t) = 0$ states there is no moment at the free end of the beam.

order equation is

$$U(x, s) = e^{-x\sqrt{s/(2a)}}\left[A(s)\cos x\sqrt{\frac{s}{2a}} + B(s)\sin x\sqrt{\frac{s}{2a}}\right]$$

$$+ e^{x\sqrt{s/(2a)}}\left[C(s)\cos x\sqrt{\frac{s}{2a}} + D(s)\sin x\sqrt{\frac{s}{2a}}\right]$$

where $A(s)$, $B(s)$, $C(s)$, and $D(s)$ are all arbitrary functions. To obtain a bounded solution, we must set $C(s) = D(s) = 0$. The remaining boundary conditions in (3.30) lead to $A(s) = b/s$ and $B(s) = 0$; thus, our solution of the transformed problem reduces to

$$U(x, s) = \frac{b}{s}e^{-x\sqrt{s/(2a)}}\cos x\sqrt{\frac{s}{2a}} \tag{3.31}$$

Next, by use of Euler's formula $\cos z = \frac{1}{2}(e^{iz} + e^{-iz})$ we may rewrite (3.31) as*

$$U(x, s) = \frac{b}{2s}(e^{-x\sqrt{-is/a}} + e^{-x\sqrt{is/a}}) \tag{3.32}$$

and using the inverse transform relation (recall Example 1, which is the special case $a = 2$)

$$\mathcal{L}^{-1}\left\{\frac{1}{s}e^{-a\sqrt{s}}; s \to t\right\} = \mathrm{erfc}\,\frac{a}{2\sqrt{t}} \tag{3.33}$$

we deduce that

$$u(x, t) = \frac{b}{2}\left[\mathrm{erfc}\left(\frac{x}{2}\sqrt{\frac{-i}{at}}\right) + \mathrm{erfc}\left(\frac{x}{2}\sqrt{\frac{i}{at}}\right)\right]$$

$$= b\left[1 - \frac{1}{2}\mathrm{erf}\left(\frac{x}{2}\sqrt{\frac{-i}{at}}\right) - \frac{1}{2}\mathrm{erf}\left(\frac{x}{2}\sqrt{\frac{i}{at}}\right)\right] \tag{3.34}$$

Finally, with the aid of the identity (see problem 9 in Exercises 3.3)

$$\mathrm{erf}\sqrt{ix} + \mathrm{erf}\sqrt{-ix} = 2\left[C\left(x\sqrt{\frac{2}{\pi}}\right) + S\left(x\sqrt{\frac{2}{\pi}}\right)\right] \tag{3.35}$$

we can express our solution in terms of the Fresnel integrals

$$u(x, t) = b\left[1 - C\left(\frac{x}{\sqrt{2a\pi t}}\right) - S\left(\frac{x}{\sqrt{2a\pi t}}\right)\right] \tag{3.36}$$

---

* Note that $1 \pm i = \sqrt{\pm 2i}$.

Exercises 3.3

1. Derive an expression for the cumulative gamma distribution function in terms of the error function, where

$$p_x(x) = \frac{1}{\Gamma(\alpha)} x^{\alpha-1} e^{-x} \qquad x > 0$$

and $p_x(x) = 0$, $x < 0$ (a) for $\alpha = \frac{1}{2}$ and (b) for $\alpha = \frac{3}{2}$.

2. The input to a *limiter* is a normal random voltage **x** with mean zero and variance σ^2. The output is a random voltage **y** described by

$$y = \begin{cases} -1 & x < -1 \\ x & -1 < x < 1 \\ 1 & x > 1 \end{cases}$$

Calculate the variance σ_y^2 in terms of the error function.

3. The temperature distribution in a uniform long slender rod is described by the boundary-value problem

$$\frac{\partial^2 u}{\partial x^2} = a^{-2} \frac{\partial u}{\partial t} \qquad -\infty < x < \infty$$

$$u(x, 0) = f(x) \qquad -\infty < x < \infty$$

Use the Fourier transform to find the formal solution

$$u(x, t) = \frac{1}{2a\sqrt{\pi t}} \int_{-\infty}^{\infty} f(\xi) e^{-(x-\xi)^2/(4a^2 t)} \, d\xi$$

4. (a) Show that the formal solution in problem 3 for the special case

$$f(x) = \begin{cases} T_0 & |x| < 1 \\ 0 & |x| > 1 \end{cases}$$

reduces to

$$u(x, t) = \frac{1}{2} T_0 \left[\operatorname{erf}\left(\frac{x+1}{2a\sqrt{t}}\right) - \operatorname{erf}\left(\frac{x-1}{2a\sqrt{t}}\right) \right]$$

(b) Find a solution for the special case

$$f(x) = \begin{cases} 0 & x < 0 \\ T_0 & x > 0 \end{cases}$$

5. (a) Show that

$$\int_u^{\infty} e^{-z^2 - bz} \, dz = \frac{\sqrt{\pi}}{2} e^{b^2/4} \operatorname{erfc}\left(u + \frac{b}{2}\right)$$

(b) Use the result of (a) to solve problem 3 when

$$f(x) = e^{-|x|}$$

6. (a) If the boundary condition in problem (3.21) is

$$u(0, t) = f(t) = \begin{cases} T_1 & 0 < t < b \\ 0 & t \geq b \end{cases}$$

show that

$$u(x, t) = \begin{cases} T_1 \operatorname{erfc}\left(\dfrac{x}{2a\sqrt{t}}\right) & 0 < t < b \\[4mm] T_1\left[\operatorname{erf}\left(\dfrac{x}{2a\sqrt{t-b}}\right) - \operatorname{erf}\left(\dfrac{x}{2a\sqrt{t}}\right)\right] & t \geq b \end{cases}$$

(b) Verify that $u(x, t)$ is continuous at $t = b$, though $f(t)$ is not.

7. If the boundary condition in (3.21) is modified to

$$\frac{\partial u}{\partial x}(0, t) = -f(t)$$

the formal solution becomes

$$u(x, t) = \frac{a}{\sqrt{\pi}} \int_0^t \frac{f(\tau)}{\sqrt{t-\tau}} \exp\left[-\frac{x^2}{4a^2(t-\tau)}\right] d\tau$$

(a) For the special case $f(t) = K$ (constant), show that

$$u(x, t) = K\left[2a\left(\frac{t}{\pi}\right)^{1/2} e^{-x^2/(4a^2 t)} - x\operatorname{erfc}\left(\frac{x}{2a\sqrt{t}}\right)\right]$$

(b) What is the temperature in the rod at the end $x = 0$ as a function of time?

8. (a) Show that

$$\mathscr{L}\left\{\frac{1}{\sqrt{t}} e^{-k/t}; t \to s\right\} = \sqrt{\frac{\pi}{s}} e^{-2\sqrt{ks}} \qquad k \geq 0$$

Hint: Use the result of problem 18 in Exercises 3.2.

(b) By formally differentiating both sides of (a) with respect to the parameter k, deduce that

$$\mathscr{L}\left\{\frac{1}{t\sqrt{t}} e^{-k/t}; t \to s\right\} = \sqrt{\frac{\pi}{k}} e^{-2\sqrt{ks}} \qquad k > 0$$

9. Using the definition of the error function, show that
   (a) $\operatorname{erf}\sqrt{ix} = (1 + i)[C(x\sqrt{2/\pi}) - iS(x\sqrt{2/\pi})]$
   (b) $\operatorname{erf}\sqrt{-ix} = (1 - i)[C(x\sqrt{2/\pi}) + iS(x\sqrt{2/\pi})]$
   (c) $\operatorname{erf}\sqrt{ix} + \operatorname{erf}\sqrt{-ix} = 2[C(x\sqrt{2/\pi}) + S(x\sqrt{2/\pi})]$

10. The free vibrations $u(x, t)$ of a uniform beam are governed by the boundary-value problem

$$\frac{\partial^4 u}{\partial x^4} + \left(\frac{1}{a^2}\right)\frac{\partial^2 u}{\partial t^2} = 0 \qquad 0 < x < \infty \qquad t > 0$$

$$u(0, t) = f(t) \qquad \frac{\partial^2 u}{\partial x^2}(0, t) = 0 \qquad u(x, t) \to 0 \qquad \text{as} \qquad x \to \infty$$

$$u(x, 0) = 0 \qquad \frac{\partial u}{\partial t}(x, 0) = 0 \qquad 0 < x < \infty$$

Given that the formal solution can be expressed in the form

$$u(x, t) = \frac{1}{\sqrt{\pi}} \int_{x/\sqrt{2t}}^{\infty} f\left(t - \frac{x^2}{2v^2}\right)\left(\cos\frac{v^2}{2} + \sin\frac{v^2}{2}\right) dv$$

show that the special case $f(t) = b$ reduces to that given by Eq. (3.36) with $a = 1$.

11. (*Stefan problem*) Imagine the half-plane $x > 0$ filled with ice at 0°C with the wall at $x = 0$ kept at constant temperature T_0. As the ice begins to melt, the interface boundary between ice and water is described by $x = s(t)$, and the temperature $u(x, t)$ in the water satisfies the boundary-value problem

$$\frac{\partial^2 u}{\partial x^2} = \frac{\partial u}{\partial t} \qquad 0 < x < s \qquad t > 0$$

$$u(0, t) = T_0 \qquad u[s(t), t] = 0 \qquad \frac{\partial u}{\partial x}[s(t), t] = -\frac{ds}{dt}$$

Assume a solution of the form

$$u(x, t) = A + B \operatorname{erf}\frac{x}{2\sqrt{t}}$$

where A and B are constants to be determined.
(a) Show that $s(t) = 2k\sqrt{t}$, where k is unknown.
(b) Show that $A = T_0$ and $B = -t_0/\operatorname{erf} k$.

3.4   The Exponential Integral and Related Functions

The **exponential integral** is defined by*

$$\operatorname{Ei}(x) = \int_{-\infty}^{x} \frac{e^t}{t}\, dt \qquad x \neq 0 \tag{3.37}$$

---

*Technically, Eq. (3.37) does not define Ei (x) for $x > 0$ unless we interpret the integral as its Cauchy principal value, i.e.,

$$\int_{-\infty}^{x} \frac{e^t}{t}\, dt = \lim_{\epsilon \to 0^+}\left(\int_{-\infty}^{-\epsilon} \frac{e^t}{t}\, dt + \int_{\epsilon}^{x} \frac{e^t}{t}\, dt\right)$$

Another definition that is often given results from the replacement of x by $-x$ and t by t, which leads to

$$-\text{Ei}\,(-x) \equiv E_1(x) = \int_x^\infty \frac{e^{-t}}{t}\,dt \qquad x > 0 \qquad (3.38)$$

The exponential integral (3.37) or (3.38) is encountered in several areas, including antenna theory and some astrophysical problems. Also many integrals of a more complicated nature can be expressed in terms of the exponential integrals.

Comparison of (3.38) with Eq. (2.43) in Sec. 2.5 reveals that $E_1(x)$ is related to the incomplete gamma functions according to

$$E_1(x) = \Gamma(0, x) = \lim_{a \to 0} [\Gamma(a) - \gamma(a, x)] \qquad (3.39)$$

Thus, properties of $E_1(x)$ can be deduced from those of the incomplete gamma functions. For example, from the series for $\gamma(a, x)$ [see Eq. (2.45) in Sec. 2.5], we have

$$E_1(x) = \lim_{a \to 0} \left[\Gamma(a) - x^a \sum_{n=0}^\infty \frac{(-1)^n x^n}{n!(n+a)} \right]$$

$$= \lim_{a \to 0} \left[\frac{a\Gamma(a) - x^a}{a} \right] - \sum_{n-1}^\infty \frac{(-1)^n x^n}{n!n} \qquad (3.40)$$

Using the recurrence formula for the gamma function and L'Hôpital's rule, it follows that*

$$\lim_{a \to 0} \left[\frac{a\Gamma(a) - x^a}{a} \right] = \lim_{a \to 0} [\Gamma'(a+1) - x^a \ln x]$$

$$= -\gamma - \ln x \qquad (3.41)$$

Hence we have derived the series representation

$$E_1(x) = -\gamma - \ln x - \sum_{n=1}^\infty \frac{(-1)^n x^n}{n!n} \qquad x > 0 \qquad (3.42)$$

Equation (3.42) illustrates the logarithmic behavior of $E_1(x)$ for small arguments, i.e.,

$$E_1(x) \sim -\ln x \qquad x \to 0^+ \qquad (3.43)$$

---

* Recall from Eq. (2.58) in Sec. 2.6 that $\Gamma'(1) = -\gamma$.

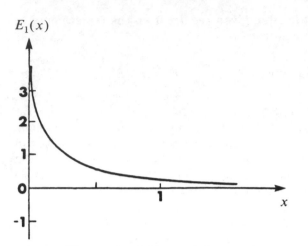

Figure 3.4 The exponential integral $E_1(x)$.

For large arguments, we can use Eq. (2.49) in Sec. 2.5 to deduce that*

$$E_1(x) \sim \frac{e^{-x}}{x} \qquad x \to \infty \tag{3.44}$$

The graph of $E_1(x)$ for positive x is shown in Fig. 3.4.

3.4.1 Logarithmic integral

Closely related to the exponential integral is the **logarithmic integral**

$$\text{li}\,(x) = \int_0^x \frac{dt}{\ln t} \qquad x \neq 1 \tag{3.45}$$

Setting $u = \ln t$, we see that (3.45) becomes

$$\text{li}\,(x) = \int_{-\infty}^{\ln x} \frac{e^u}{u}\,du$$

and thus deduce that

$$\text{li}\,(x) = \text{Ei}\,(\ln x) = -E_1(-\ln x) \qquad 0 < x < 1 \tag{3.46}$$

---

* The complete asymptotic series for $E_1(x)$ was developed in Example 19 in Sec. 1.6.2.

By using (3.46), we can immediately deduce properties of $\mathrm{li}\,(x)$ from those developed for the exponential integrals In particular, Eq. (3.42) leads to

$$\mathrm{li}\,(x) = \gamma + \ln\,(-\ln x) + \sum_{n=1}^{\infty} \frac{(\ln x)^n}{n!n} \qquad 0 < x < 1 \qquad (3.47)$$

The graph of $\mathrm{li}\,(x)$ is shown in Fig. 3.5.

3.4.2 Sine and cosine integrals

Another oot of special functions that are related to the exponential integral includes the **sine integral** and **cosine integral** defined, respectively, by

$$\mathrm{Si}\,(x) = \int_0^x \frac{\sin t}{t}\,dt \qquad x > 0 \qquad (3.48)$$

and

$$\mathrm{Ci}\,(x) = \int_\infty^x \frac{\cos t}{t}\,dt \qquad x > 0 \qquad (3.49)$$

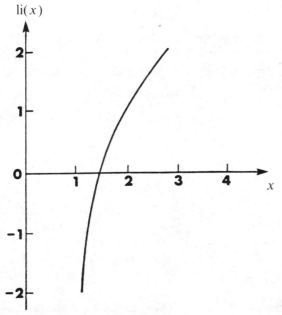

Figure 3.5 The logarithmic integral.

To relate these integrals to the exponential integral requires complex variable theory, and thus we omit the derivation.*

It is convenient in some applications to introduce another sine integral defined by

$$\text{si}\,(x) = -\int_x^\infty \frac{\sin t}{t}\,dt \tag{3.50}$$

which is related to (3.48) by (see problem 6 in Exercises 3.4)

$$\text{Si}\,(x) = \frac{\pi}{2} + \text{si}\,(x) \tag{3.51}$$

Special values of these functions include (see problem 7 in Exercises 3.4)

$$\text{Si}\,(0) = 0 \qquad \text{Si}\,(\infty) = \frac{\pi}{2} \tag{3.52a}$$

$$\text{Ci}\,(0^+) = -\infty \qquad \text{Ci}\,(\infty) = 0 \tag{3.52b}$$

Taking derivatives, we obtain

$$\text{Si}'\,(x) = \frac{\sin x}{x} \qquad \text{Ci}'\,(x) = \frac{\cos x}{x} \tag{3.53}$$

which shows that both functions are oscillatory. We observe that $\text{Si}\,(x)$ has extrema at $x = n\pi$ $(n = 0, 1, 2, \ldots)$, while $\text{Ci}\,(x)$ has extrema at $x = (n + \frac{1}{2})\pi$ $(n = 0, 1, 2, \ldots)$. The graphs of these functions are shown in Fig. 3.6.

Exercises 3.4

1. Show that $(x > 0)$

$$E_1(x) = e^{-x} \int_0^\infty \frac{e^{-xt}}{1+t}\,dt$$

2. Derive the asymptotic formula

$$\text{Ei}\,(x) \sim \frac{e^x}{x} \sum_{n=0}^\infty \frac{n!}{x^n} \qquad x \to \infty$$

---

*See N. N. Lebedev, *Special Functions and Their Applications*, Dover, New York, 1972, pp. 33–37.

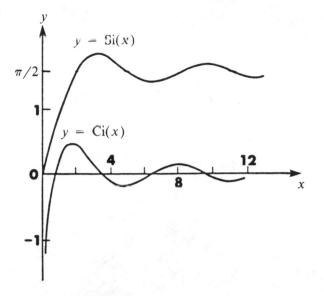

Figure 3.6 The sine and cosine integrals.

3. From the result of problem 2, show that

$$E_1(x) \sim \frac{e^{-x}}{x} \sum_{n=0}^{\infty} \frac{(-1)^n n!}{x^n} \qquad x \to \infty$$

4. Derive the asymptotic formula

$$\text{li}\,(x) \sim \frac{x}{\ln x} \sum_{n=0}^{\infty} \frac{n!}{(\ln x)^n} \qquad x \to \infty$$

5. Let

$$f(t) = \int_0^{\infty} \frac{\sin tx}{x}\, dx \qquad t > 0$$

(a) By taking the Laplace transform of both sides, show that

$$\mathscr{L}\{f(t); s\} = \int_0^{\infty} (x^2 + s^2)^{-1}\, dx$$

(b) Evaluate the integral in (a), and by taking the inverse Laplace transform, deduce the value $f(t) = \pi/2$.

6. Using the result of problem 5, show that

$$\text{Si}\,(x) = \frac{\pi}{2} + \text{si}\,(x)$$

7. Show that

    (a) $\text{Si}(\infty) = \dfrac{\pi}{2}$ (c) $\text{Si}(0) = 0$

    (b) $\text{Ci}(\infty) = 0$ (d) $\text{Ci}(0^+) = -\infty$

8. Derive the series representation

$$\text{Si}(x) = \sum_{n=0}^{\infty} \frac{(-1)^n x^{2n+1}}{(2n+1)(2n+1)!}$$

In problems 9 to 14, derive the integral relation.

9. $\displaystyle\int_0^\infty e^{-st} E_1(t)\, dt = \frac{1}{s}\ln(1+s),\ s>0$

10. $\displaystyle\int_0^\infty e^{-st}\,\text{Si}(t)\, dt = \frac{1}{s}\tan^{-1}\frac{1}{s},\ s>0$

11. $\displaystyle\int_0^\infty e^{-st}\,\text{si}(t)\, dt = -\frac{1}{s}\tan^{-1}s,\ s>0$

12. $\displaystyle\int_0^\infty e^{-st}\,\text{Ci}(t)\, dt = -\frac{1}{2s}\ln(1+s^2),\ s>0$

13. $\displaystyle\int_0^\infty \cos x\,\text{Ci}(x)\, dx = \int_0^\infty \sin x\,\text{si}(x)\, dx = -\frac{\pi}{4}$

14. $\displaystyle\int_0^\infty [\text{Ci}(x)]^2\, dx = \int_0^\infty [\text{si}(x)]^2\, dx = \frac{\pi}{2}$

In problems 15 to 20, express the given integral in terms of $\text{Si}(x)$ and/or $\text{Ci}(x)$.

15. $\displaystyle\int_a^b \frac{\sin t}{t}\, dt$ 18. $\displaystyle\int_a^b \frac{\sin t}{t^3}\, dt$

                                      Hint: Use integration by parts.

16. $\displaystyle\int_a^b \frac{\sin t^2}{t}\, dt$ 19. $\displaystyle\int_0^b \frac{1 - \cos t}{t}\sin at\, dt$

    Hint: Let $t^2 = u$.

17. $\displaystyle\int_a^b \frac{\cos t^2}{t}\, dt$ 20. $\displaystyle\int_2^3 \frac{\cos 2t}{1 - t^2}\, dt$

                                      Hint: Start with partial fractions.

3.5   Elliptic Integrals

The parametric equations for an elliptic arc are given by $(b > a)$

$$x = a \cos \theta \qquad 0 \le \theta \le \phi \qquad (3.54)$$
$$y = b \sin \theta$$

Using the formula for arclength from calculus, we find the length of the elliptic arc (3.54) leads to the integral

$$L - \int_0^\phi \sqrt{a^2 \sin^2 \theta + b^2 \cos^2 \theta}\, d\theta \qquad (3.55)$$

which can also be expressed in the form

$$L = b \int_0^\phi \sqrt{1 - e^2 \sin^2 \theta}\, d\theta \qquad (3.56)$$

where e is the eccentricity of the ellipse, defined by

$$e = \frac{1}{b} \sqrt{b^2 - a^2} \qquad (3.57)$$

The integral in (3.56) cannot be evaluated in terms of elementary functions. Because of its origin, it is called an *elliptic integral*.

There are three classifications of elliptic integrals, called **elliptic integrals of the first, second, and third kinds** and defined, respectively, by

$$F(m, \phi) = \int_0^\phi \frac{d\theta}{\sqrt{1 - m^2 \sin^2 \theta}} \qquad 0 < m < 1 \qquad (3.58)$$

$$E(m, \phi) = \int_0^\phi \sqrt{1 - m^2 \sin^2 \theta}\, d\theta \qquad 0 < m < 1 \qquad (3.59)$$

and

$$\Pi(m, \phi, a) = \int_0^\phi \frac{d\theta}{\sqrt{1 - m^2 \sin^2 \theta}\,(1 + a^2 \sin^2 \theta)} \qquad (3.60)$$

$$0 < m < 1 \qquad a \ne m, 0$$

The parameter ϕ is called the *amplitude*, and m is the *modulus*. When $\phi = \pi/2$, we refer to (3.58) to (3.60) as **complete elliptic**

integrals, and they are often given the special designations

$$K(m) = \int_0^{\pi/2} \frac{d\theta}{\sqrt{1 - m^2 \sin^2 \theta}} \qquad 0 < m < 1 \qquad (3.61)$$

$$E(m) = \int_0^{\pi/2} \sqrt{1 - m^2 \sin^2 \theta}\, d\theta \qquad 0 < m < 1 \qquad (3.62)$$

and

$$\Pi(m, a) = \int_0^{\pi/2} \frac{d\theta}{\sqrt{1 - m^2 \sin^2 \theta}\,(1 + a^2 \sin^2 \theta)} \qquad (3.63)$$

$$0 < m < 1 \qquad a \neq m, 0$$

Sometimes these integrals are designated simply by the letters K, E, and Π.

Some of the importance connected with these integrals lies in the following theorem, which we state without proof.*

Theorem 3.1.   If $R(x, y)$ is a rational function in x and y and $P(x)$ is a polynomial of degree 4 or less, then the integral

$$\int R(x, \sqrt{P(x)})\, dx$$

can always be expressed in terms of elliptic integrals.

3.5.1 Limiting values and series representations

For the limiting case $m \to 0$, we find that (3.61) leads to

$$K(0) = \int_0^{\pi/2} d\theta = \frac{\pi}{2} \qquad (3.64)$$

and similarly,

$$E(0) = \frac{\pi}{2} \qquad (3.65)$$

---

* For a proof of Theorem 3.1, see F. Bowman, *Introduction to Elliptic Functions with Applications*, Dover, New York, 1961.

In the other limiting case where $m \to 1$, we obtain the results (see problem 1 in Exercises 3.5)

$$K(1) = \infty \tag{3.66}$$

$$E(1) = 1 \tag{3.67}$$

We can generate an infinite series representation for K and E by first expanding the integrands in (3.61) and (3.62) in binomial series and then using termwise integration. For example,

$$(1 - m^2 \sin^2 \theta)^{-1/2} = \sum_{n=0}^{\infty} \binom{-1/2}{n} (-1)^n m^{2n} \sin^{2n} \theta \tag{3.68}$$

and hence, by using the integral formula (see problem 19 in Exercises 3.5)

$$\int_0^{\pi/2} \sin^{2n} \theta \, d\theta = \frac{\pi}{2} (-1)^n \binom{-1/2}{n} \tag{3.69}$$

we deduce the series representation

$$K(m) = \frac{\pi}{2} \sum_{n=0}^{\infty} \binom{-1/2}{n}^2 m^{2n} \tag{3.70}$$

In the same fashion, it follows that

$$E(m) = \frac{\pi}{2} \sum_{n=0}^{\infty} \binom{1/2}{n}\binom{-1/2}{n} m^{2n} \tag{3.71}$$

3.5.2   The pendulum problem

A mass μ is suspended from the end of a rod of constant length b (whose weight is negligible). Summing forces (see Fig. 3.7) makes it clear that the weight component $\mu g \cos \phi$ acting in the normal direction to the path is offset by the force of restraint in the rod.* Therefore, the only weight component contributing to the motion is $\mu g \sin \phi$, which acts in the direction of the tangent to the path. If we denote the arclength of the path by s, then Newton's second law of motion ($F = ma$) leads to

$$\mu \frac{d^2 s}{dt^2} = -\mu g \sin \phi$$

---

* Here g is the gravitational constant.

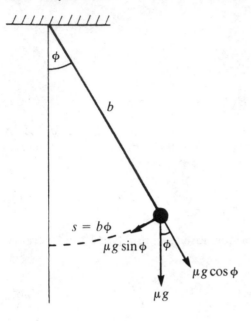

Figure 3.7 A swinging pendulum.

where the minus sign signifies that the tangential force component opposes the motion for increasing s. The arclength s of a circle of radius b is related to the central angle ϕ through the formula $s = b\phi$, and so the equation of motion (after simplification) becomes

$$\frac{d^2\phi}{dt^2} + k^2 \sin \phi = 0 \tag{3.72}$$

where $k^2 = g/b$.

Equation (3.72) is nonlinear and cannot be solved in terms of elementary functions. To solve it, first we note that it is equivalent to

$$\frac{1}{2}\left(\frac{d\phi}{dt}\right)^2 - k^2 \cos \phi = C \tag{3.73}$$

i.e., (3.72) is the derivative of (3.73). (The constant C is proportional to the energy of the system.) Solving (3.73) for $(d\phi/dt)^2$, we have

$$\left(\frac{d\phi}{dt}\right)^2 = 2C + 2k^2 \cos \phi = 2C + 2k^2 - 4k^2 \sin^2 \tfrac{1}{2}\phi$$

or

$$\left(\frac{d\phi}{dt}\right)^2 = 2(C + k^2)\left[1 - \frac{2k^2}{C + k^2} \sin^2 \frac{1}{2}\phi\right] \tag{3.74}$$

If we now introduce the parameter

$$m^2 = \frac{C + k^2}{2k^2} \tag{3.75}$$

and make the change of dependent variable

$$y = \frac{1}{m} \sin \frac{1}{2} \phi \tag{3.76}$$

the chain rule demands that

$$\frac{d\phi}{dt} = 2m \frac{dy}{dt} (1 - m^2 y)^{-1/2} \tag{3.77}$$

Upon making these replacements and taking the (positive) square root, (3.74) becomes

$$\frac{dy}{dt} = k\sqrt{(1 - y^2)(1 - m^2 y^2)} \tag{3.78}$$

If we assume the position of the pendulum is $\phi = 0$ at time $t = 0$ and position $\phi = \Phi$ at time $t = T$, then the separation of variables applied to (3.78) leads to $[Y = (1/m) \sin \frac{1}{2}\Phi]$

$$kT = \int_0^Y \frac{dy}{\sqrt{(1 - y^2)(1 - m^2 y^2)}} \tag{3.79}$$

This integral is another form of the elliptic integral of the first kind $F(m, Y)$, as can be verified by making the substitution $y = \sin x$ (see problem 4 in Exercises 3.5).

Equation (3.79) gives the total time of motion of the pendulum in terms of an elliptic integral. If we wish to solve explicitly for the angle of motion $\Phi = 2 \sin^{-1} mY$, we need to define an inverse function for $F(m, Y)$. Such a function exists and is called a **Jacobian elliptic function**. If in general we set

$$u = F(m, \phi) \tag{3.80}$$

then we can define three elliptic functions by the relations

$$\text{sn } u = \sin \phi \tag{3.81}$$

$$\text{cn } u = \cos \phi \tag{3.82}$$

and
$$\text{dn } u = \sqrt{1 - m^2 \sin^2 \phi} \tag{3.83}$$

These elliptic functions belong to the class of *doubly periodic functions* with one real period and one imaginary period. In this respect, they have characteristics of both the circular and hyperbolic functions. Much of the theory of elliptic functions is couched in the language of complex variables, and thus we do not pursue their general theory. Some elementary properties, however, are taken up in the exercises.

Exercises 3.5

1. Show that
   (a) $K(1) = \infty$ (b) $E(1) = 1$

2. Show that

$$\lim_{m \to 0} \frac{K - E}{m^2} = \frac{\pi}{4}$$

3. Verify that

$$F(m, \phi + \pi) - F(m, \phi) = 2K$$

In problems 4 to 12, derive the given integral relation.

4. $\displaystyle\int_0^{\pi/2} \frac{dx}{\sqrt{\sin x}} = \sqrt{2}\, K\!\left(\frac{1}{\sqrt{2}}\right)$

5. $\displaystyle\int_0^{\pi/2} \sqrt{1 + 4\sin^2 x}\; dx = \sqrt{5}\, E\big(\sqrt{4/5}\,\big)$

6. $\displaystyle\int_1^{\infty} (x^2 - 1)^{-1/2}(x^2 + 3)^{-1/2}\, dx = \frac{1}{2} K\!\left(\frac{\sqrt{3}}{2}\right)$

7. $\displaystyle\int_0^x \frac{dy}{\sqrt{(1 - y^2)(1 - m^2 y^2)}} = F(m, x)$

8. $\displaystyle\int_0^x \sqrt{\frac{1 - m^2 t^2}{1 - t^2}}\; dt = E(m, x)$

9. $\displaystyle\int_0^{\phi} \frac{\sin^2 x}{\sqrt{1 - m^2 \sin^2 x}}\, dx = \frac{1}{m^2}[F(m, \phi) - E(m, \phi)]$

10. $\displaystyle\int_0^{\pi/2} \frac{dx}{\sqrt{2 - \cos x}} = \frac{2}{\sqrt{3}}\left[K\!\left(\sqrt{\frac{2}{3}}\right) - F\!\left(\sqrt{\frac{2}{3}}, \frac{\pi}{4}\right)\right]$

11. $\displaystyle\int_0^2 (4 - x^2)^{-1/2}(9 - x^2)^{-1/2}\, dx = \frac{1}{3} K(\frac{2}{3})$

12. $\int_0^1 (1+x^2)^{-1/2}(1+2x^2)^{-1/2}\,dx = \dfrac{1}{\sqrt{2}}\left[K\!\left(\dfrac{1}{\sqrt{2}}\right) - F\!\left(\dfrac{1}{\sqrt{2}},\dfrac{\pi}{4}\right)\right]$

13. Show that

$$\dfrac{1}{2\pi}\int_0^\pi \dfrac{\cos\theta\,d\theta}{(a^2+b^2+z^2-2ab\,\cos\theta)^{1/2}} = \dfrac{(ab)^{-1/2}}{k\pi}\left[\left(1-\dfrac{k^2}{2}\right)K - E\right]$$

where $\qquad\qquad k^\text{y} = \dfrac{4ab}{(a+b)^2+z^2}$

14. Find the perimeter of the ellipse $8x^2 + 9y^2 = 72$.

15. Find the area enclosed by one loop of the curve $y^2 = 1 - 4\sin^2 x$.

16. Find the arclength of the lemniscate $r^2 = \cos 2\theta$, $0 \le \theta \le \pi/2$.

17. Find the length of the curve $y = \sin x$, $0 \le x \le \pi/3$.

18. Find the surface area of a right circular cylinder of radius r intercepted by a sphere of radius a $(a > r)$ whose center lies on the cylinder.

19. Show that $(n = 0, 1, 2, \ldots)$

$$\int_0^{\pi/2} \sin^{2n}\theta\,d\theta = \dfrac{\pi}{2}(-1)^n\binom{-1/2}{n}$$

   Hint: See problem 17a in Exercises 1.2.

20. Show that
    (a) $\operatorname{sn}(0) = 0$ 　　　　　　　(b) $\operatorname{cn}(0) = \operatorname{dn}(0) = 1$

21. Verify the identities.
    (a) $\operatorname{sn}^2 u + \operatorname{cn}^2 u = 1$
    (b) $m^2 \operatorname{sn}^2 u + \operatorname{dn}^2 u = 1$
    (c) $\operatorname{dn}^2 u - m^2 \operatorname{cn}^2 u = 1 - m^2$

22. Derive the derivative relations

    (a) $\dfrac{d}{du}\operatorname{sn} u = \operatorname{cn} u\ \operatorname{dn} u$

    (b) $\dfrac{d}{du}\operatorname{cn} u = -\operatorname{sn} u\ \operatorname{dn} u$

    (c) $\dfrac{d}{du}\operatorname{dn} u = -m^2 \operatorname{sn} u\ \operatorname{cn} u$

23. Show that
    (a) $\operatorname{sn}(u + 4K) = \operatorname{sn} u$
    (b) $\operatorname{cn}(u + 4K) = \operatorname{cn} u$
    (c) $\operatorname{dn}(u + 2K) = \operatorname{dn} u$

24. Verify the addition formulas.

(a) $\operatorname{sn}(u+v) = \dfrac{\operatorname{sn} u \operatorname{cn} v \operatorname{dn} v + \operatorname{sn} v \operatorname{cn} u \operatorname{dn} u}{1 - m^2 \operatorname{sn}^2 u \operatorname{sn}^2 v}$

(b) $\operatorname{cn}(u+v) = \dfrac{\operatorname{cn} u \operatorname{cn} v - \operatorname{sn} u \operatorname{dn} u \operatorname{sn} v \operatorname{dn} v}{1 - m^2 \operatorname{sn}^2 u \operatorname{sn}^2 v}$

(c) $\operatorname{dn}(u+v) = \dfrac{\operatorname{dn} u \operatorname{dn} v - m^2 \operatorname{sn} u \operatorname{cn} u \operatorname{sn} v \operatorname{cn} v}{1 - m^2 \operatorname{sn}^2 u \operatorname{sn}^2 v}$

25. Show that

(a) $\lim\limits_{m \to 1} \operatorname{sn} u = \tanh u$

(b) $\lim\limits_{m \to 1} \operatorname{cn} u = \operatorname{sech} u$

(c) $\lim\limits_{m \to 1} \operatorname{dn} u = \operatorname{sech} u$

26. Show that
(a) $\operatorname{sn}(-u) = -\operatorname{sn} u$
(b) $\operatorname{cn}(-u) = \operatorname{cn} u$
(c) $\operatorname{dn}(-u) = \operatorname{dn} u$

27. A pendulum of length 2 ft is released from rest at an angle of 60° with the vertical. Assume that acceleration due to gravity is $g = 32$ ft/s.
(a)   Determine the period of oscillation.
(b)   Find the period of oscillation if the governing DE is approximated by the linear equation

$$\frac{d^2\phi}{dt^2} + k^2\phi = 0$$

where $k^2 = g/b$.

4

Legendre Polynomials and Related Functions

4.1 Introduction

The *Legendre polynomials* are closely associated with physical phenomena for which spherical geometry is important. In particular, these polynomials first arose in the problem of expressing the newtonian potential of a conservative force field in an infinite series involving the distance variables of two points and their included central angle (see Sec. 4.2). Other similar problems dealing with either gravitational potentials or electrostatic potentials also lead to Legendre polynomials, as do certain steady-state heat conduction problems in spherical solids, and so forth.

There exist a whole class of polynomial sets which have many properties in common and for which the Legendre polynomials represent the simplest example. Each polynomial set satisfies several recurrence formulas, is involved in numerous integral relationships, and forms the basis for series expansions resembling Fourier trigonometric series, where the sines and cosines are replaced by members of the polynomial set. Because of all the similarities in these polynomial sets and because the Legendre polynomials are the simplest such set, our development of the properties associated with the Legendre polynomials will be more extensive than similar developments in Chap. 5, where we introduce other polynomial sets.

In addition to the Legendre polynomials, we present a brief discussion of the *Legendre functions of the second kind* and *associated Legendre functions*. The Legendre functions of the second kind arise as a second solution set of Legendre's equation (independent of the Legendre polynomials), and the associated functions are related to derivatives of the Legendre polynomials.

4.2   Legendre Polynomials

Our introduction to Legendre polynomials will follow the historical discovery of this important polynomial set, i.e., through expansion of the *generating function* into a particular type of series. In turn, the generating function is used to develop many of the properties associated with these polynomials.

4.2.1   The generating function

Among other areas of application, the subject of potential theory is concerned with the forces of attraction due to the presence of a gravitational field. Central to the discussion of problems of gravitational attraction is *Newton's law of universal gravitation*:

> Every particle of matter in the universe attracts every other particle with a force whose direction is that of the line joining the two, and whose magnitude is directly as the product of their masses and inversely as the square of their distance from each other.

The force field generated by a single particle is usually considered to be *conservative*. That is, there exists a potential function V such that the gravitational force $\mathbf{F}$ at a point of free space (i.e., free of point masses) is related to the potential function according to

$$\mathbf{F} = -\nabla V \tag{4.1}$$

where the minus sign is conventional. If r denotes the distance between a point mass and a point of free space, the potential function can be shown to have the form*

$$V(r) = \frac{k}{r} \tag{4.2}$$

where k is a constant whose numerical value does not concern us. Because of spherical symmetry of the gravitational field, the potential function V depends on only the radial distance r.

Valuable information on the properties of potentials like (4.2) may be inferred from developments of the potential function into power series of certain types. In 1785, A. M. Legendre published his "Sur l'attraction des sphéroïdes," in which he developed the gravitational potential (4.2) in a power series involving the ratio of two distance variables. He found that the coefficients appearing in this expansion were polynomials that exhibited interesting properties.

---

*See O. D. Kellogg, *Foundations of Potential Theory*, Dover, New York, 1953, chap. 3.

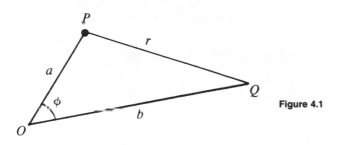

Figure 4.1

To obtain Legendre's results, let us suppose that a particle of mass m is located at point P, which is a units from the origin of our coordinate system (see Fig. 4.1). Let the point Q represent a point of free space r units from P and b units from the origin O. For the sake of definiteness, let us assume $b > a$. Then, from the law of cosines, we find the relation

$$r^2 = a^2 + b^2 - 2ab \cos \phi \qquad (4.3)$$

where ϕ is the central angle between the rays $\overline{OP}$ and $\overline{OQ}$. By rearranging the terms and factoring out b^2, it follows that

$$r^2 = b^2 \left[1 - 2\frac{a}{b} \cos \phi + \left(\frac{a}{b}\right)^2 \right] \qquad a < b \qquad (4.4)$$

For notational simplicity, we introduce the parameters

$$t = \frac{a}{b} \qquad x = \cos \phi \qquad (4.5)$$

and thus, upon taking the square root,

$$r = b(1 - 2xt + t^2)^{1/2} \qquad (4.6)$$

Finally, the substitution of (4.6) into (4.2) leads to the expression

$$V = \frac{k}{b}(1 - 2xt + t^2)^{-1/2} \qquad 0 < t < 1 \qquad (4.7)$$

for the potential function. For reasons that will soon be clear, we refer to the function $w(x, t) = (1 - 2xt + t^2)^{-1/2}$ as the **generating function** of the Legendre polynomials. Our task at this point is to develop $w(x, t)$ in a power series in the variable t.

From Example 11 in Sec. 1.3.2, we recall the binomial series

$$(1-u)^{-1/2} = \sum_{n=0}^{\infty} \binom{-1/2}{n}(-1)^n u^n \qquad |u|<1 \qquad (4.8)$$

Hence, by setting $u = t(2x - t)$, we find that

$$w(x, t) = (1 - 2xt + t^2)^{-1/2}$$

$$= \sum_{n=0}^{\infty} \binom{-1/2}{n}(-1)^n t^n (2x - t)^n \qquad (4.9)$$

which is valid for $|2xt - t^2| < 1$. For $|t| < 1$, it follows that $|x| \leq 1$. The factor $(2x - t)^n$ is simply a finite binomial series, and thus (4.9) can be further expressed as

$$w(x, t) = \sum_{n=0}^{\infty} \binom{-1/2}{n}(-1)^n t^n \sum_{k=0}^{n} \binom{n}{k}(-1)^k (2x)^{n-k} t^k$$

or

$$w(x, t) = \sum_{n=0}^{\infty} \sum_{k=0}^{n} \binom{-1/2}{n}\binom{n}{k}(-1)^{n+k}(2x)^{n-k} t^{n+k} \qquad (4.10)$$

Since our goal is to obtain a power series involving powers of t to a single index, the change of index $n \to n - k$ is suggested. Thus, recalling Eq. (1.15) in Sec. 1.2.3,

$$\sum_{n=0}^{\infty} \sum_{k=0}^{n} A_{n-k,k} = \sum_{n=0}^{\infty} \sum_{k=0}^{[n/2]} A_{n-2k,k}$$

we see that (4.10) can be written in the equivalent form

$$w(x, t) = \sum_{n=0}^{\infty} \left[\sum_{k=0}^{[n/2]} \binom{-1/2}{n-k}\binom{n-k}{k}(-1)^n (2x)^{n-2k} \right] t^n \qquad (4.11)$$

The innermost summation in (4.11) is of finite length and therefore represents a polynomial in x, which happens to be of degree n. If we denote this polynomial by the symbol

$$P_n(x) = \sum_{k=0}^{[n/2]} \binom{-1/2}{n-k}\binom{n-k}{k}(-1)^n (2x)^{n-2k} \qquad (4.12)$$

then (4.11) leads to the intended result

$$w(x, t) = (1 - 2xt + t^2)^{-1/2} = \sum_{n=0}^{\infty} P_n(x) t^n \qquad |x| \leq 1 \qquad |t| < 1$$

$$(4.13)$$

The polynomials $P_n(x)$ are called the **Legendre polynomials** in honor of their discoverer. By recognizing that [see Eq. (1.31) in Sec. 1.2.4 and Eq. (2.25) in Sec. 2.2.2]

$$\binom{-\frac{1}{2}}{n} = (-1)^n \binom{n-\frac{1}{2}}{n}$$

$$= (-1)^n \frac{\Gamma(n+\frac{1}{2})}{n!\Gamma(\frac{1}{2})}$$

$$= \frac{(-1)^n (2n)!}{2^{2n}(n!)^2} \tag{4.14}$$

it follows that the product of binomial coefficients in (4.12) is

$$\binom{-\frac{1}{2}}{n-k}\binom{n-k}{k} = \frac{(-1)^{n-k}(2n-2k)!}{2^{2n-2k}(n-k)!k!(n-2k)!} \tag{4.15}$$

and hence, (4.12) becomes

$$P_n(x) = \sum_{k=0}^{[n/2]} \frac{(-1)^k (2n-2k)! x^{n-2k}}{2^n k!(n-k)!(n-2k)!} \tag{4.16}$$

The first few Legendre polynomials are listed in Table 4.1.

Making an observation, we note that when n is an even number, the polynomial $P_n(x)$ is an even function; and when n is odd, the polynomial is an odd function. Therefore,

$$P_n(-x) = (-1)^n P_n(x) \qquad n = 0, 1, 2, \ldots \tag{4.17}$$

The graphs of $P_n(x)$, $n = 0, 1, 2, 3, 4$, are sketched in Fig. 4.2 over the interval $-1 \le x \le 1$.

Returning now to Eq. (4.7) with $x = \cos \phi$ and $t = a/b$, we find that the potential function has the series expansion

$$V = \frac{k}{b} \sum_{n=0}^{\infty} P_n(\cos \phi)\left(\frac{a}{b}\right)^n \qquad a < b \tag{4.18}$$

TABLE 4.1   Legendre Polynomials

$P_0(x) = 1$
$P_1(x) = x$
$P_2(x) = \frac{1}{2}(3x^2 - 1)$
$P_3(x) = \frac{1}{2}(5x^3 - 3x)$
$P_4(x) = \frac{1}{8}(35x^4 - 30x^2 + 3)$
$P_5(x) = \frac{1}{8}(63x^5 - 70x^3 + 15x)$

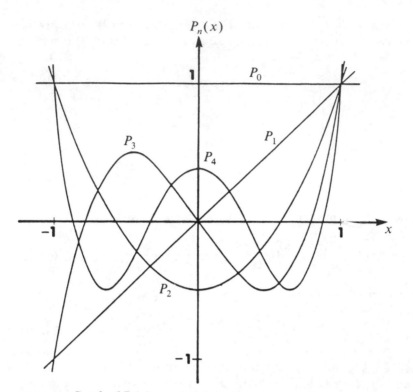

Figure 4.2 Graph of $P_n(x)$, $n = 0, 1, 2, 3, 4$.

In terms of the argument $\cos \phi$, the Legendre polynomials can be expressed as trigonometric polynomials of the form shown in Table 4.2 (see problem 3 in Exercises 4.2).

In Fig. 4.3 the first few polynomials $P_n(\cos \phi)$ are plotted as a function of the angle ϕ.

4.2.2. Special values and recurrence formulas

The Legendre polynomials are rich in recurrence relations and identities. Central to the development of many of these is the

TABLE 4.2 Legendre Trigonometric Polynomials

$$P_0(\cos \phi) = 1$$
$$P_1(\cos \phi) = \cos \phi$$
$$P_2(\cos \phi) = \tfrac{1}{2}(3 \cos^2 \phi - 1)$$
$$= \tfrac{1}{4}(3 \cos 2\phi + 1)$$
$$P_3(\cos \phi) = \tfrac{1}{2}(5 \cos^3 \phi - 3 \cos \phi)$$
$$= \tfrac{1}{8}(5 \cos 3\phi + 3 \cos \phi)$$

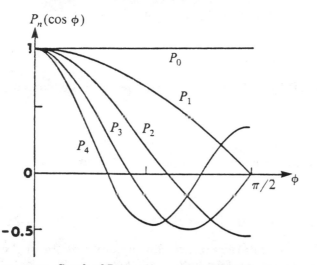

Figure 4.3 Graph of $P_n(\cos \phi)$, $n = 0, 1, 2, 3, 4$.

generating-function relation

$$(1 - 2xt + t^2)^{-1/2} = \sum_{n=0}^{\infty} P_n(x)t^n \qquad |x| \le 1 \qquad |t| < 1 \qquad (4.19)$$

Special values of the Legendre polynomials can be derived directly from (4.19) by substituting particular values for x. For example, the substitution of $x = 1$ yields

$$(1 - 2t + t^2)^{-1/2} = (1 - t)^{-1} = \sum_{n=0}^{\infty} P_n(1)t^n \qquad (4.20)$$

However, we recognize that $(1 - t)^{-1}$ is the sum of a geometric series, so (4.20) is equivalent to

$$\sum_{n=0}^{\infty} t^n = \sum_{n=0}^{\infty} P_n(1)t^n \qquad (4.21)$$

Hence, from the uniqueness theorem of the power series (Theorem 1.12), we can compare like coefficients of t^n in (4.21) to deduce the result

$$P_n(1) = 1 \qquad n = 0, 1, 2, \ldots \qquad (4.22)$$

Also from (4.17) we see that

$$P_n(-1) = (-1)^n \qquad n = 0, 1, 2, \ldots \qquad (4.23)$$

The substitution of $x = 0$ into (4.19) leads to

$$(1+t^2)^{-1/2} = \sum_{n=0}^{\infty} P_n(0)t^n \qquad (4.24)$$

but the term on the left-hand side has the binomial series expansion

$$(1+t^2)^{-1/2} = \sum_{n=0}^{\infty} \binom{-1/2}{n} t^{2n} \qquad (4.25)$$

Comparing terms of the series on the right in (4.24) and (4.25), we note that (4.25) has only *even* powers of t. Thus we conclude that $P_n(0) = 0$ for $n = 1, 3, 5, \ldots$, or equivalently,

$$P_{2n+1}(0) = 0 \qquad n = 0, 1, 2, \ldots \qquad (4.26)$$

Since all odd terms in (4.24) are zero, we can replace n by $2n$ in the series and compare with (4.25), from which we deduce

$$P_{2n}(0) = \binom{-1/2}{n} = \frac{(-1)^n (2n)!}{2^{2n}(n!)^2} \qquad n = 0, 1, 2, \ldots \qquad (4.27)$$

where we recall (4.14).

Remark: Actually, (4.26) could have been deduced from the fact that $P_{2n+1}(x)$ is an odd (continuous) function and therefore must necessarily pass through the origin. (Why?)

To obtain the desired recurrence relations, first we make the observation that the function $w(x, t) = (1 - 2xt + t^2)^{-1/2}$ satisfies the derivative relation

$$(1 - 2xt + t^2)\frac{\partial w}{\partial t} + (t - x)w = 0 \qquad (4.28)$$

Direct substitution of the series (4.13) for $w(x, t)$ into (4.28) yields

$$(1 - 2xt + t^2)\sum_{n=0}^{\infty} nP_n(x)t^{n-1} + (t - x)\sum_{n=0}^{\infty} P_n(x)t^n = 0$$

Carrying out the indicated multiplications and simplifying give us

$$\sum_{n=0}^{\infty} nP_n(x)t^{n-1} - 2x\underbrace{\sum_{n=0}^{\infty} nP_n(x)t^n}_{n \to n-1} + \underbrace{\sum_{n=0}^{\infty} nP_n(x)t^{n+1}}_{n \to n-2}$$

$$+ \underbrace{\sum_{n=0}^{\infty} P_n(x)t^{n+1}}_{n \to n-2} - x\underbrace{\sum_{n=0}^{\infty} P_n(x)t^n}_{n \to n-1} = 0 \qquad (4.29)$$

We now wish to change indices so that powers of t are the same in each summation. We accomplish this by leaving the first sum in (4.29) as it is, replacing n with $n-1$ in the second and last sums, and replacing n with $n-2$ in the remaining sums. Thus, (4.29) becomes

$$\sum_{n=0}^{\infty} nP_n(x)t^{n-1} - 2x \sum_{n=1}^{\infty} (n-1)P_{n-1}(x)t^{n-1}$$

$$+ \sum_{n=2}^{\infty} (n-2)P_{n-2}(x)t^{n-1} + \sum_{n=2}^{\infty} P_{n-2}(x)t^{n-1} - x \sum_{n=1}^{\infty} P_{n-1}(x)t^{n-1} = 0$$

Finally, combining all summations, we have

$$\sum_{n=2}^{\infty} [nP_n(x) - 2x(n-1)P_{n-1}(x) + (n-2)P_{n-2}(x)$$

$$+ P_{n-2}(x) - xP_{n-1}(x)]t^{n-1} + P_1(x) - xP_0(x) = 0 \quad (4.30)$$

But $P_1(x) - xP_0(x) = x - x = 0$, and the validity of (4.30) demands that the coefficient of t^{n-1} be zero for all x. Hence, after simplification we arrive at

$$nP_n(x) - (2n-1)xP_{n-1}(x) + (n-1)P_{n-2}(x) = 0 \qquad n = 2, 3, 4, \ldots$$

or, replacing n by $n+1$, we obtain the more conventional form

$$(n+1)P_{n+1}(x) - (2n+1)xP_n(x) + nP_{n-1}(x) = 0 \qquad (4.31)$$

where $n = 1, 2, 3, \ldots$.

We refer to (4.31) as a three-term **recurrence formula**, since it forms a connecting relation between three successive Legendre polynomials. One of the primary uses of (4.31) in computations is to produce higher-order Legendre polynomials from lower-order ones by expressing them in the form

$$P_{n+1}(x) = \frac{2n+1}{n+1} xP_n(x) - \frac{n}{n+1} P_{n-1}(x) \qquad (4.32)$$

where $n = 1, 2, 3, \ldots$. In practice, (4.32) is generally preferred to (4.16) in making computer calculations when several polynomials are involved.*

---

*Actually, to avoid excessive roundoff error in making computer calculations, Eq. (4.32) should be rewritten in the form

$$P_{n+1}(x) = 2xP_n(x) - P_{n-1}(x) - \frac{xP_n(x) - P_{n-1}(x)}{n+1}$$

A relation similar to (4.31) involving derivatives of the Legendre polynomials can be derived in the same fashion by first making the observation that $w(x, t)$ satisfies

$$(1 - 2xt + t^2)\frac{\partial w}{\partial x} - tw = 0 \qquad (4.33)$$

where this time the differentiation is with respect to x. Substituting the series for $w(x, t)$ directly into (4.33) leads to

$$(1 - 2xt + t^2)\sum_{n=0}^{\infty} P_n'(x)t^n - \sum_{n=0}^{\infty} P_n(x)t^{n+1} = 0$$

or, after carrying out the multiplications,

$$\sum_{n=0}^{\infty} P_n'(x)t^n$$

$$- 2x\underbrace{\sum_{n=0}^{\infty} P_n'(x)t^{n+1}}_{n \to n-1} + \underbrace{\sum_{n=0}^{\infty} P_n'(x)t^{n+2}}_{n \to n-2} - \underbrace{\sum_{n=0}^{\infty} P_n(x)t^{n+1}}_{n \to n-1} = 0 \quad (4.34)$$

Next, making an appropriate change of index in each summation, we get

$$\sum_{n=2}^{\infty} [P_n'(x) - 2xP_{n-1}'(x) + P_{n-2}'(x) - P_{n-1}(x)]t^n = 0 \qquad (4.35)$$

where all terms outside this summation add to zero. Thus, by equating the coefficient of t^n to zero in (4.35), we find

$$P_n'(x) - 2xP_{n-1}'(x) + P_{n-2}'(x) - P_{n-1}(x) = 0 \qquad n = 2, 3, 4, \ldots$$

or, by a change of index,

$$P_{n+1}'(x) - 2xP_n'(x) + P_{n-1}'(x) - P_n(x) = 0 \qquad (4.36)$$

for $n = 1, 2, 3, \ldots$.

Certain combinations of (4.21) and (4.36) can lead to further recurrence relations. For example, suppose we first differentiate (4.31), that is,

$$(n + 1)P_{n+1}'(x) - (2n + 1)P_n(x) - (2n + 1)xP_n'(x) + nP_{n-1}'(x) = 0 \qquad (4.37)$$

From (4.36) we find

$$P'_{n-1}(x) = P_n(x) + 2xP'_n(x) - P'_{n+1}(x) \qquad (4.38a)$$

$$P'_{n+1}(x) = P_n(x) + 2xP'_n(x) - P'_{n-1}(x) \qquad (4.38b)$$

and the successive replacement of $P'_{n-1}(x)$ and $P'_{n+1}(x)$ in (4.37) by (4.38a) and (4.38b) leads to the two relations

$$P'_{n+1}(x) - xP'_n(x) = (n+1)P_n(x) \qquad (4.39a)$$

$$xP'_n(x) - P'_{n-1}(x) = nP_n(x) \qquad (4.39b)$$

The addition of (4.39a) and (4.39b) yields the more symmetric formula

$$P'_{n+1}(x) - P'_{n-1}(x) = (2n+1)P_n(x) \qquad (4.40)$$

Finally, replacing n by $n-1$ in (4.39a) and then eliminating the term $P'_{n-1}(x)$ by use of (4.39b), we obtain

$$(1-x^2)P'_n(x) = nP_{n-1}(x) - nxP_n(x) \qquad (4.41)$$

This last relation allows us to express the *derivative* of a Legendre polynomial in terms of Legendre polynomials.

4.2.3 Legendre's differential equation

All the recurrence relations that we have derived thus far involve successive Legendre polynomials. We may well wonder if any relation exists between derivatives of the Legendre polynomials and Legendre polynomials of the same index. The answer is yes, but to derive this relation, we must consider second derivatives of the polynomials.

By taking the derivative of both sides of (4.41), we get

$$\frac{d}{dx}[(1-x^2)P'_n(x)] = nP'_{n-1}(x) - nP_n(x) - nxP'_n(x)$$

and then, using (4.39b) to eliminate $P'_{n-1}(x)$, we arrive at the derivative relation

$$\frac{d}{dx}[(1-x^2)P'_n(x)] + n(n+1)P_n(x) = 0 \qquad (4.42)$$

which holds for $n = 0, 1, 2, \ldots$. Expanding the product term in (4.42) yields

$$(1-x^2)P''_n(x) - 2xP'_n(x) + n(n+1)P_n(x) = 0 \qquad (4.43)$$

and thus we deduce that the Legendre polynomial $y = P_n(x)$ $(n = 0, 1, 2, \ldots)$ is a solution of the linear second-order differential equation (DE)

$$(1 - x^2)y'' - 2xy' + n(n + 1)y = 0 \qquad (4.44)$$

called **Legendre's differential equation.**[*]
Perhaps the most natural way in which Legendre polynomials arise in practice is as solutions of Legendre's equation. In such problems the basic model is generally a partial differential equation. Solving the partial DE by the separation-of-variables technique leads to a system of ordinary DEs, and sometimes one of these is Legendre's DE. This is precisely the case, for example, in solving for the steady-state temperature distribution (independent of the azimuthal angle) in a solid sphere. We delay any further discussion of such problems, however, until Sec. 4.8.

Remark: Any function $f_n(x)$ that satisfies Legendre's equation, i.e.,

$$(1 - x^2)f_n''(x) - 2xf_n'(x) + n(n + 1)f_n(x) = 0$$

will also satisfy *all* previous recurrence formulas given above, provided that $f_n(x)$ is properly normalized. Consequently, any further solutions of Legendre's equation can be selected in such a way that they automatically satisfy the whole set of recurrence relations already derived. The set of solutions $Q_n(x)$ introduced in Sec. 4.6 is a case in point.

Exercises 4.2

1. Use the series (4.16) to determine $P_n(x)$ directly for the specific cases $n = 0, 1, 2, 3, 4,$ and 5.

2. Given that $P_0(x) = 1$ and $P_1(x) = x$, use the recurrence formula (4.32) to determine $P_2(x)$, $P_3(x)$, and $P_4(x)$.

3. Verify that
   (a) $P_0(\cos \phi) = 1$
   (b) $P_1(\cos \phi) = \cos \phi$
   (c) $P_2(\cos \phi) = \frac{1}{4}(3 \cos 2\phi + 1)$
   (d) $P_3(\cos \phi) = \frac{1}{8}(5 \cos 3\phi + 3 \cos \phi)$

---

[*] In Sec. 4.6 we discuss other solutions of Legendre's equation.

4. The function $w(x, t) = (1 - 2xt + t^2)^{-1/2}$ is given.
   (a) Show that $w(-x, -t) = w(x, t)$.
   (b) Use the result in (a) and the generating function relation
       (4.19) to deduce that (for $n = 0, 1, 2, \ldots$)
       $$P_n(-x) = (-1)^n P_n(x)$$

5. Verify the special values ($n = 0, 1, 2, \ldots$)
   (a) $P_n'(1) = \frac{1}{2}n(n + 1)$ (b) $P_n'(-1) = (-1)^{n-1}[\frac{1}{2}n(n + 1)]$

6. Verify the special values ($n = 0, 1, 2, \ldots$)

   (a) $P_{2n}'(0) = 0$ (b) $P_{2n+1}'(0) = \dfrac{(-1)^n (2n + 1)}{2^{2n}} \dbinom{2n}{n}$

7. Establish the generating-function relation

   $$(1 - 2xt + t^2)^{-1} = \sum_{n=0}^{\infty} U_n(x)t^n \qquad |t| < 1 \qquad |x| \leq 1$$

   where $U_n(x)$ is the nth *Chebyshev polynomial of the second
   kind,** defined by

   $$U_n(x) = \sum_{k=0}^{[n/2]} \frac{(-1)^k (n - k)!}{k!(n - 2k)!} (2x)^{n-2k}$$

8. The generating function $w(x, t) = (1 - 2xt + t^2)^{-1}$ is given.
   (a) Show that it satisfies the identity

   $$(1 - 2xt + t^2)\frac{\partial w}{\partial t} + 2(t - x)w = 0$$

   (b) Substitute the series in problem 7 into the identity in (a)
       and derive the recurrence formula (for $n = 1, 2, 3, \ldots$)
       $$U_{n+1}(x) - 2xU_n(x) + U_{n-1}(x) = 0$$

9. (a) Show that the generating function in problem 8 also
       satisfies the identity

   $$(1 - 2xt + t^2)\frac{\partial w}{\partial x} - 2tw = 0$$

   (b) From (a), deduce the relation (for $n = 1, 2, 3, \ldots$)
       $$U_{n+1}'(x) - 2xU_n'(x) + U_{n-1}'(x) - 2U_n(x) = 0$$
   (c) Show that (b) can be obtained directly from problem 8b by
       differentiation.

10. Using the results of problems 7 to 9, show that
    (a) $(1 - x^2)U_n'(x) = -nxU_n(x) + (n + 1)U_{n-1}(x)$
    (b) $(1 - x^2)U_n''(x) - 3xU_n'(x) + n(n + 2)U_n(x) = 0$

---

* We discuss these polynomials further in Sec. 5.4.2.

11. Using the Cauchy product of two power series (Sec. 1.3.3), show that

$$\frac{e^{xt}}{1-t} = \sum_{n=0}^{\infty} e_n(x)t^n \qquad |t| < 1$$

where $e_n(x)$ is the polynomial equal to the first $n+1$ terms of the Maclaurin series for e^x, that is,

$$e_n(x) = \sum_{k=0}^{n} \frac{x^k}{k!}$$

12. The generating function $w(x,t) = e^{xt}/(1-t)$ is given.
    (a) Show that it satisfies the identity

$$(1-t)\frac{\partial w}{\partial t} - [x(1-t)+1]w = 0$$

    (b) Substitute the series in problem 11 into the identity in (a) and derive the recurrence formula $(n = 1, 2, 3, \ldots)$

$$(n+1)e_{n+1}(x) - (n+1+x)e_n(x) + xe_{n-1}(x) = 0$$

    (c) Show directly from the series definition of $e_n(x)$ that

$$e_n'(x) = e_{n-1}(x) \qquad n = 1, 2, 3, \ldots$$

13. Using the results of problems 11 and 12, show that $y = e_n(x)$ is a solution of the second-order linear DE

$$xy'' - (x+n)y' + ny = 0$$

14. Make the change of variable $x = \cos \phi$ in the DE

$$\frac{1}{\sin \phi}\frac{d}{d\phi}\left(\sin \phi \frac{dy}{d\phi}\right) + n(n+1)y = 0$$

and show that it reduces to Legendre's DE (4.44).

15. Determine the values of n for which $y = P_n(x)$ is a solution of
    (a) $(1-x^2)y'' - 2xy' + n(n+1)y = 0$, $y(0) = 0$, $y(1) = 1$
    (b) $(1-x^2)y'' - 2xy' + n(n+1)y = 0$, $y'(0) = 0$, $y(1) = 1$

16. When a tightly stretched string is rotating with uniform angular speed ω about its rest position along the x axis, the DE governing the displacement of the string in the vertical plane is approximately

$$\frac{d}{dx}[T(x)y'] + \rho\omega^2 y = 0$$

where $T(x)$ is the tension in the string and ρ is the linear density (constant) of the string. If $T(x) = 1 - x^2$ and the boundary condition $y(-1)=y(1)$ is prescribed, determine the two lowest possible critical speeds ω. What shape does the string assume in the vertical plane in each case?

Hint: Assume that $\rho\omega^2 = n(n+1)$.

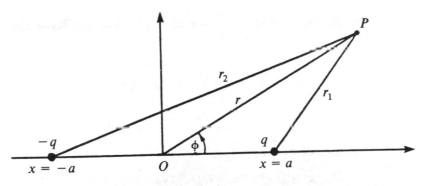

17. An electric dipole consists of electric charges q and $-q$ located along the x axis as shown in the figure above. The potential induced at point P due to the charges is known to be $(r>a)$

$$V = kq\left(\frac{1}{r_1} - \frac{1}{r_2}\right)$$

where k is a constant. Express the potential in terms of the coordinates r and ϕ, and show that it leads to an infinite series involving Legendre polynomials. Also show that if only the first nonzero term of the series is retained, the dipole potential is

$$V \simeq \frac{2akq}{r^2}\cos\phi \qquad r \gg a$$

18. The electrostatic potential induced at point P for the array of charges shown in the figure below is given by $(r>a)$

$$V = kq\left(\frac{1}{r_1} + \frac{1}{r_2} - \frac{2}{r}\right)$$

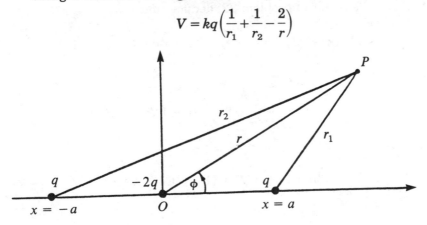

where k is constant. Expressing V entirely in terms of r and ϕ, show that the first nonzero term of the resulting series yields

$$V \simeq \frac{kqa^2}{2r^3}(3\cos 2\phi + 1) \qquad r \gg a$$

19. Show that the even and odd Legendre polynomials have the series representations (for $n = 0, 1, 2, \ldots$)

    (a) $P_{2n}(x) = \dfrac{(-1)^n}{2^{2n-1}} \displaystyle\sum_{k=0}^{n} \dfrac{(-1)^k(2n+2k-1)!}{(2k)!(n+k-1)!(n-k)!}x^{2k}$

    (b) $P_{2n+1}(x) = \dfrac{(-1)^n}{2^{2n}} \displaystyle\sum_{k=0}^{n} \dfrac{(-1)^k(2n+2k+1)!}{(2k+1)!(n+k)!(n-k)!}x^{2k+1}$

20. Derive the identity ($n = 0, 1, 2, \ldots$)

    $$(1-x^2)P_n'(x) = (n+1)[xP_n(x) - P_{n+1}(x)]$$

21. Show that

    (a) $\displaystyle\sum_{k=0}^{n}(2k+1)P_k(x) = P_{n+1}'(x) + P_n'(x)$

    (b) $(1-x)\displaystyle\sum_{k=0}^{n}(2k+1)P_k(x) = (n+1)[P_n(x) - P_{n+1}(x)]$

22. Show that

    (a) $\displaystyle\sum_{n=0}^{\infty}[xP_n'(x) - nP_n(x)]t^n = t^2(1 - 2xt + t^2)^{-3/2}$

    (b) $\displaystyle\sum_{n=0}^{\infty}\sum_{k=0}^{[n/2]}(2n-4k+1)P_{n-2k}(x)t^n = (1 - 2xt + t^2)^{-3/2}$

23. Using the result of problem 22, deduce that

    $$xP_n'(x) - nP_n(x) = \sum_{k=0}^{[(n-2)/2]}(2n-4k-3)P_{n-2-2k}(x)$$

24. Show that

    $$P_n'(x) = \sum_{k=0}^{[(n-2)/2]}(2n-4k-1)P_{n-1-2k}(x)$$

25. Show that

    $$\sum_{n=0}^{\infty}(2n+1)P_n(x)t^n = (1-t^2)(1-2xt+t^2)^{-3/2}$$

4.3   Other Representations of the Legendre Polynomials

For each n, the Legendre polynomials can be defined by either the series

$$P_n(x) = \sum_{h=0}^{[n/2]} \frac{(-1)^k(2n-2k)!}{2^n k!(n-k)!(n-2k)!} x^{n-2k} \qquad (4.45)$$

or the recurrence formula

$$P_{n+1}(x) = \frac{2n+1}{n+1} x P_n(x) - \frac{n}{n+1} P_{n-1}(x) \qquad (4.46)$$

where $P_0(x) = 1$ and $P_1(x) = x$. In some situations, however, it is advantageous to have other representations from which further properties of the polynomials are more readily found.

4.3.1   Rodrigues' formula

A representation of the Legendre polynomials involving differentiation is given by the **Rodrigues formula**

$$P_n(x) = \frac{1}{2^n n!} \frac{d^n}{dx^n} [(x^2-1)^n] \qquad n = 0, 1, 2, \ldots \qquad (4.47)$$

To verify (4.47), we start with the binomial series

$$(x^2-1)^n = \sum_{k=0}^{n} \frac{(-1)^k n!}{k!(n-k)!} x^{2n-2k}$$

and differentiate n times. Noting that

$$\frac{d^n}{dx^n} x^m = \begin{cases} \dfrac{m!}{(m-n)!} x^{m-n} & n \leq m \\ 0 & n > m \end{cases}$$

we infer

$$\frac{d^n}{dx^n} [(x^2-1)^n] = \sum_{k=0}^{[n/2]} \frac{(-1)^k n!(2n-2k)!}{k!(n-k)!(n-2k)!} x^{n-2k}$$

$$= 2^n n! P_n(x)$$

from which (4.47) now follows.

4.3.2 Laplace integral formula

An integral representation of $P_n(x)$ is given by

$$P_n(x) = \frac{1}{\pi} \int_0^\pi [x + (x^2 - 1)^{1/2} \cos \phi]^n \, d\phi \qquad n = 0, 1, 2, \ldots \quad (4.48)$$

which is called the **Laplace integral formula**. This relation is easily verified for $n = 0$ and $n = 1$, but more difficult to prove in the general case.

Let us call the integral I and expand the integrand in a finite binomial series to get

$$I = \frac{1}{\pi} \int_0^\pi [x + (x^2 - 1)^{1/2} \cos \phi]^n \, d\phi$$

$$= \frac{1}{\pi} \int_0^\pi \sum_{k=0}^n \binom{n}{k} x^{n-k} (x^2 - 1)^{k/2} \cos^k \phi \, d\phi$$

$$= \sum_{k=0}^n \binom{n}{k} x^{n-k} (x^2 - 1)^{k/2} \frac{1}{\pi} \int_0^\pi \cos^k \phi \, d\phi \qquad (4.49)$$

The residual integral in (4.49) can be shown to satisfy

$$\frac{1}{\pi} \int_0^\pi \cos^k \phi \, d\phi = 0 \qquad k = 1, 3, 5, \ldots \quad (4.50)$$

and for even values of k we set $k = 2j$ to find

$$\frac{1}{\pi} \int_0^\pi \cos^k \phi \, d\phi = \frac{2}{\pi} \int_0^{\pi/2} \cos^{2j} \phi \, d\phi$$

$$= \frac{(2j)!}{2^{2j}(j!)^2} \qquad j = 0, 1, 2, \ldots \quad (4.51)$$

The verification of (4.50) and (4.51) is left to the exercises (see problems 5 and 6 in Exercises 4.3). Thus, all odd terms in (4.49) are zero, and by setting $k = 2j$ and using (4.51), we see that

$$I = \sum_{j=0}^{[n/2]} \frac{n! x^{n-2j} (x^2 - 1)^j}{2^{2j}(n - 2j)!(j!)^2} \qquad 4.52)$$

What remains now is to show that (4.52) is a series representation of $P_n(x)$, and this we leave also to the exercises (see problem 7 in Exercises 4.3).

4.3.3 Some bounds on $P_n(x)$

One of the uses of the Laplace integral formula (4.40) is to establish some inequalities for the Legendre polynomials which furnish certain bounds on them. Of particular interest is the interval $|x| \leq 1$, but since the integrand in (4.48) is not real for this restriction on x, we first rewrite (4.48) in the form $(i^2 = -1)$

$$P_n(x) = \frac{1}{\pi} \int_0^\pi [x + i(1-x^2)^{1/2} \cos \phi]^n \, d\phi \qquad |x| \leq 1 \qquad (4.53)$$

Now, using the fact that the absolute value of an integral is less than or equal to the integral of the absolute value of the integrand, we get

$$|P_n(x)| \leq \frac{1}{\pi} \int_0^\pi |x + i(1-x^2)^{1/2} \cos \phi|^n \, d\phi \qquad (4.54)$$

From the algebra of complex numbers, it is known that $|a + ib| = (a^2 + b^2)^{1/2}$, and thus for $|x| \leq 1$ it follows that

$$|x + i(1-x^2)^{1/2} \cos \phi|^n = [x^2 + (1-x^2) \cos^2 \phi]^{n/2}$$

$$= (\cos^2 \phi + x^2 \sin^2 \phi)^{n/2}$$

$$\leq (\cos^2 \phi + \sin^2 \phi)^{n/2}$$

$$\leq 1$$

Returning now to (4.54), we have shown that

$$|P_n(x)| \leq \frac{1}{\pi} \int_0^\pi d\phi$$

or $\qquad\qquad |P_n(x)| \leq 1 \qquad |x| \leq 1 \qquad n = 0, 1, 2, \ldots \qquad (4.55)$

which is our intended result. The equality in (4.55) holds only when $x = \pm 1$.

Another inequality, less obvious and more difficult to prove, is given by

$$|P_n(x)| < \sqrt{\frac{\pi}{2n(1-x^2)}} \qquad |x| < 1 \qquad n = 1, 2, 3, \ldots \qquad (4.56)$$

Again the Legendre integral representation is used to derive this inequality, although we will not do so here (see problem 10 in Exercises 4.3).

Exercises 4.3

1. Using Rodrigues' formula (4.47), derive the identities ($n = 1, 2, 3, \ldots$)

   (a) $(n+1)P_{n+1}(x) = (2n+1)xP_n(x) - nP_{n-1}(x)$

   (b) $P_n'(x) = xP_{n-1}'(x) + nP_{n-1}(x)$

   (c) $xP_n'(x) = nP_n(x) + P_{n-1}'(x)$

   (d) $P_{n+1}'(x) - P_{n-1}'(x) = (2n+1)P_n(x)$

2. Representing $P_n(x)$ by Rodrigues' formula (4.47), show that

$$\int_{-1}^{1} P_n(x)\, dx = 0 \qquad n = 1, 2, 3, \ldots$$

3. Using Rodrigues' formula (4.47) and integration by parts, show that

$$\int_{-1}^{1} [P_n(x)]^2\, dx = \frac{2}{2n+1} \qquad n = 0, 1, 2, \ldots$$

4. Define $v = (x^2 - 1)^n$.

   (a) Show that

$$(1 - x^2)\frac{dv}{dx} - 2nxv = 0$$

   (b) Differentiating the result in (a) $n+1$ times and defining $u = v^{(n)}$, show that u satisfies Legendre's equation

$$(1 - x^2)u'' - 2xu' + n(n+1)u = 0$$

5. Verify that

$$\frac{1}{\pi}\int_0^\pi \cos^{2n+1}\theta\, d\theta = 0 \qquad n = 0, 1, 2, \ldots$$

6. (a) Verify that

$$\frac{1}{\pi}\int_0^\pi \cos^{2n}\theta\, d\theta = \frac{2}{\pi}\int_0^{\pi/2}\cos^{2n}\theta\, d\theta$$

   (b) Using properties of the gamma function, show that

$$\frac{2}{\pi}\int_0^{\pi/2}\cos^{2n}\theta\, d\theta = \frac{(2n)!}{2^{2n}(n!)^2} \qquad n = 0, 1, 2, \ldots$$

7. (a) Show that the generating function for the Legendre polynomials can be written in the form

$$(1 - 2xt + t^2)^{-1/2} = (1 - xt)^{-1}\left[1 - \frac{t^2(x^2 - 1)}{(1 - xt)^2}\right]^{-1/2}$$

(b)   Using the result in (a), expand the expression on the right in powers of t. Then, by comparing your result with Eq. (4.19) in Sec. 4.2.1, deduce that

$$P_n(x) = \sum_{k=0}^{[n/2]} \frac{n!x^{n-2k}(x^2-1)^k}{2^{2k}(n-2k)!(k!)^2}$$

8.   (*Jordan inequality*) If $0 \le \phi < \pi/2$, show that

$$\sin \phi \ge \frac{2\phi}{\pi}$$

Hint: Prove that $(\sin \phi)/\phi$ is a decreasing function on the given interval by showing its derivative is always negative. Hence, the minimum value occurs at $\phi = \pi/2$.

9.   Derive the inequality

$$1 - y < e^{-y} \qquad y > 0$$

10.   Use the Laplace integral formula (4.48).
(a)   Show that for $|x| < 1$,

$$|P_n(x)| \le \frac{2}{\pi} \int_0^{\pi/2} [1 - (1-x^2)\sin^2\phi]^{n/2} d\phi$$

(b)   Show that application of the Jordan inequality (problem 8) reduces (a) to

$$|P_n(x)| \le \frac{2}{\pi} \int_0^{\pi/2} \left[1 - \frac{4\phi^2(1-x^2)}{\pi^2}\right]^{n/2} d\phi$$

(c)   Making use of the inequality in problem 9 together with an appropriate change of variables, show that

$$|P_n(x)| < \frac{2}{\sqrt{2n(1-x^2)}} \int_0^\infty e^{-t^2} dt$$

and from this result, deduce that $(n = 1, 2, 3, \dots)$

$$|P_n(x)| < \sqrt{\frac{\pi}{2n(1-x^2)}} \qquad |x| < 1$$

11.   Starting with the identity

$$(1-x^2)P_n'(x) = nP_{n-1}(x) - nxP_n(x)$$

show that

$$|P_n'(x)| \le \frac{n}{1-|x|} \qquad |x| < 1 \qquad n = 1, 2, 3, \dots$$

12. Start with the identities

$$P_n(x) = xP_{n-1}(x) + \frac{x^2-1}{n}P'_{n-1}(x)$$

$$P'_n(x) = xP'_{n-1}(x) + nP_{n-1}(x)$$

(a)   Show that (for $n = 1, 2, 3, \ldots$)

$$\frac{1-x^2}{n^2}[P'_n(x)]^2 + [P_n(x)]^2 = \frac{1-x^2}{n^2}[P'_{n-1}(x)]^2 + [P_{n-1}(x)]^2$$

(b)   From (a), establish the inequality

$$\frac{1-x^2}{n^2}[P'_n(x)]^2 + [P_n(x)]^2 \le 1 \qquad |x| \le 1$$

(c)   From (b), deduce that

$$|P_n(x)| \le 1 \qquad |x| \le 1$$

4.4   Legendre Series

In this section we wish to show how to represent certain functions by series of Legendre polynomials, called *Legendre series*. Because the general term in such series is a polynomial, we can interpret a Legendre series as some generalization of a power series for which the general term is also a polynomial, that is, $(x-a)^n$. However, to develop a given function f in a power series requires that the function f be at least continuous and differentiable in the interval of convergence. In the case of Legendre series we make no such requirement. In fact, many functions of practical interest exhibiting (finite) discontinuities may be represented by convergent Legendre series. Legendre series are only one member of a fairly large and special class of series collectively referred to as *generalized Fourier series,* all of which have many properties in common. In Sec. 1.4 we encountered *Fourier trigonometric series,* which are perhaps the best known members of this class, and in the following chapters we will come across several other members of this general class. Besides their obvious mathematical interest, it turns out that the applications of generalized Fourier series are very extensive—so much so, in fact, that they involve almost every facet of applied mathematics.

4.4.1   Orthogonality of the polynomials

Although we have already derived many identities associated with the Legendre polynomials, none of these is so fundamental and far-reaching in practice as the **orthogonality property**

$$\int_{-1}^{1} P_n(x)P_k(x)\, dx = 0 \qquad k \neq n \qquad (4.57)$$

Remark: It is sometimes helpful to think of (4.57) as a generalization of the scalar (dot) product of vector analysis. In fact, much of the following discussion has a vector analog in three-dimensional vector space.

To prove (4.57), we first take note of the fact that both $P_k(x)$ and $P_n(x)$ satisfy Legendre's DE (4.42), and thus we write

$$\frac{d}{dx}[(1-x^2)P_k'(x)] + k(k+1)P_k(x) = 0 \qquad (4.58a)$$

$$\frac{d}{dx}[(1-x^2)P_n'(x)] + n(n+1)P_n(x) = 0 \qquad (4.58b)$$

If we multiply the first of these equations by $P_n(x)$ and the second by $P_k(x)$, subtract the results, and integrate from -1 to 1, we find

$$\int_{-1}^{1} P_n(x)\frac{d}{dx}[(1-x^2)P_k'(x)]\, dx$$

$$-\int_{-1}^{1} P_k(x)\frac{d}{dx}[(1-x^2)P_n'(x)]\, dx$$

$$+ [k(k+1) - n(n+1)]\int_{-1}^{1} P_n(x)P_k(x)\, dx = 0 \quad (4.59)$$

On integrating the first integral above by parts, we have

$$\int_{-1}^{1} P_n(x)\frac{d}{dx}[(1-x^2)P_k'(x)]\, dx$$

$$= P_n(x)(1-x^2)P_k'(x)\big|_{-1}^{1} - \int_{-1}^{1}(1-x^2)P_n'(x)P_k'(x)\, dx \quad (4.60a)$$

and similarly for the second integral,

$$\int_{-1}^{1} P_k(x)\frac{d}{dx}[(1-x^2)P_n'(x)]\, dx = -\int_{-1}^{1}(1-x^2)P_n'(x)P_k'(x)\, dx$$

$$(4.60b)$$

and therefore the difference of these two integrals is clearly zero. Hence, (4.59) reduces to

$$[k(k+1) - n(n+1)] \int_{-1}^{1} P_n(x)P_k(x)\,dx = 0$$

and since $k \neq n$ by hypothesis, the result (4.57) follows immediately. When $k = n$, the situation is different. Let us define

$$A_n = \int_{-1}^{1} [P_n(x)]^2\,dx \qquad (4.61)$$

and replace one of the $P_n(x)$ in (4.61) by the identity [replace n with $n-1$ in (4.32)]

$$P_n(x) = \frac{2n-1}{n}xP_{n-1}(x) - \frac{n-1}{n}P_{n-2}(x) \qquad (4.62)$$

to get

$$A_n = \int_{-1}^{1} P_n(x)\left[\frac{2n-1}{n}xP_{n-1}(x) - \frac{n-1}{n}P_{n-2}(x)\right]dx$$

$$= \frac{2n-1}{n}\int_{-1}^{1} xP_n(x)P_{n-1}(x)\,dx$$

$$-\frac{n-1}{n}\int_{-1}^{1} P_n(x)P_{n-2}(x)\,dx \qquad (4.63)$$

The second integral above vanishes because of the orthogonality property (4.57). To further simplify (4.63), we rewrite (4.62) in the form

$$xP_n(x) = \frac{1}{2n+1}[(n+1)P_{n+1}(x) + nP_{n-1}(x)]$$

and substitute it into (4.63), from which we deduce

$$A_n = \frac{2n-1}{n}\frac{n+1}{2n+1}\int_{-1}^{1} P_{n+1}(x)P_{n-1}(x)\,dx$$

$$+ \frac{2n-1}{2n+1}\int_{-1}^{1} [P_{n-1}(x)]^2\,dx$$

or $\qquad A_n = \frac{2n-1}{2n+1}A_{n-1} \qquad n = 2, 3, 4, \ldots \qquad (4.64)$

Equation (4.64) is simply a recurrence formula for A_n. Using the fact that

$$A_0 = \int_{-1}^{1} [P_0(x)]^2\, dx = \int_{-1}^{1} dx = 2$$

and

$$A_1 = \int_{-1}^{1} [P_1(x)]^2\, dx = \int_{-1}^{1} x^2\, dx = \frac{2}{3}$$

Eq. (4.64) yields

$$A_2 = \frac{3}{5} \times \frac{1}{0} \times 2 = \frac{2}{5}$$

$$A_3 = \frac{5}{7} \times \frac{3}{5} \times \frac{1}{3} \times 2 = \frac{2}{7}$$

while in general it can be verified by mathematical induction that

$$A_n = \frac{2n-1}{2n+1} \times \frac{2n-3}{2n-1} \times \frac{2n-5}{2n-3} \times \cdots \times \frac{1}{3} \times 2$$

$$= \frac{2}{2n+1} \qquad n = 0, 1, 2, \ldots \tag{4.65}$$

Thus, we have derived the important result

$$\int_{-1}^{1} [P_n(x)]^2\, dx = \frac{2}{2n+1} \qquad n = 0, 1, 2, \ldots \tag{4.66}$$

4.4.2 Finite Legendre series

Because of the special properties associated with Legendre polynomials, it may be useful in certain situations to represent arbitrary polynomials as linear combinations of Legendre polynomials. For example, if $q_m(x)$ denotes an arbitrary polynomial of degree m, then since $P_0(x), P_1(x), \ldots, P_m(x)$ are all polynomials of degree m or less, we might expect to find a representation of the form*

$$q_m(x) = c_0 P_0(x) + c_1 P_1(x) + \cdots + c_m P_m(x) \tag{4.67}$$

Let us illustrate with a simple example.

Example 1: Express x^2 in a series of Legendre polynomials.

---

* Two polynomials can be equated if and only if they are of the same degree.

Solution: We write

$$x^2 = c_0 P_0(x) + c_1 P_1(x) + c_2 P_2(x)$$
$$= c_0 + c_1 x + c_2 [\tfrac{1}{2}(3x^2 - 1)]$$
$$= (c_0 - \tfrac{1}{2}c_2) + c_1 x + \tfrac{3}{2}c_2 x^2$$

Now equating like coefficients, we see that

$$c_0 - \tfrac{1}{2}c_2 = 0 \qquad c_1 = 0 \qquad \tfrac{3}{2}c_2 = 1$$

from which we deduce $c_0 = \tfrac{1}{3}$, $c_1 = 0$, and $c_2 = \tfrac{2}{3}$. Hence,

$$x^2 = \tfrac{1}{3}P_0(x) + \tfrac{2}{3}P_2(x)$$

When the polynomial $q_m(x)$ is of a high degree, solving a system of simultaneous equations for the c's as we did in Example 1 is very tedious. A more systematic procedure can be developed by using the orthogonality property (4.57). We begin by writing (4.67) in the form

$$q_m(x) = \sum_{n=0}^{m} c_n P_n(x) \tag{4.68}$$

Next we multiply both sides of (4.68) by $P_k(x)$, $0 \le k \le m$, and we integrate the result termwise (which is justified because the series is finite) from -1 to 1 to get

$$\int_{-1}^{1} q_m(x) P_k(x)\, dx = \sum_{n=0}^{m} c_n \int_{-1}^{1} P_n(x) P_k(x)\, dx \tag{4.69}$$

$$\qquad\qquad\qquad\qquad\qquad \nearrow 0(n \ne k)$$

Because of the orthogonality property (4.57), each term of the series in (4.69) vanishes except the term corresponding to $n = k$, and here we find

$$\int_{-1}^{1} q_m(x) P_k(x)\, dx = c_k \int_{-1}^{1} [P_k(x)]^2\, dx$$

$$= c_k \frac{2}{2k + 1}$$

where the last step is a consequence of (4.66). Hence, we deduce that (changing the dummy index back to n)

$$c_n = (n + \tfrac{1}{2}) \int_{-1}^{1} q_m(x) P_n(x)\, dx \qquad n = 0, 1, 2, \ldots, m \tag{4.70}$$

Remark: If the polynomial $q_m(x)$ in (4.70) is even (odd), then only those c_n values with even (odd) suffixes are nonzero, due to the even–odd property of the Legendre polynomials (see problems 25 and 26 in Exercises 4.4).

As a consequence of the fact that a polynomial of degree m can be expressed as Legendre series involving only $P_m(x)$ and lower-order Legendre polynomials, we have the following theorem.*

Theorem 4.1. If $q_m(x)$ is a polynomial of degree m and $m < r$, then

$$\int_{-1}^{1} q_m(x)P_r(x)\, dx = 0 \qquad m < r$$

Proof: Since $q_m(x)$ is a polynomial of degree m, we can write

$$q_m(x) = \sum_{n=0}^{m} c_n P_n(x)$$

Then multiplying both sides of this expression by $P_r(x)$ and integrating from -1 to 1, we get

$$\int_{-1}^{1} q_m(x)P_r(x)\, dx = \sum_{n=0}^{m} c_n \int_{-1}^{1} P_n(x)P_r(x)\, dx \quad \overset{0}{\nearrow}$$

The largest value of n is m, and since $m < r$, the right-hand side is zero for each n [due to the orthogonality property (4.57)], and the theorem is proved.

4.4.3 Infinite Legendre series

In some applications we will find it necessary to represent a function f, other than a polynomial, as a linear combination of Legendre polynomials. Such a representation will lead to an *infinite series* of the general form

$$f(x) = \sum_{n=0}^{\infty} c_n P_n(x) \tag{4.71}$$

---

*Theorem 4.1 says that $P_r(x)$ is orthogonal to *every* polynomial of degree less than r.

where the coefficients can be formally derived by a process similar to the derivation of (4.70), leading to

$$c_n = (n + \tfrac{1}{2}) \int_{-1}^{1} f(x) P_n(x)\, dx \qquad n = 0, 1, 2, \ldots \qquad (4.72)$$

Conditions under which the representation of (4.71) and (4.72) is valid will be taken up in the next section. For now it suffices to say that for certain functions the series (4.71) will converge throughout the interval $-1 \le x \le 1$, even at points of finite discontinuities of the given function. Series of this type are called **Legendre series**, and because they belong to the larger class of generalized Fourier series, the coefficients (4.72) are commonly called the **Fourier coefficients** of the series.

In practice, the evaluation of integrals like (4.72) must be performed numerically. However, if the function f is not too complicated, we can sometimes use various properties of the Legendre polynomials to evaluate such integrals in closed form. The following example illustrates the point.

Remark: Because the interval of convergence of (4.71) is confined to $-1 \le x \le 1$, it really doesn't matter if the function f is defined outside this interval. That is, even if f is defined for *all* x, the representation will not be valid beyond the interval $-1 \le x \le 1$ (unless f is a polynomial).

Example 2: Find the Legendre series for

$$f(x) = \begin{cases} -1 & -1 \le x < 0 \\ 1 & 0 < x \le 1 \end{cases}$$

Solution: The function f is an odd function. Hence, owing to the even–odd property of the Legendre polynomials depending on the index n, we note that $f(x)P_n(x)$ is an odd function when n is even, and in this case it follows that (see problem 25 in Exercises 4.4)

$$c_n = (n + \tfrac{1}{2}) \int_{-1}^{1} f(x) P_n(x)\, dx = 0 \qquad n = 0, 2, 4, \ldots$$

For odd index n, the product $f(x)P_n(x)$ is an even function, and therefore

$$c_n = (n + \tfrac{1}{2}) \int_{-1}^{1} f(x) P_n(x)\, dx$$

$$= (2n + 1) \int_{0}^{1} P_n(x)\, dx \qquad n = 1, 3, 5, \ldots$$

Let us use the identity [see Eq. (4.40)]

$$P_n(x) = \frac{1}{2n+1} [P'_{n+1}(x) - P'_{n-1}(x)]$$

and set $n = 2k + 1$, thereby obtaining the result (for $k = 0, 1, 2, \ldots$)

$$c_{2k+1} = (4k+3) \int_0^1 P_{2k+1}(x)\, dx$$

$$= \int_0^1 [P'_{2k+2}(x) - P'_{2k}(x)]\, dx$$

$$= [P_{2k+2}(x) - P_{2k}(x)]\big|_0^1$$

$$= P_{2k}(0) - P_{2k+2}(0)$$

where we have used the property $P_n(1) = 1$ for all n. Referring to Eq. (4.27), we have

$$c_{2k+1} = \frac{(-1)^k (2k)!}{2^{2k}(k!)^2} - \frac{(-1)^{k+1}(2k+2)!}{2^{2k+2}[(k+1)!]^2}$$

$$= \frac{(-1)^k (2k)!}{2^{2k}(k!)^2} \left[1 + \frac{(2k+2)(2k+1)}{2^2(k+1)^2} \right]$$

$$= \frac{(-1)^k (2k)!}{2^{2k}(k!)^2} \left(1 + \frac{2k+1}{2k+2} \right)$$

$$= \frac{(-1)^k (2k)!(4k+3)}{2^{2k+1}k!(k+1)!}$$

and thus

$$f(x) = \sum_{k=0}^{\infty} \frac{(-1)^k (2k)!(4k+3)}{2^{2k+1}k!(k+1)!} P_{2k+1}(x) \qquad -1 \le x \le 1$$

Exercises 4.4

In problems 1 to 15, use the orthogonality property and/or any other relations to derive the integral formula.

1. $\int_{-1}^1 xP_n(x)\, dx = \begin{cases} 0 & n \ne 1 \\ 2/3 & n = 1 \end{cases}$

2. $\int_{-1}^1 xP_n(x)P_{n-1}(x)\, dx = \frac{2n}{4n^2 - 1}, \ n = 1, 2, 3, \ldots$

3. $\int_{-1}^{1} P_n(x)P'_{n+1}(x)\, dx = 2,\ n = 0, 1, 2, \ldots$

4. $\int_{-1}^{1} xP'_n(x)P_n(x)\, dx = \dfrac{2n}{2n+1},\ n = 0, 1, 2, \ldots$

5. $\int_{-1}^{1} (1-x^2)P'_n(x)P'_k(x)\, dx = 0,\ k \neq n$

6. $\int_{-1}^{1} (1 - 2xt + t^2)^{-1/2}P_n(x)\, dx = \dfrac{2t^n}{2n+1},\ n = 0, 1, 2, \ldots$

7. $\int_{-1}^{1} (1-x)^{-1/2}P_n(x)\, dx = \dfrac{2\sqrt{2}}{2n+1},\ n = 0, 1, 2, \ldots$

   Hint: Let $t \to 1$ in problem 6.

8. $\int_{-1}^{1} x^2 P_{n+1}(x)P_{n-1}(x)\, dx = \dfrac{2n(n+1)}{(4n^2-1)(2n+3)},\ n = 1, 2, 3, \ldots$

9. $\int_{-1}^{1} (x^2-1)P_{n+1}(x)P'_n(x)\, dx = \dfrac{2n(n+1)}{(2n+1)(2n+3)},\ n = 1, 2, 3, \ldots$

10. $\int_{-1}^{1} x^n P_n(x)\, dx = \dfrac{2^{n+1}(n!)^2}{(2n+1)!},\ n = 0, 1, 2, \ldots$

    Hint: Use problem 31.

11. If $k \le n$,
$$\int_{-1}^{1} P'_n(x)P'_k(x)\, dx = \begin{cases} 0 & n+k \text{ even} \\ k(k+1) & n+k \text{ odd} \end{cases}$$

12. $\displaystyle\int_{-1}^{1} P_n(x)P'_k(x)\, dx = \begin{cases} 0 & k \le n \\ 0 & k > n,\ k+n \text{ even} \\ 2 & k > n,\ k+n \text{ odd} \end{cases}$

13. $\int_{0}^{1} P_{2n}(x)\, dx = 0,\ n = 1, 2, 3, \ldots$

14. $\int_{-1}^{1} (1-x^2)[P'_n(x)]^2\, dx = \dfrac{2n(n+1)}{2n+1},\ n = 0, 1, 2, \ldots$

15. $\int_{-1}^{1} x^2[P_n(x)]^2\, dx = \dfrac{2}{(2n+1)^2}\left[\dfrac{(n+1)^2}{2n+3} + \dfrac{n^2}{2n-1}\right],\quad n = 0, 1, 2, \ldots$

16. Show that the orthogonality relation (4.57) for the functions $P_n(\cos\phi)$ is
$$\int_{0}^{\pi} P_n(\cos\phi)P_k(\cos\phi)\sin\phi\, d\phi = 0 \qquad k \neq n$$

In problems 17 to 21, derive the given integral formula.

17. $\displaystyle \int_0^\pi P_{2n}(\cos\phi)\,d\phi = \frac{\pi}{2^{4n}}\binom{2n}{n}^2$, $n = 0, 1, 2, \ldots$

18. $\displaystyle \int_0^{2\pi} P_{2n}(\cos\phi)\,d\phi = \frac{\pi}{2^{4n-1}}\binom{2n}{n}^2$, $n = 0, 1, 2, \ldots$

19. $\displaystyle \int_0^{\Omega\pi} P_{2n}(\cos\phi)\cos\phi\,d\phi = \frac{1}{2^{4n+1}}\binom{2n}{n}\binom{2n+2}{n+1}$, $n = 1, 2, 3, \ldots$

20. $\displaystyle \int_0^{\pi/2} P_{2n}(\cos\phi)\sin\phi\,d\phi = 0$, $n = 1, 2, 3, \ldots$

21. $\displaystyle \int_0^\pi P_n(\cos\phi)\cos n\phi\,d\phi = B(n+\tfrac{1}{2}, \tfrac{1}{2})$, $n = 0, 1, 2, \ldots$

22. Use Rodrigues' formula (4.47) for $P_n(x)$.

(a) Show that integration by parts leads to

$$\int_{-1}^1 P_n(x)P_k(x)\,dx = -\frac{1}{2^n n!}\int_{-1}^1 P_k'(x)\frac{d^{n-1}}{dx^{n-1}}[(x^2-1)^n]\,dx$$

(b) Show, by continued integration by parts, that

$$\int_{-1}^1 P_n(x)P_k(x)\,dx = \frac{(-1)^n}{2^n n!}\int_{-1}^1 \frac{d^n}{dx^n}[P_k(x)](x^2-1)^n\,dx$$

(c) For $k \neq n$, show that the integral on the right in (b) is zero.

23. (a) For $k = n$, show that problem 22b leads to ($n = 0, 1, 2, \ldots$)

$$\int_{-1}^1 [P_n(x)]^2\,dx = \frac{(2n)!}{2^{2n}(n!)^2}\int_{-1}^1 (1-x^2)^n\,dx$$

(b) By making an appropriate change of variable, evaluate the integral in (a) through use of the gamma function and hence derive Eq. (4.66).

24. Starting with the expression

$$(1-2xt+t^2)^{-1} = \sum_{n=0}^\infty \sum_{k=0}^\infty P_n(x)P_k(x)t^{n+k}$$

use the orthogonality property (4.57) to deduce Eq. (4.66).

Hint: $\ln\dfrac{1+t}{1-t} = 2\displaystyle\sum_{n=0}^\infty \frac{t^{2n+1}}{2n+1}$

25. Show that if f is an odd function,

(a) $\displaystyle \int_{-1}^1 f(x)P_n(x)\,dx = 0$, $n = 0, 2, 4, \ldots$

(b) $\displaystyle \int_{-1}^1 f(x)P_n(x)\,dx = 2\int_0^1 f(x)P_n(x)\,dx$, $n = 1, 3, 5, \ldots$

26. Show that if f is an even function,

(a) $\displaystyle\int_{-1}^{1} f(x)P_n(x)\,dx = 2\int_{0}^{1} f(x)P_n(x)\,dx, \; n = 0, 2, 4, \ldots$

(b) $\displaystyle\int_{-1}^{1} f(x)P_n(x)\,dx = 0, \; n = 1, 3, 5, \ldots$

In problems 27 to 30, find the Legendre series for the given polynomial.

27. $q(x) = x^3$

29. $q(x) = 12x^4 - 8x^2 + 7$

28. $q(x) = 9x^3 - 8x^2 + 7x - 6$

30. $q(x) = 1 + x + \dfrac{x^2}{2!} + \dfrac{x^3}{3!} + \dfrac{x^4}{4!}$

31. Using Rodrigues' formula (4.47) and integration by parts, show that

$$\int_{-1}^{1} f(x)P_n(x)\,dx = \frac{(-1)^n}{2^n n!}\int_{-1}^{1} f^{(n)}(x)(x^2 - 1)^n\,dx$$

Hint: See problem 22.

32. From the result of problem 31, deduce that

$$\int_{-1}^{1} x^m P_n(x)\,dx = 0 \qquad \text{if} \qquad m < n$$

33. From the result of problem 31, deduce that

$$\int_{-1}^{1} x^{n+2k} P_n(x)\,dx = \frac{(n + 2k)!\Gamma(k + \tfrac{1}{2})}{2^n (2k)!\Gamma(n + k + \tfrac{3}{2})} \qquad k = 0, 1, 2, \ldots$$

34. Show that

(a) $\displaystyle x^{2m} = \sum_{n=0}^{m} \frac{2^{2n}(4n + 1)(2m)!(m + n)!}{(2m + 2n + 1)!(m - n)!} P_{2n}(x)$

(b) $\displaystyle x^{2m+1} = \sum_{n=0}^{m} \frac{2^{2n+1}(4n + 3)(2m + 1)!(m + n + 1)!}{(2m + 2n + 3)!(m - n)!} P_{2n+1}(x)$

Hint: Use problem 33.

In problems 35 to 40, develop the Legendre series for the given function.

35. $f(x) = P_6(x)$

38. $f(x) = \begin{cases} 1 & -1 \le x < 0 \\ 0 & 0 < x \le 1 \end{cases}$

36. $f(x) = |x|, \; -1 \le x \le 1$

39. $f(x) = \begin{cases} 0 & -1 \le x < 0 \\ x & 0 < x \le 1 \end{cases}$

37. $f(x) = \begin{cases} 0 & -1 \le x < 0 \\ 1 & 0 < x \le 1 \end{cases}$

40. $f(x) = \begin{cases} 1 & -1 \le x < 0 \\ x & 0 < x \le 1 \end{cases}$

41. Show that the Legendre series of a function f defined in the interval $-a \leqslant x \leqslant a$ is given by

$$f(x) = \sum_{n=0}^{\infty} c_n P_n\left(\frac{x}{a}\right) \qquad -a \leq x \leq a$$

where

$$c_n = \frac{2n+1}{2a} \int_{-a}^{a} f(x) P_n\left(\frac{x}{a}\right) dx \qquad n = 0, 1, 2, \ldots$$

42. Making the change of variable $x = \cos \phi$, show that the Legendre series for a function $f(\phi)$ is given by

$$f(\phi) = \sum_{n=0}^{\infty} c_n P_n(\cos \phi) \qquad 0 \leq \phi \leq \pi$$

where

$$c_n = (n + {}^1\!/_2) \int_0^{\pi} f(\phi) P_n(\cos \phi) \sin \phi \, d\phi \qquad n = 0, 1, 2, \ldots$$

Hint: See problem 16.

43. Using the result of problem 42, find the Legendre series for

(a) $f(\phi) = \begin{cases} 0 & 0 \leq \phi < \pi/2 \\ 1 & \pi/2 < \phi \leq \pi \end{cases}$ (b) $f(\phi) = \cos^2 \phi, \; 0 \leq \phi \leq \pi$

44. (a) Show that

$$(1 - x)^n P_n\left(\frac{1+x}{1-x}\right) = \sum_{k=0}^{n} \binom{n}{k}^2 x^k$$

(b) Letting $x \to 1$, use part (a) to derive the identity

$$\sum_{k=0}^{n} \binom{n}{k}^2 = \binom{2n}{n}$$

4.5 Convergence of the Series

Given the Legendre series of some function f, we now wish to discuss the validity of such a representation. What we mean is—if a value of x is selected in the chosen interval and each term of the series is evaluated for this value of x, will the sum of the series be $f(x)$? If so, we say the series *converges pointwise* to $f(x)$.* To establish pointwise convergence of the series, we need to obtain an expression for the

---

* See also the discussion in Sec. 1.3.

partial sum*

$$S_n(x) = \sum_{k=0}^{n} c_k P_k(x) \tag{4.73}$$

and then for a fixed value of x show that

$$\lim_{n \to \infty} S_n(x) = f(x) \tag{4.74}$$

4.5.1   Piecewise continuous and piecewise smooth functions

To be sure that the Legendre series converges to the function which generates the series, it is essential to place certain restrictions on the function f. From a practical point of view, such conditions should be broad enough to cover most situations of concern and still simple enough to be easily checked for the given function.

Definition 4.1.   A function f is said to be *piecewise continuous* in the interval $a \le x \le b$ provided that

(*a*) $f(x)$ is defined and continuous at all but a finite number of points in the interval.

(*b*)   The left- and right-hand limits exist at each point in the interval.

   Remark: The left- and right-hand limits are defined, respectively, by

$$\lim_{\epsilon \to 0^+} f(x - \epsilon) = f(x^-) \qquad \lim_{\epsilon \to 0^+} f(x + \epsilon) = f(x^+)$$

Furthermore, when x is a point of continuity, $f(x^-) = f(x^+) = f(x)$.

   It is not essential that a piecewise continuous function f be defined at every point in the interval of interest. In particular, it is often not defined at a point of discontinutiy, and even when it is, it really doesn't matter what functional value is assigned at such a point. Also the interval of interest may be open or closed, or open at one end and closed at the other (see Fig. 4.4).

---

   *Although (4.73) has $n + 1$ terms, we still designate it by the symbol $S_n(x)$.

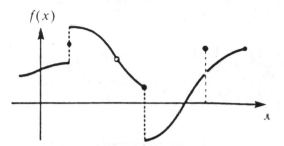

Figure 4.4 A piecewise continuous function.

Definition 4.2. A function f is said to be *smooth* in the interval $a \leq x \leq b$ if it has a continuous derivative there. We say the function is *piecewise smooth* if f and/or its derivative f' is only piecewise continuous in $a \leq x \leq b$.

Example 3: Classify the following functions as smooth, piecewise smooth, or neither in $-1 \leq x \leq 1$: (*a*) $f(x) = x$, (*b*) $f(x) = |x|$, (*c*) $f(x) = |x|^{1/2}$.

Solution: In (*a*), the function $f(x) = x$ and its derivative $f'(x) = 1$ are both continuous, and thus f is *smooth*. The function in (*b*) is also continuous, but because the derivative is discontinuous, i.e.,

$$f'(x) = \begin{cases} -1 & -1 < x < 0 \\ 1 & 0 < x < 1 \end{cases}$$

it is not smooth but only *piecewise smooth*. In (*c*), the function is once again continuous, but $|f'(x)| \to \infty$ as $x \to 0$, so it is *neither* smooth nor piecewise smooth.

4.5.2 Pointwise convergence

Before stating and proving our main theorem on convergence, we must establish two lemmas.

Lemma 4.1 (Riemann). If the function f is piecewise continuous in the closed interval $-1 \leq x \leq 1$, then

$$\lim_{n \to \infty} \sqrt{n + \tfrac{1}{2}} \int_{-1}^{1} f(x) P_n(x)\, dx = 0$$

Proof: Let the nth partial sum be denoted by

$$S_n(x) = \sum_{k=0}^{n} c_k P_k(x)$$

and consider the nonnegative quantity

$$\int_{-1}^{1} [f(x) - S_n(x)]^2 \, dx \geq 0$$

or
$$\int_{-1}^{1} f^2(x) \, dx - 2 \int_{-1}^{1} f(x)S_n(x) \, dx + \int_{-1}^{1} S_n^2(x) \, dx \geq 0$$

Now
$$\int_{-1}^{1} f(x)S_n(x) \, dx = \sum_{k=0}^{n} c_k \int_{-1}^{1} f(x)P_k(x) \, dx$$

$$= \sum_{k=0}^{n} \frac{c_k^2}{k + \frac{1}{2}}$$

and
$$\int_{-1}^{1} S_n^2(x) \, dx = \sum_{j=0}^{n} \sum_{k=0}^{n} c_j c_k \overset{0(j \neq k)}{\int_{-1}^{1} P_j(x)P_k(x) \, dx}$$

$$= \sum_{k=0}^{n} c_k^2 \int_{-1}^{1} [P_k(x)]^2 \, dx$$

$$= \sum_{k=0}^{n} \frac{c_k^2}{k + \frac{1}{2}}$$

Accordingly, we have

$$\int_{-1}^{1} f^2(x) \, dx - 2 \sum_{k=0}^{n} \frac{c_k^2}{k + \frac{1}{2}} + \sum_{k=0}^{n} \frac{c_k^2}{k + \frac{1}{2}} \geq 0$$

from which we deduce

$$\sum_{k=0}^{n} \frac{c_k^2}{k + \frac{1}{2}} \leq \int_{-1}^{1} f^2(x) \, dx$$

Because this last inequality is valid for all n, we simply pass to the limit to get

$$\sum_{k=0}^{\infty} \frac{c_k^2}{k + \frac{1}{2}} \leq \int_{-1}^{1} f^2(x) \, dx$$

The integral on the right is necessarily bounded, since f is assumed to be piecewise continuous in the closed interval of integration. Hence, the series on the left is a convergent series (because its sum is finite), and therefore it follows that

$$\lim_{k \to \infty} \frac{c_k^2}{k + \frac{1}{2}} = 0$$

or equivalently (changing the index back to n)

$$\lim_{n \to \infty} \sqrt{n + \tfrac{1}{2}} \int_{-1}^{1} f(x) P_n(x) \, dx = 0$$

Lemma 4.2 (Christoffel-Darboux). The Legendre polynomials satisfy the identity

$$\sum_{k=0}^{n} (2k + 1) P_k(t) P_k(x) = \frac{n + 1}{t - x} [P_{n+1}(t) P_n(x) - P_n(t) P_{n+1}(x)]$$

Proof: We begin by multiplying the recurrence relation (4.31) by $P_k(t)$ to get

$$(2k + 1) x P_k(t) P_k(x) = (k + 1) P_k(t) P_{k+1}(x) + k P_k(t) P_{k-1}(x)$$

If we now interchange the roles of x and t in this expression and subtract the two results, we obtain

$$(2k + 1)(t - x) P_k(t) P_k(x) = (k + 1)[P_{k+1}(t) P_k(x) - P_k(t) P_{k+1}(x)]$$
$$- k[P_k(t) P_{k-1}(x) - P_{k-1}(t) P_k(x)]$$

Finally, summing both sides of this identity as k runs from 0 to n and setting $P_{-1}(x) = 0$, we find

$$(t - x) \sum_{k=0}^{n} (2k + 1) P_k(t) P_k(x) = (n + 1)[P_{n+1}(t) P_n(x) - P_n(t) P_{n+1}(x)]$$

and the lemma is proved.

We note that integration of the Christoffel-Darboux formula leads to

$$\sum_{k=0}^{n} (2k + 1) P_k(x) \int_{-1}^{1} P_k(t) \, dt$$

$$= (n + 1) \int_{-1}^{1} \frac{P_{n+1}(t) P_n(x) - P_n(t) P_{n+1}(x)}{t - x} \, dt$$

from which we deduce

$$(n + 1) \int_{-1}^{1} \frac{P_{n+1}(t) P_n(x) - P_n(t) P_{n+1}(x)}{t - x} \, dt = 2 \qquad (4.75)$$

where we are using the orthogonality property

$$\int_{-1}^{1} P_k(t)\, dt = \begin{cases} 0 & k \neq 0 \\ 2 & k = 0 \end{cases} \tag{4.76}$$

We are now prepared to state and prove our main result.

Theorem 4.2. If the function f is piecewise smooth in the closed interval $-1 \leq x \leq 1$, then the Legendre series

$$f(x) = \sum_{n=0}^{\infty} c_n P_n(x)$$

where $c_n = (n + \frac{1}{2}) \int_{-1}^{1} f(x) P_n(x)\, dx \qquad n = 0, 1, 2, \ldots$

converges pointwise to $f(x)$ at every continuity point of the function f in the interval $-1 < x < 1$. At points of discontinuity of f in the interval $-1 < x < 1$, the series converges to the average value $\frac{1}{2}[f(x^+) + f(x^-)]$. Finally, at $x = -1$ the series converges to $f(-1^+)$, and at $x = 1$ it converges to $f(1^-)$.

Proof (for a point of continuity): Let us assume that x is a point of continuity of the function f, and consider the partial sum $(-1 < x < 1)$

$$S_n(x) = \sum_{k=0}^{n} c_k P_k(x)$$

$$= \sum_{k=0}^{n} \left[(k + \frac{1}{2}) \int_{-1}^{1} f(t) P_k(t)\, dt \right] P_k(x)$$

where we have replaced the constants c_k by their integral representation. Interchanging the order of summation and integration and recalling the Christoffel-Darboux formula (Lemma 4.2), we obtain

$$S_n(x) = \frac{1}{2} \int_{-1}^{1} f(t) \sum_{k=0}^{n} (2k + 1) P_k(t) P_k(x)\, dt$$

$$= \frac{1}{2}(n + 1) \int_{-1}^{1} f(t) \frac{P_{n+1}(t) P_n(x) - P_n(t) P_{n+1}(x)}{t - x}\, dt$$

If we add and subtract the function $f(x)$ (which is independent of the variable of integration), we get

$$S_n(x) = \frac{1}{2}(n+1)f(x)\int_{-1}^{1}\frac{P_{n+1}(t)P_n(x) - P_n(t)P_{n+1}(x)}{t-x}\,dt$$

$$+ \frac{1}{2}(n+1)\int_{-1}^{1}\frac{f(t)-f(x)}{t-x}[P_{n+1}(t)P_n(x) - P_n(t)P_{n+1}(x)]\,dt$$

For notational convenience we introduce the function

$$g(t) = \frac{f(t)-f(x)}{t-x}$$

and use (4.75) to obtain

$$S_n(x) = f(x) + \tfrac{1}{2}(n+1)P_n(x)\int_{-1}^{1} g(t)P_{n+1}(t)\,dt$$

$$+ \tfrac{1}{2}(n+1)P_{n+1}(x)\int_{-1}^{1} g(t)P_n(t)\,dt$$

At this point we wish to show that g satisfies the conditions of Riemann's lemma, i.e., that g is at least piecewise continuous. Because f is at least piecewise smooth, it follows that g is also piecewise smooth for all $t \neq x$. To investigate the behavior of g at $t = x$, we consider the limit (remembering that x is a point of continuity of f)

$$g(x) = \lim_{t \to x}\frac{f(t)-f(x)}{t-x} = f'(x)$$

Since by hypothesis f' is at least piecewise continuous (why?), we see that g is indeed a piecewise continuous function.

Letting

$$b_n = \sqrt{n+\tfrac{1}{2}}\int_{-1}^{1} g(t)P_n(t)\,dt$$

we can express the partial sum in the form

$$S_n(x) = f(x) + \frac{(n+1)P_n(x)}{2\sqrt{n+\tfrac{3}{2}}}\,b_{n+1} - \frac{(n+1)P_{n+1}(x)}{2\sqrt{n+\tfrac{1}{2}}}\,b_n$$

By recognizing that the Legendre polynomials are bounded on the interval $-1 < x < 1$ [see Eq. (4.56)] and applying Riemann's lemma, it can now be shown that the last two terms in the expression for

$S_n(x)$ vanish in the limit as $n \to \infty$ (see problem 10 in Exercises 4.5), and hence we deduce our intended result

$$\lim_{n \to \infty} S_n(x) = f(x)$$

at a point of continuity of f.

To prove that*

$$\lim_{n \to \infty} S_n(x) = \tfrac{1}{2}[f(x^+) + f(x^-)]$$

at a point of discontinuity of f requires only a slight modification of the above proof. Similar comments can be made about the points $x = \pm 1$.

Exercises 4.5

In problems 1 to 8, discuss whether the function is piecewise continuous, continuous, piecewise smooth, smooth, or none of these in the interval $-1 \le x \le 1$.

1. $f(x) = \tan 2x$

2. $f(x) = \sin x$

3. $f(x) = \dfrac{x^2 - 1}{x - 1}$, $x \ne 1$

4. $f(x) = \begin{cases} 1 & \text{if } x \text{ is rational} \\ 0 & \text{if } x \text{ is irrational} \end{cases}$

5. $f(x) = \dfrac{\sin x}{x}$, $x \ne 0$, $f(0) = 1$

6. $f(x) = \dfrac{\sin x}{x}$, $x \ne 0$

7. $f(x) = \sin \dfrac{1}{x}$, $x \ne 0$

8. $f(x) = xe^{-1/x}$, $x \ne 0$

9. Suppose that a piecewise smooth function f is to be approximated on the interval $-1 \le x \le 1$ by the finite sum

$$S_n(x) = \sum_{k=0}^{n} b_k P_k(x) \qquad -1 \le x \le 1$$

Determine the constants b_k so that the *mean square error* is minimized, i.e., minimize

$$E_n = \int_{-1}^{1} [f(x) - S_n(x)]^2 \, dx$$

Hint: Set $\partial E_n / \partial b_k = 0$, $k = 1, 2, \ldots, n$.

---

*For details, see D. Jackson, *Fourier Series and Orthogonal Polynomials*, Carus Math. Monogr. **6**, Math. Assoc. Amer., Open Court Publ. Co., LaSalle, Ill., 1941.

10. Given that

$$|P_n(x)| < \sqrt{\frac{\pi}{2n(1-x^2)}} \qquad |x| < 1$$

and

$$b_n = \sqrt{n + \tfrac{1}{2}} \int_{-1}^{1} g(t)P_n(t)\,dt$$

where $g(t)$ is piecewise continuous, deduce that

(a) $\displaystyle \lim_{n \to \infty} \frac{(n+1)P_n(x)}{2\sqrt{n + \tfrac{3}{2}}} b_{n+1} = 0$

(b) $\displaystyle \lim_{n \to \infty} \frac{(n+1)P_{n+1}(x)}{2\sqrt{n + \tfrac{1}{2}}} b_n = 0$

4.6 Legendre Functions of the Second Kind

The Legendre polynomial $P_n(x)$ represents only one solution of Legendre's equation

$$(1 - x^2)y'' - 2xy' + n(n+1)y = 0 \qquad (4.77)$$

Because the equation is second-order, we know from the general theory of differential equations that there exists a second linearly independent solution $Q_n(x)$ such that the combination

$$y = C_1 P_n(x) + C_2 Q_n(x) \qquad (4.78)$$

where C_1 and C_2 are arbitrary constants, is a general solution of (4.77).

Also from the theory of second-order linear DEs it is well known that if $y_1(x)$ is a nontrivial solution of

$$y'' + a(x)y' + b(x)y = 0 \qquad (4.79)$$

then a second linearly independent solution can be defined by*

$$y_2(x) = y_1(x) \int \frac{\exp\left[-\int a(x)\,dx\right]}{y_1^2(x)}\,dx \qquad (4.80)$$

---

* See Theorem 4.6 in L. C. Andrews, *Introduction to Differential Equations with Boundary Value Problems,* HarperCollins, New York, 1991.

Hence, if we express (4.77) in the form

$$y'' - \frac{2x}{1-x^2}y' + \frac{n(n+1)}{1-x^2}y = 0$$

and let $y_1(x) = P_n(x)$, it follows that

$$y_2(x) = P_n(x) \int \frac{dx}{(1-x^2)[P_n(x)]^2} \tag{4.81}$$

is a second solution, linearly independent of $P_n(x)$. Because any linear combination of solutions is also a solution of a homogeneous DE, it has become customary to define the second solution of (4.77), not by (4.81), but by

$$Q_n(x) = P_n(x)\left\{A_n + B_n \int \frac{dx}{(1-x^2)[P_n(x)]^2}\right\} \tag{4.82}$$

where A_n and B_n are constants to be chosen for each n. We refer to $Q_n(x)$ as the **Legendre function of the second kind** of integral order.

Accordingly, when $n = 0$, we choose $A_0 = 0$ and $B_0 = 1$, and hence

$$Q_0(x) = \int \frac{dx}{1-x^2}$$

which leads to

$$Q_0(x) = \frac{1}{2}\ln\frac{1+x}{1-x} \qquad |x| < 1 \tag{4.83}$$

For $n = 1$, we set $A_1 = 0$ and $B_1 = 1$, from which we obtain

$$Q_1(x) = x \int \frac{dx}{(1-x^2)x^2}$$

$$= x \int \left(\frac{1}{1-x^2} + \frac{1}{x^2}\right) dx$$

$$= \frac{1}{2}x \ln\frac{1+x}{1-x} - 1 \tag{4.84}$$

or
$$Q_1(x) = xQ_0(x) - 1 \qquad |x| < 1 \tag{4.85}$$

Rather than continuing in this fashion, which leads to more difficult integrals to evaluate, we recall the Remark made at the end of Sec. 4.2.3 which stated that all (properly normalized) solutions of Legendre's equation automatically satisfy the recurrence formulas for $P_n(x)$. Hence, we select the Legendre functions $Q_n(x)$ so that necessarily

$$Q_{n+1}(x) = \frac{2n+1}{n+1} x Q_n(x) - \frac{n}{n+1} Q_{n-1}(x) \qquad (4.86)$$

for $n = 1, 2, 3, \ldots$. With $Q_0(x)$ and $Q_1(x)$ already defined, the substitution of $n = 1$ into (4.86) yields

$$Q_2(x) = \tfrac{3}{2} x Q_1(x) - \tfrac{1}{2} Q_0(x)$$
$$= \tfrac{1}{2}(3x^2 - 1)Q_0(x) - \tfrac{3}{2}x$$

which we recognize as

$$Q_2(x) = P_2(x)Q_0(x) - \tfrac{3}{2}x \qquad |x| < 1 \qquad (4.87)$$

For $n = 2$, we find

$$Q_3(x) = P_3(x)Q_0(x) - \tfrac{5}{2}x^2 + \tfrac{2}{3} \qquad |x| < 1 \qquad (4.88)$$

whereas in general it has been shown that*

$$Q_n(x) = P_n(x)Q_0(x) - \sum_{k=0}^{[(n-1)/2]} \frac{(2n - 4k - 1)}{(2k+1)(n-k)} P_{n-2k-1}(x)$$

$$|x| < 1 \qquad (4.89)$$

for $n = 1, 2, 3, \ldots$.

Because of the logarithm term in $Q_0(x)$, it becomes clear that all $Q_n(x)$ have infinite discontinuities at $x = \pm 1$. However, within the interval $-1 < x < 1$ these functions are well defined. The first few Legendre functions of the second kind are sketched in Fig. 4.5 for the interval $0 \le x < 1$.

In some applications it is important to consider $Q_n(x)$ defined on the interval $x > 1$. While Eq. (4.89) is not valid for $x > 1$, the functions $Q_n(x)$ can be expanded in a convergent asymptotic series (see problem 16 in Exercises 4.6). Based on this series, it can then be shown that all $Q_n(x)$ approach zero as $x \to \infty$. Such behavior for large x is quite distinct from that of the Legendre polynomials $P_n(x)$, which become unbounded as $x \to \infty$ except for $P_0(x) = 1$.

---

* See W. W. Bell, *Special Functions for Scientists and Engineers*, Van Nostrand, London, 1968, pp. 71–77.

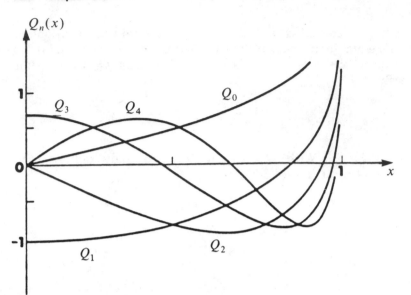

Figure 4.5　Graph of $Q_n(x)$, $n = 0, 1, 2, 3, 4$.

4.6.1　Basic properties

We have already mentioned that the Legendre functions $Q_n(x)$ satisfy all recurrence relations given in Sec. 4.2.2 for $P_n(x)$. In addition, there are several relations that directly involve both $P_n(x)$ and $Q_n(x)$. For example, if $|t| < |x|$, then*

$$\frac{1}{x - t} = \sum_{n=0}^{\infty} (2n + 1)P_n(t)Q_n(x) \tag{4.90}$$

From this result, it is easily shown that (see problem 13 in Exercises 4.6)

$$Q_n(x) = \frac{1}{2} \int_{-1}^{1} \frac{P_n(t)}{x - t}\,dt \qquad n = 0, 1, 2, \ldots \tag{4.91}$$

which is called the *Neumann formula*. Other properties are taken up in the exercises.

Exercises 4.6

In problems 1 to 4, find a general solution of the DE in terms of $P_n(x)$ and $Q_n(x)$.

1. $(1 - x^2)y'' - 2xy' = 0$
2. $(1 - x^2)y'' - 2xy' + 2y = 0$
3. $(1 - x^2)y'' - 2xy' + 12y = 0$
4. $(1 - x^2)y'' - 2xy' + 30y = 0$

---

*See E. T. Whittaker and G. N. Watson, *A Course of Modern Analysis*, Cambridge University Press, London, 1965, pp. 321–322.

5. Given $P_0(x) = 1$ and $Q_0(x) = \frac{1}{2}\ln[(1+x)/(1-x)]$, verify directly that their wronskian* satisfies

$$W(P_0, Q_0)(x) = \frac{1}{1-x^2}$$

6. Use Eq. (4.82) for $Q_n(x)$ to deduce that, in general, the wronskian of $P_n(x)$ and $Q_n(x)$ is given by

$$W(P_n, Q_n)(x) = \frac{B_n}{1-x^2} \qquad n = 0, 1, 2, \ldots$$

7. Show that $Q_n(x)$ satisfies the relations ($n = 1, 2, 3, \ldots$)
   (a) $Q'_{n+1}(x) - 2xQ'_n(x) + Q'_{n-1}(x) - Q_n(x) = 0$
   (b) $Q'_{n+1}(x) - xQ'_n(x) - (n+1)Q_n(x) = 0$
   (c) $xQ'_n(x) - Q'_{n-1}(x) - nQ_n(x) = 0$
   (d) $Q'_{n+1}(x) - Q'_{n-1}(x) = (2n+1)Q_n(x)$
   (e) $(1-x^2)Q'_n(x) = n[Q_{n-1}(x) - xQ_n(x)]$

8. Show that
   (a) $Q_0(-x) = -Q_0(x)$
   (b) $Q_n(-x) = (-1)^{n+1}Q_n(x)$, $n = 1, 2, 3, \ldots$

9. Show that (for $n = 1, 2, 3, \ldots$)
   $$n[Q_n(x)P_{n-1}(x) - Q_{n-1}(x)P_n(x)]$$
   $$= (n-1)[Q_{n-1}(x)P_{n-2}(x) - Q_{n-2}(x)P_{n-1}(x)]$$

10. From the result of problem 9, deduce that ($n = 1, 2, 3, \ldots$)
    $$Q_n(x)P_{n-1}(x) - Q_{n-1}(x)P_n(x) = -\frac{1}{n}$$

11. Deduce the result of problem 10 by using the wronskian relation in problem 6 and appropriate recurrence relations.

12. Show that $Q_n(x)$ satisfies the Christoffel-Darboux formula
    $$\sum_{k=0}^{n}(2k+1)Q_k(t)Q_k(x) = \frac{n+1}{t-x}[Q_{n+1}(t)Q_n(x) - Q_n(t)Q_{n+1}(x)]$$

13. Use the result of Eq. (4.90) to deduce the Neumann formula
    $$Q_n(x) = \frac{1}{2}\int_{-1}^{1}\frac{P_n(t)}{x-t}dt \qquad |x| > 1$$

14. For $x > 1$, use the Neumann formula in problem 13 to show that
    $$Q_n(x) = \frac{1}{2^{n+1}}\int_{-1}^{1}\frac{(1-t^2)^n}{(x-t)^{n+1}}dt$$

---

* Recall that the wronskian is defined by $W(y_1, y_2) = y_1 y'_2 - y'_1 y_2$.

15. Using the result of problem 14, deduce that $(x > 1)$

(a) $Q_n(x) = \int_0^\infty \dfrac{d\theta}{[x + (x^2 - 1)^{1/2} \cosh \theta]^{n+1}}$

   Hint: Set $t = \dfrac{e^\theta (x + 1)^{1/2} - (x - 1)^{1/2}}{e^\theta (x + 1)^{1/2} + (x - 1)^{1/2}}$

(b) $Q_n(x) \sim \dfrac{2^n}{x^{n+1}} \displaystyle\sum_{k=0}^\infty \dfrac{(n + k)!(n + 2k)!}{k!(2n + 2k + 1)!} \dfrac{1}{x^{2k}}, \quad x \to \infty$

16. Solve Legendre's equation

$$(1 - x^2)y'' - 2xy' + n(n + 1)y = 0$$

by assuming a power series solution of the form $y = \sum_{m=0}^\infty c_m x^m$.
(a)   Show that the general solution is

$$y = Ay_1(x) + By_2(x)$$

where A and B are any constants and

$$y_1(x) = 1 - \frac{n(n + 1)}{2!} x^2 + \frac{(n - 2)n(n + 1)(n + 3)}{4!} x^4 - \cdots$$

and

$$y_2(x) = x - \frac{(n - 1)(n + 2)}{3!} x^3$$

$$+ \frac{(n - 3)(n - 1)(n + 2)(n + 4)}{5!} x^5 - \cdots$$

(b)   For $n = 0$ show that

$$P_0(x) = \frac{y_1(x)}{y_1(1)} \qquad Q_0(x) = y_1(1)y_2(x)$$

(c)   For $n = 1$, show that

$$P_1(x) = \frac{y_2(x)}{y_2(1)} \qquad Q_1(x) = -y_2(1)y_1(x)$$

4.7   Associated Legendre Functions

In applications involving either the Laplace or the Helmholtz equation in spherical, oblate spheroidal, or prolate spheroidal coordinates, it is not Legendre's equation (4.44) that ordinarily arises but rather the **associated Legendre equation**

$$(1 - x^2)y'' - 2xy' + \left[n(n + 1) - \frac{m^2}{1 - x^2} \right] y = 0 \qquad (4.92)$$

Observe that for $m = 0$, (4.92) reduces to Legendre's equation (4.44). The DE (4.92) and its solutions, called *associated Legendre functions*, can be developed directly from Legendre's equation and its solutions. To show this, we will need the *Leibniz formula* for the mth derivative of a product

$$\frac{d^m}{dx^m}(fg) = \sum_{k=0}^{m} \binom{m}{k} \frac{d^{m-k}f}{dx^{m-k}} \frac{d^k g}{dx^k} \qquad m = 1, 2, 3, \dots \qquad (4.93)$$

If z is a solution of Legendre's equation, that is, if

$$(1 - x^2)z'' - 2xz' + n(n + 1)z = 0 \qquad (4.94)$$

we wish to show that

$$y = (1 - x^2)^{m/2} \frac{d^m z}{dx^m} \qquad (4.95)$$

is then a solution of (4.92). By taking m derivatives of (4.94), we get

$$\frac{d^m}{dx^m}[(1 - x^2)z''] - 2\frac{d^m}{dx^m}(xz') + n(n + 1)\frac{d^m z}{dx^m} = 0$$

which, by applying the Leibniz formula (4.93), becomes

$$(1 - x^2)\frac{d^{m+2}z}{dx^{m+2}} - 2mx\frac{d^{m+1}z}{dx^{m+1}} - m(m - 1)\frac{d^m z}{dx^m}$$

$$- 2\left(x\frac{d^{m+1}z}{dx^{m+1}} + m\frac{d^m z}{dx^m}\right) + n(n + 1)\frac{d^m z}{dx^m} = 0$$

Collecting like terms gives us

$$(1 - x^2)\frac{d^2 u}{dx^2} - 2(m + 1)x\frac{du}{dx} + [n(n + 1) - m(m + 1)]u = 0 \quad (4.96)$$

where for notational convenience we have set $u = d^m z/dx^m$. Next, by introducing the new variable $y = (1 - x^2)^{m/2}u$, or equivalently,

$$u = y(1 - x^2)^{-m/2}$$

we find that (4.96) takes the form

$$(1 - x^2)\frac{d^2}{dx^2}[y(1 - x^2)^{-m/2}] - 2(m + 1)x\frac{d}{dx}[y(1 - x^2)^{-m/2}]$$

$$+ [n(n + 1) - m(m + 1)]y(1 - x^2)^{-m/2} = 0 \quad (4.97)$$

Carrying out the indicated derivatives in (4.97) leads to

$$\frac{d}{dx}[y(1-x^2)^{-m/2}] = y'(1-x^2)^{-m/2} + mxy(1-x^2)^{-1-m/2}$$

$$= \left(y' + \frac{mxy}{1-x^2}\right)(1-x^2)^{-m/2} \qquad (4.98)$$

and similarly

$$\frac{d^2}{dx^2}[y(1-x^2)^{-m/2}]$$

$$= \left[y'' + \frac{m(2xy'+y)}{1-x^2} + \frac{m(m+2)x^2y}{(1-x^2)^2}\right](1-x^2)^{-m/2} \quad (4.99)$$

Finally, the substitution of (4.98) and (4.99) into (4.97) and cancellation of the common factor $(1-x^2)^{-m/2}$ then yield

$$(1-x^2)\left[y'' + \frac{m(2xy'+y)}{1-x^2} + \frac{m(m+2)x^2y}{(1-x^2)^2}\right]$$

$$- 2(m+1)x\left(y' + \frac{mxy}{1-x^2}\right) + [n(n+1) - m(m+1)]y = 0$$

which reduces to (4.92) upon algebraic simplification.

We define the **associated Legendre functions of the first and second kinds**, respectively, by ($m = 0, 1, 2, \ldots, n$)

$$P_n^m(x) = (1-x^2)^{m/2}\frac{d^m}{dx^m}P_n(x) \qquad (4.100)$$

and
$$Q_n^m(x) = (1-x^2)^{m/2}\frac{d^m}{dx^m}Q_n(x) \qquad (4.101)$$

Since $P_n(x)$ and $Q_n(x)$ are solutions of Legendre's equation, it follows from (4.95) that $P_n^m(x)$ and $Q_n^m(x)$ are solutions of the associated Legendre equation (4.92).

The associated Legendre functions have many properties in common with the simpler Legendre polynomials $P_n(x)$ and Legendre functions of the second kind $Q_n(x)$. Many of these properties can be developed directly from the corresponding relation involving either $P_n(x)$ or $Q_n(x)$ by taking derivatives and applying the definitions (4.100) and (4.101).

4.7.1   Basic properties of $P_n^m(x)$

Using the Rodrigues formula (4.47), we can write (4.100) in the form

$$P_n^m(x) = \frac{1}{2^n n!}(1-x^2)^{m/2}\frac{d^{n+m}}{dx^{n+m}}[(x^2-1)^n] \qquad (4.102)$$

Here we make the interesting observation that the right-hand side of (4.102) is well defined for all values of m such that $n+m \ge 0$, i.e., for $m \ge -n$, whereas (4.100) is valid only for $m \ge 0$. Thus, (4.102) may be used to extend the definition of $P_n^m(x)$ to include all integer values of m such that $-n \le m \le n$. [If $m > n$, then necessarily $P_n^m(x) \equiv 0$, which we leave to the reader to prove.] Moreover, using the Leibniz formula (4.93) once again, it can be shown that (see problem 5 in Exercises 4.7)

$$P_n^{-m}(x) = (-1)^m \frac{(n-m)!}{(n+m)!}P_n^m(x) \qquad (4.103)$$

Last, we note that for $m=0$ we get the special case

$$P_n^0(x) = P_n(x) \qquad (4.104)$$

The associated Legendre functions $P_n^m(x)$ satisfy many recurrence relations, several of which are generalizations of the recurrence formulas for $P_n(x)$. But because $P_n^m(x)$ has two indices instead of just one, there exist a wider variety of possible relations than for $P_n(x)$.

To derive the *three-term recurrence formula* for $P_n^m(x)$, we start with the known relation [see Eq. (4.31)]

$$(n+1)P_{n+1}(x) - (2n+1)xP_n(x) + nP_{n-1}(x) = 0 \qquad (4.105)$$

and differentiate it m times to obtain

$$(n+1)\frac{d^m}{dx^m}P_{n+1}(x) - (2n+1)x\frac{d^m}{dx^m}P_n(x)$$

$$-m(2n+1)\frac{d^{m-1}}{dx^{m-1}}P_n(x) + n\frac{d^m}{dx^m}P_{n-1}(x) = 0 \quad (4.106)$$

Now recalling [Eq. (4.40)]

$$(2n+1)P_n(x) = P'_{n+1}(x) - P'_{n-1}(x)$$

we find that taking $m-1$ derivatives leads to

$$m(2n+1)\frac{d^{m-1}}{dx^{m-1}}P_n(x) = m\frac{d^m}{dx^m}P_{n+1}(x) - m\frac{d^m}{dx^m}P_{n-1}(x) \quad (4.107)$$

and using this result, (4.106) becomes

$$(n - m + 1)\frac{d^m}{dx^m}P_{n+1}(x) - (2n + 1)x\frac{d^m}{dx^m}P_n(x)$$

$$+ (n + m)\frac{d^m}{dx^m}P_{n-1}(x) = 0$$

Finally, multiplication of this last result by $(1 - x^2)^{m/2}$ yields the desired **recurrence formula**

$$(n - m + 1)P_{n+1}^m(x) - (2n + 1)xP_n^m(x) + (n + m)P_{n-1}^m(x) = 0 \quad (4.108)$$

Additional recurrence relations, which are left to the exercises for verification, include the following:

$$(1 - x^2)^{1/2}P_n^m(x) = \frac{1}{2n + 1}[P_{n+1}^{m+1}(x) - P_{n-1}^{m+1}(x)] \qquad (4.109)$$

$$(1 - x^2)^{1/2}P_n^m(x) = \frac{1}{2n + 1}[(n + m)(n + m - 1)P_{n-1}^{m-1}(x)$$

$$- (n - m + 1)(n - m + 2)P_{n+1}^{m-1}(x)] \quad (4.110)$$

$$P_n^{m+1}(x) = 2mx(1 - x^2)^{-1/2}P_n^m(x)$$

$$- [n(n + 1) - m(m - 1)]P_n^{m-1}(x) \qquad (4.111)$$

By constructing a proof exactly analogous to the proof of orthogonality of the Legendre polynomials, it can be shown that

$$\int_{-1}^{1} P_n^m(x)P_k^m(x)\, dx = 0 \qquad k \neq n \qquad (4.112)$$

Also the evaluation of

$$\int_{-1}^{1} [P_n^m(x)]^2\, dx = \frac{2(n + m)!}{(2n + 1)(n - m)!} \qquad (4.113)$$

follows exactly our derivation of (4.66) given in Sec. 4.4.1. The details of proving (4.112) and (4.113) are left for the exercises.

As a final comment we mention that although it is essentially only a mathematical curiosity, there is another orthogonality relation for the associated Legendre functions given by

$$\int_{-1}^{1} P_n^m(x)P_n^k(x)(1 - x^2)^{-1}\, dx = \begin{cases} 0 & k \neq m \\ \dfrac{(n + m)!}{m(n - m)!} & k = m \end{cases} \qquad (4.114)$$

Exercises 4.7

1. Directly from Eq. (4.100), show that

   (a) $P_1^1(x) = (1 - x^2)^{1/2}$ (d) $P_3^1(x) = \frac{3}{2}(5x^2 - 1)(1 - x^2)^{1/2}$

   (b) $P_2^1(x) = 3x(1 - x^2)^{1/2}$ (e) $P_3^2(x) = 15x(1 - x^2)$

   (c) $P_2^2(x) = 3(1 - x^2)$

2. Show that

   (a) $P_n^m(-x) = (-1)^{n+m} P_n^m(x)$

   (b) $P_n^m(\pm 1) = 0, \; m > 0$

3. Show that (for $n = 0, 1, 2, \ldots$)

   (a) $P_{2n}^1(0) = 0$ (b) $P_{2n+1}^1(0) = \dfrac{(-1)^n (2n + 1)!}{2^{2n} (n!)^2}$

4. Show that

   (a) $P_n^m(0) = 0, \; n + m$ odd

   (b) $P_n^m(0) = (-1)^{(n-m)/2} \dfrac{(n + m)!}{2^n [(n - m)/2]! [(n + m)/2]!}, \; n + m$ even

5. Applying the Leibniz formula (4.93) to the product $(x + 1)^n (x - 1)^n$ and using (4.102), verify that

$$P_n^{-m}(x) = (-1)^m \frac{(n - m)!}{(n + m)!} P_n^m(x)$$

6. Derive the generating function relation

$$\frac{(2m)!(1 - x^2)^{m/2}}{2^m m!(1 - 2xt + t^2)^{m+1/2}} = \sum_{n=0}^{\infty} P_{n+m}^m(x) t^n$$

In problems 7 to 11, derive the given recurrence formula.

7. $(1 - x^2)P_n^{m\prime}(x) = (n + m)P_{n-1}^m(x) - nx P_n^m(x)$

8. $(1 - x^2)P_n^{m\prime}(x) = (n + 1)x P_n^m(x) - (n - m + 1)P_{n+1}^m(x)$

9. $(1 - x^2)^{1/2} P_n^m(x) = \dfrac{1}{2n + 1}[P_{n+1}^{m+1}(x) - P_{n-1}^{m+1}(x)]$

10. $(1 - x^2)^{1/2} P_n^m(x) = \dfrac{1}{2n + 1}[(n + m)(n + m - 1)P_{n-1}^{m-1}(x)$

$$- (n - m + 1)(n - m + 2)P_{n+1}^{m-1}(x)]$$

11. $P_n^{m+1}(x) = 2mx(1 - x^2)^{-1/2} P_n^m(x) - [n(n + 1) - m(m - 1)]P_n^{m-1}(x)$

12. Prove the orthogonality relation

$$\int_{-1}^{1} P_n^m(x) P_k^m(x) \, dx = 0 \qquad k \neq n$$

13. Prove the orthogonality relation

$$\int_{-1}^{1} P_n^m(x)P_n^k(x)(1-x^2)^{-1}\,dx = 0 \qquad k \neq m$$

14. (a) By defining

$$A_n = \int_{-1}^{1} [P_n^m(x)]^2\,dx \qquad n = 0, 1, 2, \ldots$$

show that

$$A_n = \frac{(2n-1)(n+m)}{(2n+1)(n-m)}A_{n-1} \qquad n = 2, 3, 4, \ldots$$

(b) Evaluate A_0 and A_1 directly and use (a) to deduce that

$$A_n = \frac{2(n+m)!}{(2n+1)(n-m)!} \qquad n = 0, 1, 2, \ldots$$

15. Show that

$$\int_{-1}^{1} [P_n^m(x)]^2(1-x^2)^{-1}\,dx = \frac{(n+m)!}{m(n-m)!}$$

4.8 Applications

It was during the nineteenth century that problems of heat conduction and electromagnetic theory were first formulated in terms of partial DEs, the solutions of which often led to one or more special functions. Since that time the study of special functions has been closely linked with the study of DEs. It turns out that the geometry of the problem, rather than the DE itself, has the greatest influence on which special function will arise in the solution process. For example, Legendre polynomials and the associated Legendre functions are prominent in applications featuring spherical geometry, such as finding the electric potential inside a spherical shell or the steady-state temperature inside a homogeneous solid sphere. On the other hand, Bessel functions (see Chaps. 6 to 8) are common in the solution of problems featuring circular or cylindrical domains, and Hermite polynomials (Chap. 5) arise in parabolic cylindrical domains.

The most important partial DE (or PDE) in mathematical physics is **Laplace's equation**, also known as the **potential equation**. Its general form is given by

$$\nabla^2 u = 0 \tag{4.115}$$

where the *laplacian* $\nabla^2 u$ in rectangular coordinates is defined by

$$\nabla^2 u = \frac{\partial^2 u}{\partial x^2} + \frac{\partial^2 u}{\partial y^2} + \frac{\partial^2 u}{\partial z^2} \qquad (4.116)$$

Laplace's equation arises in *steady-state heat conduction* problems involving homogeneous solids. This same equation is satisfied by the *gravitational potential* in free space, the *electrostatic potential* in a uniform dielectric, the *magnetic potential* in free space, and the *velocity potential* of inviscid, irrotational flows.

4.8.1 Electric potential due to a sphere

Problems involving spherical domains are usually formulated in spherical coordinates (r, θ, ϕ), as shown in Fig. 4.6. Recall that the relation between a point (x, y, z) in rectangular coordinates and that in spherical coordinates is

$$x = r \cos \theta \sin \phi \qquad y = r \sin \theta \sin \phi \qquad z = r \cos \phi$$

Suppose that on the surface of a hollow sphere of unit radius a fixed distribution of electric potential is maintained in such a way that it is independent of the polar azimuthal angle θ, shown in Fig. 4.6. In the absence of any further charges within the sphere, we wish to find the potential distribution within the sphere. Laplace's equation (4.115) is the governing equation for this problem which, formulated in spherical coordinates independent of θ, becomes (see problems 11 and 12 in Exercises 4.8)

$$\frac{\partial}{\partial r}\left(r^2 \frac{\partial u}{\partial r}\right) + \frac{1}{\sin \phi}\frac{\partial}{\partial \phi}\left(\sin \phi \frac{\partial u}{\partial \phi}\right) = 0 \qquad (4.117)$$

If the electric potential on the spherical shell is described by $f(\phi)$, then we impose the boundary condition

$$u(1, \phi) = f(\phi) \qquad 0 < \phi < \pi \qquad (4.118)$$

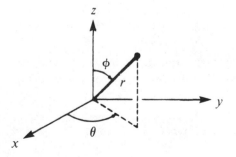

Figure 4.6 Spherical coordinates.

The solution of (4.117) subject to the boundary condition (4.118) is known as a **Dirichlet problem**, or **boundary-value problem of the first kind**.

To solve (4.117), we start with the assumption that the solution can be expressed in the product form

$$u(r, \phi) = R(r)\Phi(\phi) \tag{4.119}$$

(this is called the method of **separation of variables**). The direct substitution of (4.119) into Laplace's equation (4.117) leads to*

$$\frac{d}{dr}[r^2R'(r)\Phi(\phi)] + \frac{1}{\sin \phi \, d\phi} \frac{d}{d\phi}[\sin \phi \, R(r)\Phi'(\phi)] = 0$$

which, by rearranging and dividing by the product $R(r)\Phi(\phi)$, becomes

$$\frac{\dfrac{d}{dr}[r^2R'(r)]}{R(r)} = -\frac{\dfrac{1}{\sin \phi \, d\phi} \dfrac{d}{d\phi}[\sin \phi \, \Phi'(\phi)]}{\Phi(\phi)} \tag{4.120}$$

We now make the observation that the left-hand side of (4.120) involves only functions of r and the right-hand side only functions of ϕ. Thus we have "separated the variables." Because r and ϕ are independent variables, it follows that the only way (4.120) can be valid is if both sides are constant. Equating each side of (4.120) to the constant λ and simplifying, we obtain two ordinary DEs:

$$r^2R''(r) + 2rR'(r) - \lambda R(r) = 0 \qquad 0 < r < 1 \tag{4.121}$$

and

$$\frac{1}{\sin \phi \, d\phi} \frac{d}{d\phi}[\sin \phi \, \Phi'(\phi)] + \lambda\Phi(\phi) = 0 \qquad 0 < \phi < \pi \tag{4.122}$$

Our problem has now been reduced to solving (4.121) and (4.122).

By setting $x = \cos \phi$ in (4.122), we get the more recognizable form (see problem 14 in Exercises 4.2)

$$\frac{d}{dx}\left[(1-x^2)\frac{d\Phi}{dx}\right] + \lambda\Phi = 0 \qquad -1 < x < 1 \tag{4.123}$$

Physical considerations demand that the potential u everywhere on and inside the sphere remain bounded. The only bounded solutions of

---

*All partial derivatives become ordinary derivatives under the assumption (4.119), and thus we can resort to the prime notation for derivatives when convenient.

(4.123) occur when λ assumes one of the values (called *eigenvalues*)

$$\lambda_n = n(n+1) \qquad n = 0, 1, 2, \ldots \qquad (4.124)$$

and in this case we recognize (4.123) as Legendre's equation. Hence, the bounded solutions are given by the Legendre polynomials*

$$\Phi_n(\phi) \equiv P_n(x) = P_n(\cos \phi) \qquad n = 0, 1, 2, \ldots \qquad (4.125)$$

With the separation constant λ defined by (4.124), Eq. (4.121) becomes

$$r^2 R''(r) + 2rR'(r) - n(n+1)R(r) = 0 \qquad (4.126)$$

This DE is a *Cauchy-Euler equation* with general solution (see problems 4 and 5 in Exercises 4.8)

$$R_n(r) = a_n r^n + b_n r^{-(n+1)} \qquad n = 0, 1, 2, \ldots \qquad (4.127)$$

where a_n and b_n denote arbitrary constants. To avoid infinite values of $R_n(r)$ at $r = 0$, we must select $b_n = 0$ for all n. Therefore,

$$R_n(r) = a_n r^n \qquad n = 0, 1, 2, \ldots \qquad (4.128)$$

and by forming the product of (4.125) and (4.128) we generate the family of solutions

$$u_n(r, \phi) = a_n r^n P_n(\cos \phi) \qquad n = 0, 1, 2, \ldots$$

Finally, summing over all possible values of n (*superposition principle*†), we get

$$u(r, \phi) = \sum_{n=0}^{\infty} a_n r^n P_n(\cos \phi) \qquad (4.129)$$

Equation (4.129) represents a bounded solution of Laplace's equation (4.117) for any choice of the constants a_n. To satisfy the boundary condition (4.118), however, we must select constants a_n such that

$$u(1, \phi) = f(\phi) = \sum_{n=0}^{\infty} a_n P_n(\cos \phi) \qquad 0 < \phi < \pi \qquad (4.130)$$

---

* Recall that the Legendre functions $Q_n(x)$ are not bounded at $x = \pm 1$, that is, for $\phi = 0$ or $\phi = \pi$ (see Sec. 4.6).

† The *superposition principle* states that if $u_1, u_2, \ldots, u_n, \ldots$, are all solutions of a homogeneous linear PDE, then $u = \sum u_n$ is also a solution.

This last expression is recognized as a *Legendre series* for $f(\phi)$, and therefore the Fourier coefficients of (4.130) are given by (see problem 42 in Exercises 4.4)

$$a_n = (n + \tfrac{1}{2}) \int_0^\pi f(\phi) P_n(\cos \phi) \sin \phi \, d\phi \qquad n = 0, 1, 2, \ldots \quad (4.131)$$

Example 4: Find the electric potential inside a unit sphere when the boundary potential is prescribed by

$$u(1, \phi) = f(\phi) = \begin{cases} U_0 & 0 \le \phi < \dfrac{\pi}{2} \\[2mm] -U_0 & \dfrac{\pi}{2} < \phi \le \pi \end{cases}$$

Solution: The solution is given by Eq. (4.129), where the constants are determined from (4.131), which yields

$$a_n = \left(n + \frac{1}{2}\right) U_0 \left[\int_0^{\pi/2} P_n(\cos \phi) \sin \phi \, d\phi - \int_{\pi/2}^\pi P_n(\cos \phi) \sin \phi \, d\phi \right]$$

$$= \left(n + \frac{1}{2}\right) U_0 \left[\int_0^1 P_n(x) \, dx - \int_{-1}^0 P_n(x) \, dx \right]$$

The last step follows from the change of variables $x = \cos \phi$. Owing to the even–odd character of the Legendre polynomials, we see that the replacement of x by $-x$ in the last integral leads to the conclusion

$$a_n = 0 \qquad n = 0, 2, 4, \ldots$$

and
$$a_n = (2n + 1) U_0 \int_0^1 P_n(x) \, dx \qquad n = 1, 3, 5, \ldots$$

Recalling Example 2 in Sec. 4.4.3, the evaluation of this integral yields (setting $n = 2k + 1$)

$$a_{2k+1} = U_0 \frac{(-1)^k (2k)! (4k + 3)}{2^{2k+1} k! (k + 1)!} \qquad k = 0, 1, 2, \ldots$$

Hence the solution we seek becomes

$$u(r, \phi) = U_0 \sum_{k=0}^\infty \frac{(-1)^k (2k)! (4k + 3)}{2^{2k+1} k! (k + 1)!} r^{2k+1} P_{2k+1}(\cos \phi)$$

For problems involving electric potentials it is also natural to inquire about the potential outside the sphere $(r > 1)$. To determine the potential in this region, we must again solve Laplace's equation (4.117) subject to the boundary condition (4.118). In this case, however, our boundedness condition is not prescribed at $r = 0$ (which is outside the region of interest), but for $r \to \infty$. Hence, this time we set $a_n = 0$ $(n = 0, 1, 2, \dots)$ in Eq. (4.127) and obtain

$$R_n(r) = \frac{b_n}{r^{n+1}} \qquad n = 0, 1, 2, \dots \qquad (4.132)$$

Combining (4.125) and (4.132) by the superposition principle leads to

$$u(r, \phi) = \sum_{n=0}^{\infty} \frac{b_n}{r^{n+1}} P_n(\cos \phi) \qquad (4.133)$$

The determination of the constants b_n from the boundary condition (4.118) leads to the same integral as before [see (4.131)].

4.8.2 Steady-state temperatures in a sphere

Let us now consider the case where the temperature distribution on the surface of a homogeneous solid sphere of unit radius is maintained at a fixed distribution independent of time. Assuming the sphere is void of any heat sources, we wish to determine the (steady-state) temperature distribution everywhere within the sphere. The general form of Laplace's equation in spherical coordinates is given by (see problem 12 in Exercises 4.8)

$$\frac{\partial}{\partial r}\left(r^2 \frac{\partial u}{\partial r}\right) + \frac{1}{\sin \phi}\frac{\partial}{\partial \phi}\left(\sin \phi \frac{\partial u}{\partial \phi}\right) + \frac{1}{\sin^2 \phi}\frac{\partial^2 u}{\partial \theta^2} = 0 \qquad (4.134)$$

and the temperature on the surface of the sphere is prescribed by the boundary condition

$$u(1, \theta, \phi) = f(\theta, \phi) \qquad -\pi < \theta < \pi \qquad 0 < \phi < \pi \qquad (4.135)$$

To solve (4.134) by the separation-of-variables method, we initially assume the product form

$$u(r, \theta, \phi) = P(r, \phi)\Theta(\theta) \qquad (4.136)$$

The substitution of (4.136) into (4.134) and subsequent division by the product $P(r, \phi)\Theta(\theta)$ lead to*

$$\frac{\sin^2\phi\,\dfrac{\partial}{\partial r}\left(r^2\dfrac{\partial P}{\partial r}\right) + \sin\phi\,\dfrac{\partial}{\partial\phi}\left(\sin\phi\,\dfrac{\partial P}{\partial\phi}\right)}{P} = -\frac{\Theta''}{\Theta} \qquad (4.137)$$

In (4.137) we have separated the variables r and ϕ from θ, and thus by equating both sides to the constant λ, we obtain

$$\Theta'' + \lambda\Theta = 0 \qquad -\pi < \theta < \pi \qquad (4.138)$$

and
$$\frac{\partial}{\partial r}\left(r^2\frac{\partial P}{\partial r}\right) + \frac{1}{\sin\phi}\frac{\partial}{\partial\phi}\left(\sin\phi\frac{\partial P}{\partial\phi}\right) - \frac{\lambda}{\sin^2\phi}P = 0 \qquad (4.139)$$

To preserve the single-valuedness of the temperature distribution $u(r, \theta, \phi)$, we must require that $\Theta(\theta)$ be a *periodic function* with period 2π. Hence, we impose the *periodic boundary conditions*

$$\Theta(-\pi) = \Theta(\pi) \qquad \Theta'(-\pi) = \Theta'(\pi) \qquad (4.140)$$

The solution of (4.138) satisfying the periodic conditions (4.140) demands that λ be restricted to the values

$$\lambda_m = m^2 \qquad m = 0, 1, 2, \ldots \qquad (4.141)$$

and thus we obtain the solutions

$$\Theta_m(\theta) = \begin{cases} a_0 & m = 0 \\ a_m\cos m\theta + b_m\sin m\theta & m = 1, 2, 3, \ldots \end{cases} \qquad (4.142)$$

Equation (4.139) is still a PDE, and so we apply the separation-of-variables method once more in the hopes of reducing (4.139) to a system of ordinary DEs. Writing $\lambda = m^2$ and setting

$$P(r, \phi) = R(r)\Phi(\phi) \qquad (4.143)$$

we find that

$$\frac{\dfrac{d}{dr}(r^2R')}{R} = -\frac{\dfrac{1}{\sin\phi}\dfrac{d}{d\phi}(\sin\phi\,\Phi')}{\Phi} + \frac{m^2}{\sin^2\phi} = \mu$$

---

*For notational convenience, we will no longer display the arguments of the functions involved in the separation of variables.

and consequently,

$$r^2 R'' + 2r R' - \mu R = 0 \qquad 0 < r < 1 \qquad (4.144)$$

and

$$\frac{1}{\sin \phi} \frac{d}{d\phi}(\sin \phi \; \Phi') + \left(\mu - \frac{m^2}{\sin^2 \phi}\right)\Phi = 0 \qquad 0 < \phi < \pi \quad (4.145)$$

where μ is the new separation constant.

The change of variable $x = \cos \phi$ in (4.145) puts it in the form

$$\frac{d}{dx}\left[(1 - x^2)\frac{d\Phi}{dx}\right] + \left(\mu - \frac{m^2}{1 - x^2}\right)\Phi = 0 \qquad -1 < x < 1 \quad (4.146)$$

The temperature distribution throughout the sphere must remain bounded, and this condition requires that μ be restricted to the set of values

$$\mu_n = n(n + 1) \qquad n = 0, 1, 2, \ldots \qquad (4.147)$$

However, for these values of μ we see that (4.146) is the *associated Legendre equation* (Sec. 4.7), and its bounded solutions are the *associated Legendre functions* defined by

$$\Phi_{mn}(\phi) \equiv P_n^m(x) = P_n^m(\cos \phi) \qquad m, n = 0, 1, 2, \ldots \quad (4.148)$$

For $\mu = n(n + 1)$, the bounded solutions of (4.144) are of the form

$$R_n(r) = c_n r^n$$

and thus we see that $u_{mn}(r, \theta, \phi) = R_n(r)\Theta_m(\theta)\Phi_{mn}(\phi)$ gives us the family of solutions

$$u_{mn}(r, \theta, \phi)$$

$$= \begin{cases} A_{0n} r^n P_n(\cos \phi) & m = 0 \\ (A_{mn} \cos m\theta + B_{mn} \sin m\theta) r^n P_n^m(\cos \phi) & m = 1, 2, 3, \ldots \end{cases}$$

$$(4.149)$$

where $A_{0n} = a_0 c_n$, $A_{mn} = a_m c_n$, and $B_{mn} = b_m c_n$. Finally, summing over all such solutions by invoking the superposition principle, we arrive at

$$u(r, \theta, \phi) = \sum_{n=0}^{\infty} A_{0n} r^n P_n(\cos \phi)$$

$$+ \sum_{m=1}^{\infty} \sum_{n=0}^{\infty} (A_{mn} \cos m\theta + B_{mn} \sin m\theta) r^n P_n^m(\cos \phi) \quad (4.150)$$

The constants A_{0m}, A_{mn}, and B_{mn} have to be selected in such a way that the boundary condition (4.135) is satisfied, which leads to

$$f(\theta, \phi) = \sum_{n=0}^{\infty} A_{0n}P_n(\cos \phi)$$

$$+ \sum_{m=1}^{\infty} \sum_{n=0}^{\infty} (A_{mn} \cos m\theta + B_{mn} \sin m\theta)P_n^m(\cos \phi) \quad (4.151)$$

This last relation is what is known as a *generalized Fourier series in two variables*. Although the theory associated with such series follows in a natural way from the theory of one variable, it goes beyond the intended scope of this text. As a final observation here, we note that for the special case where the prescribed temperatures are independent of the angle θ, the temperatures inside the sphere will also be independent of θ. This condition necessitates that we allow only $m = 0$ in the solution (4.150), and in this case our solution (4.150) reduces to the result (4.129).

Exercises 4.8

1. Find the electric potential in the *interior* of the unit sphere, assuming the potential on the surface is
   (a) $f(\phi) = 1$ (c) $f(\phi) = \cos^2 \phi$
   (b) $f(\phi) = \cos \phi$ (d) $f(\phi) = \cos 2\phi$

2. Find the electric potential in the *exterior* of the unit sphere, assuming the potential on the surface is prescribed as given in problem 1.

3. The base $\phi = \frac{1}{2}\pi$, $r < 1$, of a solid hemisphere $r \leq 1$, $0 \leq \phi \leq \frac{1}{2}\pi$, is kept at temperature $u = 0$, while $u = T_0$ on the hemisphere surface $r = 1$, $0 < \phi < \frac{1}{2}\pi$. Show that the steady-state temperature distribution is given by

$$u(r, \phi) = T_0 \sum_{n=0}^{\infty} (-1)^n \left(\frac{4n + 3}{2n + 2}\right) \frac{(2n)!}{2^{2n}(n!)^2} r^{2n+1}P_{2n+1}(\cos \phi)$$

4. The DE

$$ax^2y'' + bxy' + cy = 0$$

   where a, b, and c are constants, is called a *Cauchy-Euler equation*.
   (a) Show that the change of variable $x = e^t$ leads to

$$xy' = Dy \qquad x^2y'' = D(D - 1)y \qquad D = \frac{d}{dt}$$

(b) Using the result of part (a), show that the Cauchy-Euler equation can be transformed to the constant-coefficient DE

$$[aD^2 + (b-a)D + c]y = 0$$

5. Use the result of problem 4 to verify that (4.127) is the general solution of the Cauchy-Euler equation

$$r^2R'' + 2rR' - n(n+1)R = 0$$

6. Solve the electric-potential problem in Sec. 4.8.1 for a sphere of radius c.

7. If the potential on the surface of a sphere of unit radius is that in Example 1, show that at points far from the sphere surface the potential is (approximately) given by

$$u(r,\phi) \simeq \frac{3U_0}{2r^2}\cos\phi \qquad r \gg 1$$

8. For a long time, the temperature $u(r,\phi)$ on the surface of a sphere of radius c has been maintained at $u(c,\phi) = T_0(1 - \cos^2\phi)$, where T_0 is a constant and ϕ is the cone angle in spherical coordinates. Find the temperature inside the sphere.

9. A spherical shell has an inner radius of 1 unit and an outer radius of 2 units. The prescribed temperatures on the inner and outer surfaces are given, respectively, by

$$u(1,\phi) = 30 + 10\cos\phi \qquad u(2,\phi) = 50 - 20\cos\phi$$

Determine the steady-state temperature everywhere within the spherical shell

10. The temperature on the surface of a solid homogeneous sphere of unit radius is prescribed by

$$u(1,\phi) = \begin{cases} T_0 & 0 < \phi < \alpha \\ 0 & \alpha < \phi < \pi \end{cases}$$

Show that the steady-state temperature distribution throughout the sphere is described by

$$u(r,\phi) = \tfrac{1}{2}T_0\Big\{1 - \cos\alpha - \sum_{n=1}^{\infty}[P_{n+1}(\cos\alpha)$$

$$- P_{n-1}(\cos\alpha)]r^nP_n(\cos\phi)\Big\}$$

11. Show that the laplacian $\nabla^2 u$ in spherical coordinates defined by

$$x = r\cos\theta\sin\phi \qquad y = r\sin\theta\sin\phi \qquad z = r\cos\phi$$

is given by

$$\nabla^2 u = \frac{\partial^2 u}{\partial r^2} + \frac{2}{r}\frac{\partial u}{\partial r} + \frac{1}{r^2}\frac{\partial^2 u}{\partial \phi^2} + \frac{\cot \phi}{r^2}\frac{\partial u}{\partial \phi} + \frac{1}{r^2 \sin^2 \phi}\frac{\partial^2 u}{\partial \theta^2}$$

12. Show that the laplacian in problem 11 can also be expressed as

$$\nabla^2 u = \frac{1}{r^2}\left[\frac{\partial}{\partial r}\left(r^2 \frac{\partial u}{\partial r}\right) + \frac{1}{\sin \phi}\frac{\partial}{\partial \phi}\left(\sin \phi \frac{\partial u}{\partial \phi}\right) + \frac{1}{\sin^2 \phi}\frac{\partial^2 u}{\partial \theta^2}\right]$$

5

Other Orthogonal
Polynomials

5.1 Introduction

A set of functions $\{\phi_n(x)\}$, $n = 0, 1, 2, \ldots$, is said to be *orthogonal* on the interval $a < x < b$, with respect to a weight function $r(x) > 0$, if*

$$\int_a^b r(x)\phi_n(x)\phi_k(x)\,dx = 0 \qquad k \neq n$$

Sets of orthogonal functions play an extremely important role in analysis, primarily because functions belonging to a very general class can be represented by series of orthogonal functions, called *generalized Fourier series*.

A special class of orthogonal functions consists of the sets of *orthogonal polynomials* $\{p_n(x)\}$, where n denotes the degree of the polynomial $p_n(x)$. The Legendre polynomials discussed in Chap. 4 are probably the simplest set of polynomials belonging to this class. Other polynomial sets which commonly occur in applications are the *Hermite, Laguerre,* and *Chebyshev polynomials*. More general polynomial sets are defined by the *Gegenbauer* and *Jacobi polynomials*, which include the others as special cases.

The study of general polynomial sets like the Jacobi polynomials facilitates the study of each polynomial set by focusing on those properties that are characteristic of all the individual sets. For example, the sets $\{p_n(x)\}$ that we will study all satisfy a second-

---

* In some cases the interval of orthogonality may be of infinite extent.

order linear DE and Rodrigues' formula, and the related set $\{(d^m/dx^m)p_n(x)\}$ (for example, the associated Legendre functions) is also orthogonal. Moreover, it can be shown that any orthogonal polynomial set satisfying these three conditions is necessarily a member of the Jacobi polynomial set, or a limiting case such as the Hermite and Laguerre polynomials.

5.2 Hermite Polynomials

The *Hermite polynomials* play an important role in problems involving Laplace's equation in parabolic coordinates, in various problems in quantum mechanics, and in probability theory.

We define the **Hermite polynomials** $H_n(x)$ by means of the generating function*

$$\exp(2xt - t^2) = \sum_{n=0}^{\infty} H_n(x)\frac{t^n}{n!} \qquad |t| < \infty, \qquad |x| > \infty \qquad (5.1)$$

By writing

$$\exp(2xt - t^2) = e^{2xt} \cdot e^{-t^2}$$

$$= \left[\sum_{m=0}^{\infty} \frac{(2xt)^m}{m!} \right]\left[\sum_{k=0}^{\infty} \frac{(-t^2)^k}{k!} \right]$$

$$= \sum_{n=0}^{\infty} \sum_{k=0}^{[n/2]} \frac{(-1)^k(2x)^{n-2k}}{k!(n-2k)!}t^n \qquad (5.2)$$

where the last step follows from the index change $m = n - 2k$ [see Eq. (1.17) in Sec. 1.2.3], we identify

$$H_n(x) = \sum_{k=0}^{[n/2]} \frac{(-1)^k n!}{k!(n-2k)!}(2x)^{n-2k} \qquad (5.3)$$

Examination of the series (5.3) reveals that $H_n(x)$ is a polynomial of degree n and, further, is an even function of x for even n and an odd function of x for odd n. Thus, it follows that

$$H_n(-x) = (-1)^n H_n(x) \qquad (5.4)$$

The first few Hermite polynomials are listed in Table 5.1 for easy reference.

------

*There is another definition of the Hermite polynomials that uses the generating function $\exp(xt - \frac{1}{2}t^2)$. This definition occurs most often in statistical applications.

TABLE 5.1 Hermite Polynomials

$H_0(x) = 1$
$H_1(x) = 2x$
$H_2(x) = 4x^2 - 2$
$H_3(x) = 8x^3 - 12x$
$H_4(x) = 16x^4 - 48x^2 + 12$
$H_5(x) = 32x^5 - 160x^3 + 120x$

In addition to the series (5.3), the Hermite polynomials can be defined by the *Rodrigues formula* (see problem 3 in Exercises 5.2)

$$H_n(x) = (-1)^n e^{x^2} \frac{d^n}{dx^n} (e^{-x^2}) \qquad n = 0, 1, 2, \ldots \qquad (5.5)$$

and the *integral representation* (see problem 5 in Exercises 5.2)

$$H_n(x) = \frac{(-i)^n 2^n e^{x^2}}{\sqrt{\pi}} \int_{-\infty}^{\infty} e^{-t^2 + 2ixt} t^n \, dt \qquad n = 0, 1, 2, \ldots \qquad (5.6)$$

The Hermite polynomials have many properties in common with the Legendre polynomials, and in fact, there are many relations connecting the two polynomial sets. For example, two of the simplest relations are given by ($n = 0, 1, 2, \ldots$)

$$\frac{2}{n!\sqrt{\pi}} \int_0^{\infty} e^{-t^2} t^n H_n(xt) \, dt = P_n(x) \qquad (5.7)$$

and

$$2^{n+1} e^{x^2} \int_x^{\infty} e^{-t^2} t^{n+1} P_n\left(\frac{x}{t}\right) dt = H_n(x) \qquad (5.8)$$

the verifications of which are left for the exercises.

Example 1: Use the generating function to derive the relation

$$x^n = \sum_{k=0}^{[n/2]} \frac{n! H_{n-2k}(x)}{2^n k! (n-2k)!} \qquad n = 0, 1, 2, \ldots$$

Solution: From (5.1) we have

$$\exp(2xt - t^2) = \sum_{k=0}^{\infty} H_k(x) \frac{t^k}{k!}$$

or

$$e^{2xt} = e^{t^2} \sum_{k=0}^{\infty} H_k(x) \frac{t^k}{k!}$$

Expressing both exponentials in power series leads to

$$\sum_{n=0}^{\infty} \frac{(2x)^n t^n}{n!} = \sum_{m=0}^{\infty} \frac{t^{2m}}{m!} \cdot \sum_{k=0}^{\infty} H_k(x) \frac{t^k}{k!}$$

$$= \sum_{n=0}^{\infty} \sum_{k=0}^{[n/2]} \frac{H_{n-2k}(x) t^n}{k!(n-2k)!}$$

where we have interchanged m and k and set $m = n - 2k$. Finally, by comparing the coefficients of t^n in the two series, we deduce that

$$x^n = \sum_{k=0}^{[n/2]} \frac{n! H_{n-2k}(x)}{2^n k!(n-2k)!}$$

5.2.1 Recurrence formulas

By substituting the series for $w(x, t) = \exp(2xt - t^2)$ into the identity

$$\frac{\partial w}{\partial t} - 2(x - t)w = 0 \qquad (5.9)$$

we obtain (after some manipulation)

$$\sum_{n=1}^{\infty} [H_{n+1}(x) - 2xH_n(x) + 2nH_{n-1}(x)] \frac{t^n}{n!} + H_1(x) - 2xH_0(x) = 0$$

$$(5.10)$$

But $H_1(x) - 2xH_0(x) = 0$, and thus we deduce the **recurrence formula**

$$H_{n+1}(x) - 2xH_n(x) + 2H_{n-1}(x) = 0 \qquad (5.11)$$

for $n = 1, 2, 3, \ldots$.

Another recurrence relation satisfied by the Hermite polynomials follows the substitution of the series for $w(x, t)$ into

$$\frac{\partial w}{\partial x} - 2tw = 0 \qquad (5.12)$$

This time we find

$$\sum_{n=1}^{\infty} [H_n'(x) - 2nH_{n-1}(x)] \frac{t^n}{n!} = 0$$

which leads to

$$H_n'(x) = 2nH_{n-1}(x) \qquad n = 1, 2, 3, \ldots \qquad (5.13)$$

The elimination of $H_{n-1}(x)$ from (5.11) and (5.13) yields

$$H_{n+1}(x) \quad 2xH_n(x) + H_n'(x) - 0 \qquad (5.14)$$

and by differentiating this expression and using (5.13) once again, we find

$$H_n''(x) - 2xH_n'(x) + 2nH_n(x) = 0 \qquad (5.15)$$

for $n = 0, 1, 2, \ldots$. Therefore we see that $y = H_n(x)$ $(n = 0, 1, 2, \ldots)$ is a solution of the linear second-order DE

$$y'' - 2xy' + 2ny = 0 \qquad (5.16)$$

called **Hermite's equation**.

5.2.2 Hermite series

The **orthogonality property** of the Hermite polynomials is given by*

$$\int_{-\infty}^{\infty} e^{-x^2} H_n(x) H_k(x)\, dx = 0 \qquad k \neq n \qquad (5.17)$$

We could construct a proof of (5.17) analogous to that given in Sec. 4.4.1 for the Legendre polynomials, but for the Hermite polynomials an interesting alternative proof exists.

Let us start with the generating-function relations

$$\sum_{n=0}^{\infty} \frac{t^n}{n!} H_n(x) = e^{2xt - t^2} \qquad (5.18a)$$

$$\sum_{k=0}^{\infty} \frac{s^k}{k!} H_k(x) = e^{2xs - s^2} \qquad (5.18b)$$

and multiply these two series to obtain

$$\sum_{n=0}^{\infty} \sum_{k=0}^{\infty} \frac{t^n s^k}{n! \, k!} H_n(x) H_k(x) = \exp\left[-(t^2 + s^2) + 2x(t+s)\right] \quad (5.19)$$

Next, we multiply both sides of (5.19) by the weight function e^{-x^2} and integrate (assuming that termwise integration is permitted), to find

$$\sum_{n=0}^{\infty} \sum_{k=0}^{\infty} \frac{t^n s^k}{n! \, k!} \int_{-\infty}^{\infty} e^{-x^2} H_n(x) H_k(x)\, dx = e^{-(t^2 + s^2)} \int_{-\infty}^{\infty} e^{-x^2 + 2x(t+s)}\, dx$$

$$= \sqrt{\pi}\, e^{2ts}$$

---

*The function e^{-x^2} in (5.17) is called a *weight function*. In the case of the Legendre polynomials, the weight function is unity.

where we have made the observation (see Example 2 below)

$$\int_{-\infty}^{\infty} e^{-x^2+2bx} \, dx = \sqrt{\pi} \, e^{b^2} \tag{5.20}$$

Finally, expanding e^{2ts} in a power series, we have

$$\sum_{n=0}^{\infty} \sum_{k=0}^{\infty} \frac{t^n s^k}{n! \, k!} \int_{-\infty}^{\infty} e^{-x^2} H_n(x) H_k(x) \, dx = \sqrt{\pi} \sum_{n=0}^{\infty} \frac{2^n t^n s^n}{n!} \tag{5.21}$$

and by comparing like coefficients of $t^n s^k$, we deduce that

$$\int_{-\infty}^{\infty} e^{-x^2} H_n(x) H_k(x) \, dx = 0 \qquad k \ne n$$

As a bonus, we find that when $k = n$ in (5.21), we get the additional important result (for $n = 0, 1, 2, \ldots$)

$$\int_{-\infty}^{\infty} e^{-x^2} [H_n(x)]^2 \, dx = 2^n n! \sqrt{\pi} \tag{5.22}$$

Based on relations (5.17) and (5.22), we can generate a theory concerning the expansion of arbitrary polynomials, or functions in general, in a series of Hermite polynomials. Specifically, if f is a suitable function defined for all x, we look for expansions of the general form

$$f(x) = \sum_{n=0}^{\infty} c_n H_n(x) \qquad -\infty < x < \infty \tag{5.23}$$

where the (Fourier) coefficients are given by*

$$c_n = \frac{1}{2^n n! \sqrt{\pi}} \int_{-\infty}^{\infty} e^{-x^2} f(x) H_n(x) \, dx \qquad n = 0, 1, 2, \ldots \tag{5.24}$$

Series of this type are called **Hermite series**. We have the following theorem for them.

Theorem 5.1. If f is piecewise smooth in every finite interval and

$$\int_{-\infty}^{\infty} e^{-x^2} f^2(x) \, dx < \infty$$

---

*The constants c_n can be formally derived through use of the orthogonality property analogous to the technique used in Sec. 4.4.2.

then the Hermite series (5.23) with constants defined by (5.24) converges pointwise to $f(x)$ at every continuity point of f. At points of discontinuity, the series converges to the average value $\frac{1}{2}[f(x^+) + f(x^-)]$.

The proof of Theorem 5.1 closely follows that of Theorem 4.1 (see N. N. Lebedev, *Special Functions and Their Applications*, Dover, New York, 1972, pp. 71–73).

Example 2: Express $f(x) = e^{2bx}$ in a Hermite series, and use this result to deduce the value of the integral

$$\int_{-\infty}^{\infty} e^{-x^2+2bx} H_n(x)\, dx$$

Solution: In this case we can obtain the series in an indirect way. We simply set $t = b$ in the generating function (5.1) to obtain

$$\exp(2bx - b^2) = \sum_{n=0}^{\infty} \frac{b^n}{n!} H_n(x)$$

and hence we have our intended series

$$e^{2bx} = e^{b^2} \sum_{n=0}^{\infty} \frac{b^n}{n!} H_n(x)$$

The direct derivation of this result from (5.24) leads to

$$c_n = \frac{1}{2^n n! \sqrt{\pi}} \int_{-\infty}^{\infty} e^{-x^2+2bx} H_n(x)\, dx \qquad n = 0, 1, 2, \ldots$$

However, we have already shown that

$$c_n = \frac{b^n}{n!} e^{b^2}$$

and thus it follows that

$$\int_{-\infty}^{\infty} e^{-x^2+2bx} H_n(x)\, dx = \sqrt{\pi}\,(2b)^n e^{b^2} \qquad n = 0, 1, 2, \ldots$$

In particular, for $n = 0$ we get the result of Eq. (5.20).

5.2.3 Simple harmonic oscillator

In wave mechanics, the basic equation for describing the (one-dimensional) location of a particle attracted by an energy potential

$V(z)$ is **Schrödinger's** (*time-independent*) **equation**

$$\frac{d^2\psi}{dz^2} + \frac{8m\pi^2}{h^2}[V(z) - E]\psi = 0 \qquad (5.25)$$

where m is the mass of the particle, E is the total energy, and h is Planck's constant. The unknown quantity ψ is called the *wave function*, i.e., the amplitude of the wave whose intensity gives the probability of finding the particle at any given point in space. A fundamental problem in wave mechanics concerns the motion of a particle bound in a potential well. It has been established that bounded solutions of Schrödinger's equation for such problems are obtainable only for certain discrete energy levels of the particle within the well. A particular example of this important class of problems is the *linear oscillator* (also called the *simple harmonic oscillator*), the solutions of which lead to Hermite polynomials.

If the restoring force on a particle displaced a distance z from its equilibrium position is $-kz$, where k can be associated with the spring constant of the classical oscillator, then its potential energy is $V(z) = \frac{1}{2}kz^2$. Substituting this expression into (5.25) and introducing the dimensionless parameters

$$x = z\left(\frac{4mk\pi^2}{h^2}\right)^{1/4} \qquad \lambda = \frac{4\pi E}{h}\sqrt{\frac{m}{k}} = \frac{4\pi E}{h\omega}$$

where $\omega = \sqrt{k/m}$ is the angular frequency of the classical oscillator, we find that Eq. (5.25) becomes

$$\psi'' + (\lambda - x^2)\psi = 0 \qquad -\infty < x < \infty \qquad (5.26)$$

where the primes denote differentiation with respect to x. In addition to (5.26), the wave function ψ must satisfy the boundary condition

$$\lim_{|x| \to \infty} \psi(x) = 0 \qquad (5.27)$$

In looking for bounded solutions of (5.26), we start with the observation that λ becomes negligible compared with x^2 for large values of x. Thus, asymptotically we expect the solution of (5.26) to behave like

$$\psi(x) \sim e^{\pm x^2/2} \qquad |x| \to \infty$$

where only the negative sign in the exponent is appropriate in order that (5.27) be satisfied. Based on this observation, we make the assumption that (5.26) has solutions of the form

$$\psi(x) = y(x)e^{-x^2/2} \qquad (5.28)$$

for suitable y. The substitution of (5.28) into (5.26) yields the DE

$$y'' - 2xy' + (\lambda - 1)y = 0 \qquad (5.29)$$

The boundary condition (5.27) suggests that whatever functional form y assumes, it must either be finite for all x or approach infinity at a rate slower than $e^{-x^2/2}$ approaches zero. It has been shown* that the only solutions of (5.29) satisfying this condition are those for which $\lambda - 1 = 2n$, or

$$\lambda \equiv \lambda_n = 2n + 1 \qquad n = 0, 1, 2, \ldots \qquad (5.30)$$

called *eigenvalues*. In terms of E, we find

$$E_n = \frac{(n + {}^1\!/_2)h\omega}{2\pi} \qquad n = 0, 1, 2, \ldots \qquad (5.31)$$

with $E_0 = h\omega/(4\pi)$ being the lowest or minimum energy level. With λ so restricted, we see that (5.29) becomes

$$y'' - 2xy' + 2ny = 0$$

which is Hermite's equation with solutions $y = H_n(x)$. (The other solutions of Hermite's equation are not appropriate in this problem.) Hence, we conclude that to each eigenvalue λ_n given by (5.30) there corresponds the solution of (5.26) (called an *eigenfunction* or *eigenstate*) given by

$$\psi_n(x) = H_n(x)e^{-x^2/2} \qquad n = 0, 1, 2, \ldots \qquad (5.32)$$

Exercises 5.2

1. Show that (for $n = 0, 1, 2, \ldots$)

   (a) $H_{2n}(0) = (-1)^n \dfrac{(2n)!}{n!}$ \qquad (c) $H'_{2n}(0) = 0$

   (b) $H_{2n+1}(0) = 0$ \qquad (d) $H'_{2n+1}(0) = (-1)^n \dfrac{(2n+2)!}{(n+1)!}$

2. Derive the generating-function relations

   (a) $e^{t^2} \cos 2xt = \displaystyle\sum_{n=0}^{\infty} (-1)^n H_{2n}(x) \frac{t^{2n}}{(2n)!}, \; |t| < \infty$

   (b) $e^{t^2} \sin 2xt = \displaystyle\sum_{n=0}^{\infty} (-1)^n H_{2n+1}(x) \frac{t^{2n+1}}{(2n+1)!}, \; |t| < \infty$

---

* See E. C. Kemble, *The Fundamental Principles of Quantum Mechanics with Elementary Applications*, Dover, New York, 1958, p. 87.

3. Derive the Rodrigues formula (for $n = 0, 1, 2, \ldots$)

$$H_n(x) = (-1)^n e^{x^2} \frac{d^n}{dx^n} (e^{-x^2})$$

4. Start with the integral formula

$$\int_{-\infty}^{\infty} e^{-t^2 + 2bt}\, dt = \sqrt{\pi}\, e^{b^2}$$

(a) Show that differentiating both sides with respect to b leads to

$$\int_{-\infty}^{\infty} te^{-t^2 + 2bt}\, dt = \sqrt{\pi}\, be^{b^2}$$

(b) For $n = 1, 2, 3, \ldots$, show that

$$\int_{-\infty}^{\infty} t^n e^{-t^2 + 2bt}\, dt = \frac{\sqrt{\pi}}{2^n} \frac{d^n}{db^n} (e^{b^2})$$

5. Set $b = ix$ in the result of problem 4b to deduce that (for $n = 0, 1, 2, \ldots$)

$$H_n(x) = \frac{(-i)^n 2^n e^{x^2}}{\sqrt{\pi}} \int_{-\infty}^{\infty} e^{-t^2 + 2ixt} t^n\, dt$$

6. Using the result of problem 5, show that (for $n = 0, 1, 2, \ldots$)

(a) $\quad H_{2n}(x) = \dfrac{(-1)^n 2^{2n+1}}{\sqrt{\pi}} e^{x^2} \displaystyle\int_0^{\infty} e^{-t^2} t^{2n} \cos 2xt\, dt$

(b) $\quad H_{2n+1}(x) = \dfrac{(-1)^n 2^{2n+2}}{\sqrt{\pi}} e^{x^2} \displaystyle\int_0^{\infty} e^{-t^2} t^{2n+1} \sin 2xt\, dt$

7. Derive the Fourier transform relations.

(a) $\quad \dfrac{1}{\sqrt{2\pi}} \displaystyle\int_{-\infty}^{\infty} e^{-t^2/2 + ixt} H_n(t)\, dt = i^n e^{-x^2/2} H_n(x)$

(b) $\quad \sqrt{\dfrac{2}{\pi}} \displaystyle\int_0^{\infty} e^{-t^2/2} H_{2n}(t) \cos xt\, dt = (-1)^n e^{-x^2/2} H_{2n}(x)$

(c) $\quad \sqrt{\dfrac{2}{\pi}} \displaystyle\int_0^{\infty} e^{-t^2/2} H_{2n+1}(t) \sin xt\, dt = (-1)^n e^{-x^2/2} H_{2n+1}(x)$

In problems 8 to 11, verify the integral relation.

8. $\displaystyle\int_{-\infty}^{\infty} x^k e^{-x^2} H_n(x)\, dx = 0, \ k = 0, 1, \ldots, n-1$

9. $\displaystyle\int_{-\infty}^{\infty} x^2 e^{-x^2} [H_n(x)]^2\, dx = \sqrt{\pi}\, 2^n n! (n + \frac{1}{2})$

10. $\displaystyle\int_0^\infty t^n e^{-t^2} H_n(xt)\,dt = \frac{\sqrt{\pi}\,n!}{2}\,P_n(x)$

11. $\displaystyle\int_x^\infty e^{-t^2} t^{n+1} P_n\!\left(\frac{x}{t}\right)dt = \frac{1}{2^{n+1}}\,e^{-x^2}H_n(x)$

12. Use the result of problem 5 to deduce that

   (a) $\displaystyle (1-t^2)^{-1/2}\exp\frac{2xyt-(x^2+y^2)t^2}{1-t^2} = \sum_{n=0}^{\infty} H_n(x)H_n(y)\frac{(t/2)^n}{n!}$

   (b) $\displaystyle (1-t^2)^{-1/2}\exp\frac{2x^2 t}{1+t} = \sum_{n=0}^{\infty} [H_n(x)]^2\frac{(t/2)^n}{n!}$

13. Use problem 12 to show that $(n = 0, 1, 2, \ldots)$

   $$\int_{-\infty}^{\infty} e^{-2x^2}[H_n(x)]^2\,dx = 2^{n-1/2}\Gamma(n+\tfrac{1}{2})$$

14. Derive the Hermite series relations.

   (a) $\displaystyle x^{2k} = \frac{(2k)!}{2^{2k}}\sum_{n=0}^{k}\frac{H_{2n}(x)}{(2n)!(k-n)!}$

   (b) $\displaystyle x^{2k+1} = \frac{(2k+1)!}{2^{2k+1}}\sum_{n=0}^{k}\frac{H_{2n+1}(x)}{(2n+1)!(k-n)!}$

15. For $m < n$, prove that

   $$\frac{d^m}{dx^m}H_n(x) = \frac{2^m n!}{(n-m)!}H_{n-m}(x)$$

In problems 16 to 19, derive the series relationship.

16. $\displaystyle [H_n(x)]^2 = 2^n(n!)^2\sum_{k=0}^{n}\frac{H_{2k}(x)}{2^k(k!)^2(n-k)!}$

17. $\displaystyle \sum_{k=0}^{n}\frac{H_k(x)H_k(y)}{2^k k!} = \frac{H_n(x)H_{n+1}(y)-H_{n+1}(x)H_n(y)}{2^{n+1}n!(y-x)}$

18. $\displaystyle H_n(x+y) = \frac{1}{2^{n/2}}\sum_{k=0}^{n}\binom{n}{k}H_{n-k}(x\sqrt{2})H_k(y\sqrt{2})$

19. $\displaystyle H_n(x)H_{n+p}(x) = 2^n n!(n+p)!\sum_{k=0}^{n}\frac{H_{2k+p}(x)}{2^k k!(k+p)!(n-k)!}$

20. Show that the functions $\psi_n(x) = e^{-x^2/2}H_n(x)$ satisfy the relations
   (a) $2n\psi_{n-1}(x) = x\psi_n(x) + \psi_n'(x)$
   (b) $2x\psi_n(x) - 2n\psi_{n-1}(x) = \psi_{n+1}(x)$
   (c) $\psi_n'(x) = x\psi_n(x) - \psi_{n+1}(x)$

21. Show that the functions $\psi_n(x) = e^{-x^2/2}H_n(x)$ satisfy

   $$\int_{-\infty}^{\infty} x^2[\psi_n(x)]^2\,dx = \sqrt{\pi}\,2^n n!(n+\tfrac{1}{2}) \qquad n = 0, 1, 2, \ldots$$

22. By writing $\psi_n(x) = c_n e^{-x^2/2} H_n(x)$, show that the *normalized eigenfunctions*, that is, $\int_{-\infty}^{\infty} [\psi_n(x)]^2 \, dx = 1$, assume the form

$$\psi_n(x) = [\sqrt{\pi} \, 2^n n!]^{-1/2} e^{-x^2/2} H_n(x) \qquad n = 0, 1, 2, \ldots$$

23. Ladder operators, called *creation* and *annihilation operators*, are defined by

$$L_+ = \frac{1}{\sqrt{2}} \left(x - \frac{d}{dx} \right) \qquad L_- = \frac{1}{\sqrt{2}} \left(x + \frac{d}{dx} \right)$$

For the normalized eigenfunctions defined in problem 22, show that

(a) $L_-[\psi_0(x)] = 0$
(b) $L_+[\psi_n(x)] = \sqrt{n+1} \, \psi_{n+1}(x), \ n = 0, 1, 2, \ldots$
(c) $L_-[\psi_n(x)] = \sqrt{n} \, \psi_{n-1}(x), \ n = 1, 2, 3, \ldots$

5.3 Laguerre Polynomials

The **generating function**

$$(1-t)^{-1} \exp\left(-\frac{xt}{1-t} \right) = \sum_{n=0}^{\infty} L_n(x) t^n \qquad |t| < 1, \qquad 0 \le x < \infty \quad (5.33)$$

leads to yet another important class of polynomials, called *Laguerre polynomials*. By expressing the exponential function in a series, we have

$$(1-t)^{-1} \exp\left(-\frac{xt}{1-t} \right) = \sum_{k=0}^{\infty} \frac{(-1)^k}{k!} (xt)^k (1-t)^{-k-1}$$

$$= \sum_{k=0}^{\infty} \frac{(-1)^k}{k!} (xt)^k \sum_{m=0}^{\infty} \binom{-k-1}{m} (-1)^m t^m \quad (5.34)$$

but since [see Eq. (1.27) in Sec. 1.2.4]

$$\binom{-k-1}{m} = (-1)^m \binom{k+m}{m}$$

it follows that (5.34) becomes

$$(1-t)^{-1} \exp\left(-\frac{xt}{1-t} \right) = \sum_{m=0}^{\infty} \sum_{k=0}^{\infty} \frac{(-1)^k (k+m)! x^k}{(k!)^2 m!} t^{k+m} \quad (5.35)$$

where we have reversed the order of summation. Finally, the change of index $m = n - k$ leads to (5.33) where the **Laguerre polynomials** are defined by

$$L_n(x) = \sum_{k=0}^{n} \frac{(-1)^k n! x^k}{(k!)^2 (n-k)!} \tag{5.36}$$

In Table 5.2 we have listed the first few polynomials $L_n(x)$.
   The **Rodrigues formula** for polynomials $L_n(x)$ is given by

$$L_n(x) = \frac{e^x}{n!} \frac{d^n}{dx^n} (x^n e^{-x}) \qquad n = 0, 1, 2, \dots \tag{5.37}$$

which can be verified by application of the Leibniz formula

$$\frac{d^n}{dx^n}(fg) = \sum_{k=0}^{n} \binom{n}{k} \frac{d^{n-k}f}{dx^{n-k}} \frac{d^k g}{dx^k} \qquad n = 1, 2, 3, \dots \tag{5.38}$$

5.3.1   Recurrence formulas

It is easily verified that the generating function

$$w(x, t) = (1-t)^{-1} \exp\left(-\frac{xt}{1-t}\right)$$

satisfies the identity

$$(1-t)^2 \frac{\partial w}{\partial t} + (x - 1 + t)w = 0 \tag{5.39}$$

By substituting the series (5.33) for $w(x, t)$ into (5.39), we find upon simplification that

$$\sum_{n=1}^{\infty} [(n+1)L_{n+1}(x) + (x - 1 - 2n)L_n(x) + nL_{n-1}(x)]t^n = 0 \tag{5.40}$$

TABLE 5.2   Laguerre Polynomials

$L_0(x) = 1$

$L_1(x) = -x + 1$

$L_2(x) = \dfrac{1}{2!}(x^2 - 4x + 2)$

$L_3(x) = \dfrac{1}{3!}(-x^3 + 9x^2 - 18x + 6)$

$L_4(x) = \dfrac{1}{4!}(x^4 - 16x^3 + 72x^2 - 96x + 24)$

Hence, equating the coefficient of t^n to zero, we obtain the **recurrence formula**

$$(n+1)L_{n+1}(x) + (x-1-2n)L_n(x) + nL_{n-1}(x) = 0 \qquad (5.41)$$

for $n = 1, 2, 3, \ldots$.

Similarly, substituting (5.33) into the identity

$$(1-t)\frac{\partial w}{\partial x} + tw = 0 \qquad (5.42)$$

leads to the derivative relation

$$L_n'(x) - L_{n-1}'(x) + L_{n-1}(x) = 0 \qquad (5.43)$$

where $n = 1, 2, 3, \ldots$.

If we now differentiate (5.41), we obtain

$$(n+1)L_{n+1}'(x) + (x-1-2n)L_n'(x) + L_n(x) + nL_{n-1}'(x) = 0 \quad (5.44)$$

and by writing (5.43) in the equivalent forms

$$L_{n+1}'(x) = L_n'(x) - L_n(x) \qquad (5.45a)$$

$$L_{n-1}'(x) = L_n'(x) + L_{n-1}(x) \qquad (5.45b)$$

we can eliminate $L_{n+1}'(x)$ and $L_{n-1}'(x)$ from (5.44), which yields

$$xL_n'(x) = nL_n(x) - nL_{n-1}(x) \qquad (5.46)$$

This last relation allows us to express the derivative of a Laguerre polynomial in terms of Laguerre polynomials.

To obtain the governing DE for the Laguerre polynomials, we begin by differentiating (5.46) and using (5.43) to get

$$xL_n''(x) + L_n'(x) = nL_n'(x) - nL_{n-1}'(x)$$

$$= -nL_{n-1}(x)$$

We can eliminate $L_{n-1}(x)$ by use of (5.46), which leads to

$$xL_n''(x) + (1-x)L_n'(x) + nL_n(x) = 0 \qquad (5.47)$$

Hence we conclude that $y = L_n(x)$ $(n = 0, 1, 2, \ldots)$ is a solution of **Laguerre's equation**

$$xy'' + (1-x)y' + ny = 0 \qquad (5.48)$$

5.3.2 Laguerre series

Like the Legendre polynomials and Hermite polynomials, various functions satisfying rather general conditions can be expanded in a series of Laguerre polynomials. Fundamental to the theory of such series is the **orthogonality property**

$$\int_0^\infty e^{-x}L_n(x)L_k(x)\,dx = 0 \qquad k \neq n \tag{5.49}$$

Our proof of (5.49) will parallel that given for the Hermite polynomials.

We begin by multiplying the two series

$$\sum_{n=0}^\infty L_n(x)t^n = (1-t)^{-1}\exp\left(-\frac{xt}{1-t}\right) \tag{5.50a}$$

$$\sum_{k=0}^\infty L_k(x)s^k = (1-s)^{-1}\exp\left(-\frac{xs}{1-s}\right) \tag{5.50b}$$

to obtain

$$\sum_{n=0}^\infty \sum_{k=0}^\infty t^n s^k L_n(x)L_k(x) = \frac{\exp\left[-x\left(\frac{t}{1-t}+\frac{s}{1-s}\right)\right]}{(1-t)(1-s)} \tag{5.51}$$

Next, multiplication of both sides of (5.51) by the weight function e^{-x} and subsequent integration lead to (see problem 29 in Exercises 5.3)

$$\sum_{n=0}^\infty \sum_{k=0}^\infty t^n s^k \int_0^\infty e^{-x}L_n(x)L_k(x)\,dx = (1-ts)^{-1}$$

$$= \sum_{n=0}^\infty t^n s^n \tag{5.52}$$

By comparing the coefficient of $t^n s^k$ on both sides of (5.52) we deduce the result (5.49), while for $k = n$, we also see that (for $n = 0, 1, 2, \ldots$)

$$\int_0^\infty e^{-x}[L_n(x)]^2\,dx = 1 \tag{5.53}$$

By a **Laguerre series**, we mean a series of the form

$$f(x) = \sum_{n=0}^\infty c_n L_n(x) \qquad 0 < x < \infty \tag{5.54}$$

where

$$c_n = \int_0^\infty e^{-x} f(x) L_n(x) \, dx \qquad n = 0, 1, 2, \ldots \qquad (5.55)$$

Without proof, we state the following theorem.

Theorem 5.2. If f is piecewise smooth in every finite interval $x_1 \leq x \leq x_2$, $0 < x_1 < x_2 < \infty$, and

$$\int_0^\infty e^{-x} f^2(x) \, dx < \infty$$

then the Laguerre series (5.54) with constants defined by (5.55) converges pointwise to $f(x)$ at every continuity point of f. At points of discontinuity, the series converges to the average value $\frac{1}{2}[f(x^+) + f(x^-)]$.

5.3.3 Associated Laguerre polynomials

In many applications, particularly in quantum mechanical problems, we need a generalization of the Laguerre polynomials called the **associated Laguerre polynomials**, i.e.,

$$L_n^{(m)}(x) = (-1)^m \frac{d^m}{dx^m} [L_{n+m}(x)] \qquad m = 0, 1, 2, \ldots \qquad (5.56)$$

By repeated differentiation of the series representation (5.36), it readily follows that (see problem 4 in Exercises 5.3)

$$L_n^{(m)}(x) = \sum_{k=0}^{n} \frac{(-1)^k (n+m)! x^k}{(n-k)!(m+k)!k!} \qquad m = 0, 1, 2, \ldots \qquad (5.57)$$

A *generating function* for the Laguerre polynomials $L_n^{(m)}(x)$ can be derived from that for $L_n(x)$. We first replace n by $n+m$ in (5.33) to get

$$(1-t)^{-1} \exp\left(-\frac{xt}{1-t}\right) = \sum_{n=-m}^{\infty} L_{n+m}(x) t^{n+m}$$

and then differentiate both sides m times with respect to x, that is,

$$(-1)^m t^m (1-t)^{-1-m} \exp\left(-\frac{xt}{1-t}\right) = \sum_{n=-m}^{\infty} \frac{d^m}{dx^m} [L_{n+m}(x)] t^{n+m}$$

The terms of the series for which $n = -1, -2, \ldots, -m$ are all zero, since the mth derivative of a polynomial of degree less than m is zero, and hence we deduce that

$$(1-t)^{-1-m} \exp\left(-\frac{xt}{1-t}\right) = \sum_{n=0}^{\infty} L_n^{(m)}(x)t^n \qquad |t| < 1 \qquad (5.58)$$

The associated Laguerre polynomials have many properties that are simple generalizations of those for the Laguerre polynomials. Among these are the **recurrence relations***

$$(n+1)L_{n+1}^{(m)}(x)$$
$$+ (x - 1 - 2n - m)L_n^{(m)}(x) + (n + m)L_{n-1}^{(m)}(x) = 0 \qquad (5.59)$$

$$xL_n^{(m)\prime}(x) - nL_n^{(m)}(x) + (n+m)L_{n-1}^{(m)}(x) = 0 \qquad (5.60)$$

and the **Rodrigues formula**

$$L_n^{(m)}(x) = \frac{1}{n!}e^x x^{-m}\frac{d^n}{dx^n}(e^{-x}x^{n+m}) \qquad (5.61)$$

The polynomials $L_n^{(m)}(x)$ also satisfy numerous relations where the upper index does not remain constant. Two such relations are given by

$$L_{n-1}^{(m)}(x) + L_n^{(m-1)}(x) - L_n^{(m)}(x) = 0 \qquad (5.62)$$

and
$$L_n^{(m)\prime}(x) = -L_{n-1}^{(m+1)}(x) \qquad (5.63)$$

The second-order DE satisfied by the polynomials $L_n^{(m)}(x)$ is the **associated Laguerre's equation**

$$xy'' + (m + 1 - x)y' + ny = 0 \qquad (5.64)$$

To show this, first we note that the polynomial $z = L_{n+m}(x)$ is a solution of Laguerre's equation

$$xz'' + (1-x)z' + (n+m)z = 0 \qquad (5.65)$$

By differentiating (5.65) m times, using the Leibniz rule (5.38), we obtain

$$x\frac{d^{m+2}z}{dx^{m+2}} + m\frac{d^{m+1}z}{dx^{m+1}} + (1-x)\frac{d^{m+1}z}{dx^{m+1}} + n\frac{d^m z}{dx^m} = 0$$

---

* Note that for $m = 0$, (5.59) reduces to (5.41).

or equivalently,

$$x \frac{d^2}{dx^2} \left(\frac{d^m z}{dx^m} \right) + (m + 1 - x) \frac{d}{dx} \left(\frac{d^m z}{dx^m} \right) + n \frac{d^m z}{dx^m} = 0 \qquad (5.66)$$

Comparing (5.64) and (5.66), we see that any function $y = C_1 (d^m z / dx^m)$ is a solution of (5.64) where C_1 is arbitrary. In particular, $y = L_n^{(m)}(x)$ is a solution.

Example 3: Prove the *addition formula*

$$L_n^{(a+b+1)}(x + y) = \sum_{k=0}^{n} L_k^{(a)}(x) L_{n-k}^{(b)}(y) \qquad a, b > -1$$

Solution: From the generating function (5.58), we have

$$\sum_{n=0}^{\infty} L_n^{(a+b+1)}(x + y) t^n = \frac{\exp\left[-(x + y) t / (1 - t) \right]}{(1 - t)^{a+b+2}}$$

$$= \frac{\exp\left[-xt / (1 - t) \right]}{(1 - t)^{a+1}} \cdot \frac{\exp\left[-yt / (1 - t) \right]}{(1 - t)^{b+1}}$$

$$= \sum_{k=0}^{\infty} L_k^{(a)}(x) t^k \cdot \sum_{m=0}^{\infty} L_m^{(b)}(y) t^m$$

$$= \sum_{m=0}^{\infty} \sum_{k=0}^{\infty} L_k^{(a)}(x) L_m^{(b)}(y) t^{m+k}$$

Next, making the change of index $m = n - k$ leads to

$$\sum_{n=0}^{\infty} L_n^{(a+b+1)}(x + y) t^n = \sum_{n=0}^{\infty} \sum_{k=0}^{n} L_k^{(a)}(x) L_{n-k}^{(b)}(y) t^n$$

and by comparing the coefficient of t^n in each series, we get our intended result.

Remark: The associated Laguerre polynomial $L_n^{(m)}(x)$ can be generalized to the case where m is not restricted to integer values by writing

$$L_n^{(a)}(x) = \sum_{k=0}^{n} \frac{(-1)^k \Gamma(n + a + 1) x^k}{(n - k)! \Gamma(k + a + 1) k!} \qquad a > -1$$

Most of the above relations are also valid for this more general polynomial.

5.3.4 The hydrogen atom

In Sec. 5.2.3 we solved the one-dimensional *Schrödinger equation* for the linear oscillator problem, the solutions of which led to Hermite polynomials. An important application involving the Laguerre polynomials is to find the wave function associated with the electron in a hydrogen atom. This problem involves the notion of central forces and leads to the form of Schrödinger's equation given by

$$\nabla^2 \psi + \frac{8\mu\pi^2}{h^2}[V(r) - E]\psi = 0 \qquad (5.67)$$

where μ denotes the mass of the electron, h is Planck's constant, $V(r)$ is the potential energy of the electron, and E is its total energy. Here we assume that $V(r) = k/r$, where k is a positive constant [recall Eq. (4.2)]. In spherical coordinates, Eq. (5.67) takes the form (see problem 12 in Exercises 4.8)

$$\frac{1}{r^2}\left[\frac{\partial}{\partial r}\left(r^2 \frac{\partial \psi}{\partial r}\right) + \frac{1}{\sin \phi}\frac{\partial}{\partial \phi}\left(\sin \phi \frac{\partial \psi}{\partial \phi}\right) + \frac{1}{\sin^2 \phi}\frac{\partial^2 \psi}{\partial \theta^2} \right]$$

$$+ \frac{8\mu\pi^2}{h^2}\left(\frac{k}{r} - E\right)\psi = 0 \quad (5.68)$$

To obtain bounded wave functions, we begin by looking for solutions of (5.68) having the product form

$$\psi(r, \theta, \phi) = R(r)\Theta(\theta)\Phi(\phi) \qquad (5.69)$$

By following the approach used in Sec. 4.8.2, it can be shown that $\Theta(\theta)$ and $\Phi(\phi)$ satisfy the DEs, respectively (see problem 31 in Exercises 5.3),

$$\Theta'' + \mu\Theta = 0 \qquad -\pi < \theta < \pi \qquad (5.70)$$

and

$$\frac{1}{\sin \phi}\frac{d}{d\phi}(\sin \phi \ \Phi') + \left(\nu - \frac{\mu}{\sin^2 \phi}\right)\Phi = 0 \qquad 0 < \phi < \pi \quad (5.71)$$

where μ and ν are separation constants. For physical reasons, we must require $\Theta(\theta)$ to be a periodic function with period 2π. This requirement leads to $\mu = m^2$, $m = 0, 1, 2, \ldots$, from which we deduce that $\Theta(\theta)$ is any multiple of*

$$\Theta_m(\theta) = e^{im\theta} \qquad m = 0, 1, 2, \ldots \qquad (5.72)$$

---

* It is customary here to use $e^{im\theta}$ rather than $\cos m\theta$ and $\sin m\theta$.

For $\mu = m^2$, it follows that Eq. (5.71) has bounded solutions only when $\nu = l(l+1)$, $l = 0, 1, 2, \ldots$, and in this case the solutions are multiples of

$$\Phi_{lm}(\phi) = P_l^m(\cos \phi) \qquad m, l = 0, 1, 2, \ldots \qquad (5.73)$$

where $P_l^m(\cos \phi)$ is an *associated Legendre function*. The products of these solutions, denoted by

$$Y_l^m(\theta, \phi) = P_l^m(\cos \phi)e^{im\theta} \qquad (5.74)$$

are called **spherical harmonics** and are important in a variety of applications beyond the hydrogen atom.

Based on the above results, the radial part $R(r)$ of the wave function then satisfies

$$\frac{1}{r^2}\frac{d}{dr}\left(r^2\frac{dR}{dr}\right) + \frac{8\mu\pi^2}{h^2}\left(\frac{k}{r} - E\right)R - \frac{l(l+1)}{r^2}R = 0 \qquad (5.75)$$

By introducing the new parameters

$$\rho = \alpha r \qquad \alpha^2 = -\frac{32\mu\pi^2 E}{h^2} \qquad (E < 0)$$

and

$$\chi(\rho) = R\left(\frac{\rho}{\alpha}\right) \qquad \lambda = \frac{8k\mu\pi^2}{\alpha h^2}$$

Eq. (5.75) becomes

$$\frac{1}{\rho^2}\frac{d}{d\rho}\left(\rho^2\frac{d\chi}{d\rho}\right) + \left[\frac{\lambda}{\rho} - \frac{1}{4} - \frac{l(l+1)}{\rho^2}\right]\chi = 0 \qquad (5.76)$$

the solution of which is given by (see problem 32 in Exercises 5.3 and problem 22 in Exercises 10.4)

$$\chi(\rho) = e^{-\rho/2}\rho^l L_{\lambda-l-1}^{(2l+1)}(\rho) \qquad (5.77)$$

However, to satisfy the physical constraint

$$\lim_{r\to\infty} R(r) = 0 \qquad (5.78)$$

we must restrict λ to integral values, that is, $\lambda = n$, $n = 1, 2, 3, \ldots$, where $n > l$. Such a restriction on λ has the effect of restricting the energy to certain discrete values described by

$$E_n = -\frac{2\mu k^2 \pi^2}{n^2 h^2} \qquad n = 1, 2, 3, \ldots (n > l) \qquad (5.79)$$

The corresponding wave functions are given by

$$\psi_{nlm}(r, \theta, \phi) = e^{-\alpha r/2}(\alpha r)^l L_{n-l-1}^{(2l+1)}(\alpha r)P_l^m(\cos \phi)e^{im\theta} \qquad (5.80)$$

Exercises 5.3

1. Show that (for $n = 0, 1, 2, \ldots$)
   (a) $L_n(0) = 1$
   (b) $L_n'(0) = -n$
   (c) $L_n''(0) = \frac{1}{2}n(n-1)$

2. Derive the Rodrigues formula.

   (a) $L_n(x) = \dfrac{e^x}{n!}\dfrac{d^n}{dx^n}(x^n e^{-x})$

   (b) $L_n^{(m)}(x) = \dfrac{1}{n!}x^{-m}e^{-x}\dfrac{d^n}{dx^n}(x^{n+m}e^{-x})$

   Hint: Use the Leibniz formula (5.38).

3. Derive the recurrence formulas.
   (a) $L_n'(x) - L_{n-1}'(x) + L_{n-1}(x) = 0$

   (b) $L_n'(x) = -\displaystyle\sum_{k=0}^{n-1} L_k(x)$

4. By repeated differentiation of the series (5.36), show that

   $$L_n^{(m)}(x) = \sum_{k=0}^{n} \frac{(-1)^k(m+n)!x^k}{(n-k)!(m+k)!k!} \qquad m = 0, 1, 2, \ldots$$

5. Show that

   $$n!\frac{d^k}{dx^k}[e^{-x}x^m L_n^{(m)}(x)] = (n+k)!e^{-x}x^{m-k}L_{n+k}^{(m-k)}(x)$$

6. Show that

   $$L_n^{(m)}(0) = \frac{(n+m)!}{n!m!}$$

In problems 7 to 10, verify the given recurrence relation.

7. $(n+1)L_{n+1}^{(m)}(x) + (x-1-2n-m)L_n^{(m)}(x) + (n+m)L_{n-1}^{(m)}(x) = 0$

8. $xL_n^{(m)'}(x) - nL_n^{(m)}(x) + (n+m)L_{n-1}^{(m)}(x) = 0$

9. $L_{n-1}^{(m)}(x) + L_n^{(m-1)}(x) - L_n^{(m)}(x) = 0$

10. $L_n^{(m)'}(x) = -L_{n-1}^{(m+1)}(x)$

In problems 11 to 18, verify the integral formula.

11. $\int_0^\infty e^{-x} x^k L_n(x)\,dx = \begin{cases} 0 & k < n \\ (-1)^n n! & k = n \end{cases}$

12. $\int_0^x L_k(t) L_n(x-t)\,dt = \int_0^x L_{n+k}(t)\,dt = L_{n+k}(x) - L_{n+k+1}(x)$

13. $\int_x^\infty e^{-t} L_n^{(m)}(t)\,dt = e^{-x}[L_n^{(m)}(x) - L_{n-1}^{(m)}(x)]$, $m = 0, 1, 2, \ldots$

14. $\int_0^x (x-t)^m L_n(t)\,dt = \dfrac{m!\,n!}{(m+n+1)!} x^{m+1} L_n^{(m+1)}(x)$, $m = 0, 1, 2, \ldots$

15. $\int_0^1 t^a (1-t)^{b-1} L_n^{(a)}(xt)\,dt = \dfrac{\Gamma(b)\Gamma(n+a+1)}{\Gamma(n+a+b+1)} L_n^{(a+b)}(x)$, $a > -1$,
$b > 0$

16. $\int_0^\infty e^{-x} x^a L_n^{(a)}(x) L_k^{(a)}(x)\,dx = 0$, $k \neq n$, $a > -1$

17. $\int_0^\infty e^{-x} x^a [L_n^{(a)}(x)]^2\,dx = \dfrac{\Gamma(n+a+1)}{n!}$, $a > -1$

18. $\int_0^\infty e^{-x} x^{a+1} [L_n^{(a)}(x)]^2\,dx = \dfrac{\Gamma(n+a+1)}{n!}(2n+a+1)$, $a > -1$

In problems 19 to 23, derive the given relation between the Hermite
and Laguerre polynomials.

19. $L_n^{(-1/2)}(x) = \dfrac{(-1)^n}{2^{2n} n!} H_{2n}(\sqrt{x})$

20. $L_n^{(1/2)}(x) = \dfrac{(-1)^n}{2^{2n+1} n!\sqrt{x}} H_{2n+1}(\sqrt{x})$

21. $\int_0^\infty e^{-t^2} [H_n(t)]^2 \cos\sqrt{2x}\,t\,dt = \sqrt{\pi}\,2^{n-1} n!\, e^{-x/2} L_n(x)$

22. $\int_{-1}^1 (1-t^2)^{a-1/2} H_{2n}(\sqrt{x}\,t)\,dt$

$= (-1)^n \sqrt{\pi}\, \dfrac{\Gamma(a+1/2)(2n)!}{\Gamma(n+a+1)} L_n^{(a)}(x)$, $a > -\frac{1}{2}$

23. $L_n(x^2+y^2) = \dfrac{(-1)^n}{2^{2n}} \sum_{k=0}^n \dfrac{H_{2k}(x) H_{2n-2k}(y)}{k!(n-k)!}$

In problems 24 and 25, derive the Laguerre series.

24. $x^p = p! \sum_{n=0}^{\#} \binom{p}{n} (-1)^n L_n(x)$

25. $e^{-ax} = (a+1)^{-1} \sum_{n=0}^{\infty} \left(\frac{a}{a+1}\right)^n L_n(x), a > -\frac{1}{2}$

Hint: Set $t - a/(a+1)$ in the generating function.

26. Show that $(x > 0)$

$$\int_0^\infty \frac{e^{-xt}}{t+1} dt = \sum_{n=0}^{\infty} \frac{L_n(x)}{n+1}$$

Hint: Use problem 25.

27. Show that $(x > 0)$

$$e^t (xt)^{-m/2} J_m(2\sqrt{xt}) = \sum_{n=0}^{\infty} \frac{L_n^{(m)}(x)}{(n+m)!} t^n \qquad m = 0, 1, 2, \ldots$$

where $J_m(x)$ is the *Bessel function* defined by (see Chap. 6)

$$J_m(x) = \sum_{k=0}^{\infty} \frac{(-1)^k (x/2)^{2k+m}}{k!(k+m)!}$$

28. Show that for $m > 1$,

$$\int_0^\infty t^{n+m/2} J_m(2\sqrt{xt}) e^{-t} dt = n! e^{-x} x^{m/2} L_n^{(m)}(x)$$

Hint: See problem 27.

29. Show that

$$\frac{1}{(1-t)(1-s)} \int_0^\infty \exp\left[-x\left(1 + \frac{t}{1-t} + \frac{s}{1-s}\right)\right] dx = \frac{1}{1-ts}$$

30. Show that the Laplace transform of $L_n(t)$ leads to

$$\int_0^\infty e^{-st} L_n(t) dt = \frac{1}{s}\left(1 - \frac{1}{s}\right)^n \qquad s > 0$$

31. Verify that by assuming $\psi(r, \theta, \phi) = R(r)\Theta(\theta)\Phi(\phi)$, Eq. (5.68) reduces to the system of equations given by (5.70), (5.71), and (5.75).

32. Assume (5.75) has solutions of the form

$$\chi(\rho) = e^{-\rho/2}\rho^l y(\rho)$$

(a) Show that $y(\rho)$ is a solution of the associated Laguerre equation

$$\rho y'' + (2l + 2 - \rho)y' + (\lambda - l - 1)y = 0$$

(b) From (a), deduce that $y(\rho) = L_{\lambda-l-1}^{(2l+1)}(\rho)$.

33. By assuming $R_{nl}(r) = C_{nl}e^{-\alpha r/2}(\alpha r)^l L_{n-l-1}^{(2l+1)}(\alpha r)$, show that the *normalized radial functions* require the choice

$$C_{nl} = \alpha^{3/2}\sqrt{\frac{(n-l-1)!}{2n(n+l)}} \qquad n > l$$

34. Using the normalized radial functions found in problem 33, show that

$$\int_0^\infty r^3[R_{nl}(r)]^2\,dr = \frac{1}{2}a_0[3n^2 - l(l+1)]$$

where $a_0 = h^2/(4\pi^2\mu k)$ is the radius of the innermost circular electron orbit known as the *Bohr radius*. (This result gives the average displacement of the electron from the nucleus.)

5.4 Generalized Polynomial Sets

The many properties that are shared by the Legendre, Hermite, and Laguerre polynomials suggest that there may exist more general polynomial sets of which these are certain specializations. Indeed, the *Gegenbauer* and *Jacobi polynomials* are two such generalizations. The Gegenbauer polynomials are closely connected with axially symmetric potentials in n dimensions and contain the Legendre, Hermite, and Chebyshev polynomials as special cases. The Jacobi polynomials are more general yet, since they contain the Gegenbauer polynomials as a special case.

5.4.1 Gegenbauer polynomials

The **Gegenbauer polynomials*** $C_n^\lambda(x)$ are defined by the generating function

$$(1 - 2xt + t^2)^{-\lambda} = \sum_{n=0}^\infty C_n^\lambda(x)t^n \qquad |t| < 1,\ |x| \le 1 \qquad (5.81)$$

where $\lambda > -\frac{1}{2}$. By expanding the function $w(x, t) = (1 - 2xt + t^2)^{-\lambda}$ in a binomial series and following our approach in Sec. 4.2.1, we find

$$w(x, t) = \sum_{n=0}^\infty \binom{-\lambda}{n}(-1)^n t^n (2x - t)^n$$

$$= \sum_{n=0}^\infty \sum_{k=0}^\infty \binom{-\lambda}{n}\binom{n}{k}(-1)^{n+k}(2x)^{n-k}t^{n+k}$$

$$= \sum_{n=0}^\infty \sum_{k=0}^{[n/2]} \binom{-\lambda}{n-k}\binom{n-k}{k}(-1)^n(2x)^{n-2k}t^n \qquad (5.82)$$

---

* The polynomials $C_n^\lambda(x)$ are also called *ultraspherical polynomials*.

and thus deduce that

$$C_n^\lambda(x) = (-1)^n \sum_{k=0}^{[n/2]} \binom{-\lambda}{n-k}\binom{n-k}{k}(2x)^{n-2k} \tag{5.83}$$

By substituting the series (5.81) into the identity

$$(1 - 2xt + t^2)\frac{\partial w}{\partial t} + 2\lambda(t - x)w = 0 \tag{5.84}$$

where $w(x, t) = (1 - 2xt + t^2)^{-\lambda}$, we obtain the three-term recurrence formula $(n = 1, 2, 3, \ldots)$

$$(n + 1)C_{n+1}^\lambda(x) - 2(\lambda + n)xC_n^\lambda(x) + (2\lambda + n - 1)C_{n-1}^\lambda(x) = 0 \tag{5.85}$$

Other recurrence formulas satisfied by the Gegenbauer polynomials include the following:

$$(n + 1)C_{n+1}^\lambda(x) - 2\lambda xC_n^{\lambda+1}(x) + 2\lambda C_{n-1}^{\lambda+1}(x) = 0 \tag{5.86}$$

$$(n + 2\lambda)C_n^\lambda(x) - 2\lambda C_n^{\lambda+1}(x) + 2\lambda xC_{n-1}^{\lambda+1}(x) = 0 \tag{5.87}$$

$$C_n^{\lambda'}(x) = 2\lambda C_{n+1}^{\lambda+1}(x) \tag{5.88}$$

The **orthogonality property** is given by (see problem 13 in Exercises 5.4)

$$\int_{-1}^{1} (1 - x^2)^{\lambda-1/2}C_n^\lambda(x)C_k^\lambda(x)\, dx = 0 \qquad k \neq n \tag{5.89}$$

and the governing DE is

$$(1 - x^2)y'' - (2\lambda + 1)xy' + n(n + 2\lambda)y = 0 \tag{5.90}$$

which can be verified by substituting the series (5.83) directly into (5.90).

One of the main advantages of developing properties of the Gegenbauer polynomials is that each recurrence formula, etc., becomes a master formula for all the polynomial sets that are generated as special cases. For example, when $\lambda = \frac{1}{2}$, we see that (5.81) is the generating function for the Legendre polynomials, and thus

$$P_n(x) = C_n^{1/2}(x) \qquad n = 0, 1, 2, \ldots \tag{5.91}$$

By setting $\lambda = \frac{1}{2}$ in (5.85), (5.89), and (5.90), we immediately obtain the recurrence formula, orthogonality property, and governing DE, respectively, for the Legendre polynomials.

The Hermite polynomials can also be generated from the Gegenbauer polynomials through the limit relation

$$H_n(x) = n!\lim_{\lambda\to\infty} \lambda^{-n/2} C_n^\lambda\left(\frac{x}{\sqrt{\lambda}}\right) \qquad n = 0, 1, 2, \ldots \qquad (5.92)$$

To show this, we start with the series representation

$$\lambda^{-n/2} C_n^\lambda\left(\frac{x}{\sqrt{\lambda}}\right) = (-1)^n \sum_{k=0}^{[n/2]} \binom{-\lambda}{n-k}\binom{n-k}{k}\frac{(2x)^{n-2k}}{\lambda^{n-k}} \qquad (5.93)$$

From Eq. (1.31) in Sec. 1.2.4, we obtain the relation

$$\frac{(-1)^n}{\lambda^{n-k}}\binom{-\lambda}{n-k} = \frac{(-1)^k}{\lambda^{n-k}}\binom{\lambda+n-k-1}{n-k}$$

$$= \frac{(-1)^k \Gamma(\lambda+n-k)}{\lambda^{n-k}\Gamma(\lambda)(n-k)!}$$

and thus establish that (see problem 3 in Exercises 5.4)

$$\lim_{\lambda\to\infty} \frac{(-1)^n}{\lambda^{n-k}}\binom{-\lambda}{n-k} = \frac{(-1)^k}{(n-k)!} \qquad (5.94)$$

Hence, from (5.93) we now deduce our intended result

$$n!\lim_{\lambda\to\infty} \lambda^{-n/2} C_n^\lambda\left(\frac{x}{\sqrt{\lambda}}\right) = \sum_{k=0}^{[n/2]} \frac{(-1)^k n!}{k!(n-2k)!}(2x)^{n-2k}$$

$$= H_n(x)$$

Properties of the Hermite polynomials can be obtained from properties of the Gegenbauer polynomials, although most such relations are more difficult to deduce than for the Legendre polynomials.

5.4.2   Chebyshev polynomials*

An important subclass of Gegenbauer polynomials is the Chebyshev polynomials of which there are two kinds. The **Chebyshev polynomials of the first kind** are defined by

$$T_0(x) = 1 \qquad T_n(x) = \frac{n}{2}\lim_{\lambda\to 0}\frac{C_n^\lambda(x)}{\lambda} \qquad n = 1, 2, 3, \ldots \qquad (5.95)$$

---

*There are numerous spellings of Chebyshev that occur throughout the literature, e.g., Tchebysheff, Tchebycheff, Tchebichef, and Chebysheff, among others.

**TABLE 5.3 Chebyshev Polynomials
of the First Kind**

$T_0(x) = 1$
$T_1(x) = x$
$T_2(x) = 2x^2 - 1$
$T_3(x) = 4x^3 - 3x$
$T_4(x) = 8x^4 - 8x^2 + 1$
$T_5(x) = 16x^5 - 20x^3 + 5x$

Because the Gegenbauer polynomials vanish when $\lambda = 0$, we cannot just simply define the polynomials $T_n(x)$ by $C_n^0(x)$. The choice $T_0(x) - 1$ is made to preserve the recurrence relation (5.99) given below. By following a procedure similar to that used to verify the relation (5.92), it can be established that (see problem 15 in Exercises 5.4)

$$T_n(x) = \frac{n}{2} \sum_{k=0}^{[n/2]} \frac{(-1)^k (n - k - 1)!}{k!(n - 2k)!} (2x)^{n-2k} \qquad (5.96)$$

The **Chebyshev polynomials of the second kind** are simply*

$$U_n(x) = C_n^1(x) \qquad n = 0, 1, 2, \ldots \qquad (5.97)$$

and thus by setting $\lambda = 1$ in (5.83) we immediately deduce that

$$U_n(x) = \sum_{k=0}^{[n/2]} \binom{n-k}{k}(-1)^k (2x)^{n-2k} \qquad (5.98)$$

The first few polynomials of each kind are tabulated in Tables 5.3 and 5.4.

By using properties previously cited for the Gegenbauer polyno-

**TABLE 5.4 Chebyshev Polynomials
of the Second Kind**

$U_0(x) = 1$
$U_1(x) = 2x$
$U_2(x) = 4x^2 - 1$
$U_3(x) = 8x^3 - 4x$
$U_4(x) = 16x^4 - 12x^2 + 1$
$U_5(x) = 32x^5 - 32x^3 + 6x$

---

*Some authors call $(1 - x^2)^{1/2} U_n(x)$ the Chebyshev functions of the second kind.

mials, we readily obtain the **recurrence formulas**

$$T_{n+1}(x) - 2xT_n(x) + T_{n-1}(x) = 0 \qquad (5.99)$$

$$U_{n+1}(x) - 2xU_n(x) + U_{n-1}(x) = 0 \qquad (5.100)$$

orthogonality properties

$$\int_{-1}^{1} (1 - x^2)^{-1/2} T_n(x) T_k(x)\, dx = 0 \qquad k \neq n \qquad (5.101)$$

$$\int_{-1}^{1} (1 - x^2)^{1/2} U_n(x) U_k(x)\, dx = 0 \qquad k \neq n \qquad (5.102)$$

and governing DE for $T_n(x)$

$$(1 - x^2)y'' - xy' + n^2 y = 0 \qquad (5.103)$$

and for $U_n(x)$

$$(1 - x^2)y'' - 3xy' + n(n+2)y = 0 \qquad (5.104)$$

There are also several recurrence-type formulas connecting the polynomials $T_n(x)$ and $U_n(x)$, such as

$$T_n(x) = U_n(x) - xU_{n-1}(x) \qquad (5.105)$$

and $\qquad (1 - x^2)U_{n-1}(x) = xT_n(x) - T_{n+1}(x) \qquad (5.106)$

the proofs of which are left for the exercises.

By making the substitution $x = \cos \phi$ in (5.103), we find it reduces to

$$\frac{d^2 y}{d\phi^2} + n^2 y = 0$$

with solutions $\cos n\phi$ and $\sin n\phi$. Thus we speculate that

$$T_n(\cos \phi) = c_n \cos n\phi$$

for some constant c_n. But since $T_n(1) = 1$ for all n (see problem 26 in Exercises 5.4), it follows that $c_n = 1$ for all n. It turns out that this speculation is correct, and in general we write

$$T_n(x) = \cos n\phi = \cos (n \cos^{-1} x) \qquad (5.107)$$

Similarly, it can be shown that

$$U_n(x) = \frac{\sin [(n+1) \cos^{-1} x]}{\sqrt{1 - x^2}} \qquad (5.108)$$

The significance of these observations is that the properties of sines and cosines can be used to establish many of the properties of the Chebyshev polynomials.

The Chebyshev polynomials have acquired great practical importance in polynomial approximation methods. Specifically, it has been shown that a series of Chebyshev polynomials converges more rapidly than any other series of Gegenbauer polynomials, and it converges much more rapidly than a power series.[*]

5.4.3 Jacobi polynomials

The **Jacobi polynomials**, which are generalizations of the Gegenbauer polynomials, are defined by the generating function

$$\frac{2^{a+b}}{R}(1-t+R)^{-a}(1+t+R)^{-b} = \sum_{n=0}^{\infty} P_n^{(a,b)}(x)t^n$$

$$a > -1, \qquad b > -1 \quad (5.109)$$

where
$$R = (1 - 2xt + t^2)^{1/2} \quad (5.110)$$

The Jacobi polynomials have the following three series representations (among others), which are somewhat involved to derive:

$$P_n^{(a,b)}(x) = \sum_{k=0}^{n} \binom{n+a}{n-k}\binom{n+b}{n-k}\left(\frac{x-1}{2}\right)^k\left(\frac{x+1}{2}\right)^{n-k} \quad (5.111)$$

$$P_n^{(a,b)}(x) = \sum_{k=0}^{n} \binom{n+a}{n-k}\binom{n+k+a+b}{k}\left(\frac{x-1}{2}\right)^k \quad (5.112)$$

$$P_n^{(a,b)}(x) = \sum_{k=0}^{n} (-1)^{n-k}\binom{n+b}{n-k}\binom{n+k+a+b}{k}\left(\frac{x+1}{2}\right)^k \quad (5.113)$$

By examination of the generating function (5.109), we observe that the Legendre polynomials are a specialization of the Jacobi polynomials for which $a = b = 0$, that is,

$$P_n(x) = P_n^{(0,0)}(x) \qquad n = 0, 1, 2, \ldots \quad (5.114)$$

whereas the associated Laguerre polynomials arise as the limit (see problem 37 in Exercises 5.4)

$$L_n^{(a)}(x) = \lim_{b\to\infty} P_n^{(a,b)}\left(1 - \frac{2x}{b}\right) \qquad n = 0, 1, 2, \ldots \quad (5.115)$$

---

[*] For theory and applications involving the Chebyshev polynomials, see L. Fox and I. B. Parker, *Chebyshev Polynomials in Numerical Analysis*, Oxford University Press, London, 1968.

In addition to the Legendre and Laguerre polynomials, the Gegenbauer polynomials are a special case of the Jacobi polynomials. To derive the relation between the Gegenbauer and Jacobi polynomials, we start with the identity

$$(1 - 2xt + t^2)^{-\lambda} = (1 - t)^{-2\lambda}\left[1 - \frac{2t(x-1)}{(1-t)^2}\right]^{-\lambda} \tag{5.116}$$

and expand the right-hand side in a series. This action leads to

$$(1 - 2xt + t^2)^{-\lambda} = \sum_{k=0}^{\infty} \binom{-\lambda}{k} \frac{(-1)^k (2t)^k (x-1)^k}{(1-t)^{2(k+\lambda)}}$$

$$= \sum_{m=0}^{\infty} \sum_{k=0}^{\infty} \binom{-\lambda}{k}\binom{-2k - 2\lambda}{m}(-1)^{m+k} 2^k (x-1)^k t^{m+k} \tag{5.117}$$

where we have expanded $(1 - t)^{-2(k+\lambda)}$ in another binomial series and interchanged the order of summation. Next, replacing the left-hand side of (5.117) by the series (5.81) and making the change of index $m = n - k$, we get

$$\sum_{n=0}^{\infty} C_n^{\lambda}(x)t^n = \sum_{n=0}^{\infty} \sum_{k=0}^{n} \binom{-\lambda}{k}\binom{-2k - 2\lambda}{n - k}(-1)^n 2^k (x-1)^k t^n$$

from which we deduce

$$C_n^{\lambda}(x) = (-1)^n \sum_{k=0}^{n} \binom{-\lambda}{k}\binom{-2k - 2\lambda}{n - k} 2^k (x-1)^k \tag{5.118}$$

Recalling Eq. (1.31) in Sec. 1.2.4 and the Legendre duplication formula, we see that

$$(-1)^n \binom{-\lambda}{k}\binom{-2k - 2\lambda}{n - k} = \binom{\lambda + k - 1}{k}\binom{n + k + 2\lambda - 1}{n - k}$$

$$= \frac{\Gamma(\lambda + k)\Gamma(n + k + 2\lambda)}{\Gamma(\lambda)k!(2\lambda + 2k)(n - k)!}$$

$$= \frac{\Gamma(\lambda + \frac{1}{2})\Gamma(n + k + 2\lambda)}{\Gamma(2\lambda)\Gamma(\lambda + k + \frac{1}{2})k!(n - k)!2^{2k}}$$

and hence (5.118) can be expressed in the form

$$C_n^{\lambda}(x) = \frac{\Gamma(\lambda + \frac{1}{2})\Gamma(n + 2\lambda)}{\Gamma(2\lambda)\Gamma(n + \lambda + \frac{1}{2})}$$

$$\times \sum_{k=0}^{n} \binom{n + \lambda - \frac{1}{2}}{n - k}\binom{n + k + 2\lambda - 1}{k}\left(\frac{x-1}{2}\right)^k \tag{5.119}$$

or, by comparing with (5.112),

$$C_n^\lambda(x) = \frac{\Gamma(\lambda + \frac{1}{2})\Gamma(n + 2\lambda)}{\Gamma(2\lambda)\Gamma(n + \lambda + \frac{1}{2})} P_n^{(\lambda - 1/2, \lambda - 1/2)}(x) \qquad (5.120)$$

The basic **recurrence formula** for the polynomials $P_n^{(a,b)}(x)$ is

$$2(n + 1)(a + b + n + 1)(a + b + 2n)P_{n+1}^{(a,b)}(x)$$
$$= (a + b + 2n + 1)[a^2 - b^2 + x(a + b + 2n + 2)(a + b + 2n)]P_n^{(a,b)}(x)$$
$$- 2(a + n)(b + n)(a + b + 2n + 2)P_{n-1}^{(a,b)}(x) \qquad (5.121)$$

for $n = 1, 2, 3, \ldots$. Also the **orthogonality property** and governing DE are given, respectively, by

$$\int_{-1}^{1} (1 - x)^a (1 + x)^b P_n^{(a,b)}(x) P_k^{(a,b)}(x)\, dx = 0 \qquad k \neq n \quad (5.122)$$

and

$$(1 - x^2)y'' + [b - a - (a + b + 2)x]y' + n(n + a + b + 1)y = 0 \quad (5.123)$$

Some additional properties concerning the Jacobi polynomials are taken up in the exercises.

Exercises 5.4

1. Show that (for $n = 0, 1, 2, \ldots$)
$$C_n^\lambda(-x) = (-1)^n C_n^\lambda(x)$$

2. Show that (for $n = 0, 1, 2, \ldots$)

(a) $C_{2n}^\lambda(0) = \begin{pmatrix} -\lambda \\ n \end{pmatrix}$

(c) $C_n^\lambda(1) = (-1)^n \begin{pmatrix} -2\lambda \\ n \end{pmatrix}$

(b) $C_{2n+1}^\lambda(0) = 0$

(d) $C_n^\lambda(-1) = \begin{pmatrix} -2\lambda \\ n \end{pmatrix}$

3. Show that
$$\lim_{\lambda \to \infty} \frac{\Gamma(\lambda + n - k)}{\lambda^{n-k}\Gamma(\lambda)} = 1$$

In problems 4 to 8, derive the given recurrence relation.

4. $xC_n^{\lambda'}(x) = nC_n^\lambda(x) + C_{n-1}^{\lambda'}(x)$
5. $2(\lambda + n)C_n^\lambda(x) = C_{n+1}^{\lambda'}(x) - C_{n-1}^{\lambda'}(x)$
6. $xC_n^{\lambda'}(x) = C_{n+1}^{\lambda'}(x) - (2\lambda + n)C_n^\lambda(x)$
7. $(x^2 - 1)C_n^{\lambda'}(x) = nxC_n^\lambda(x) - (2\lambda - 1 + n)C_{n-1}^\lambda(x)$

8. $nC_n^\lambda(x) = 2x(\lambda + n - 1)C_{n-1}^\lambda(x) - (2\lambda + n - 2)C_{n-2}^\lambda(x)$

9. Use any of the results of problems 4 to 8 and the recurrence formula (5.85) to show that $y = C_n^\lambda(x)$ is a solution of

$$(1 - x^2)y'' - (2\lambda + 1)xy' + n(n + 2\lambda)y = 0$$

10. Show that (for $k = 1, 2, 3, \ldots$)

$$\frac{d^k}{dx^k} C_n^\lambda(x) = 2^k \frac{\Gamma(\lambda + k)}{\Gamma(\lambda)} C_{n-k}^{\lambda+k}(x)$$

   Hint: Use Eq. (5.88).

11. Verify that (for $k = 1, 2, 3, \ldots$)*

$$C_{n-k}^{k+1/2}(x) = \frac{1}{(2k-1)!!} \frac{d^k}{dx^k} P_n(x)$$

12. Derive the recurrence relation

$$\sum_{k=0}^{n} (n + \lambda)C_k^\lambda(x) = \frac{(n + 2\lambda)C_n^\lambda(x) - (n + 1)C_{n+1}^\lambda(x)}{2(1 - x)}$$

13. Verify the orthogonality property

$$\int_{-1}^{1} (1 - x^2)^{\lambda - 1/2} C_n^\lambda(x) C_k^\lambda(x)\, dx = 0 \qquad k \neq n$$

14. Show that (for $n = 0, 1, 2, \ldots$)

$$\int_{-1}^{1} (1 - x^2)^{\lambda - 1/2}[C_n^\lambda(x)]^2\, dx = \frac{2^{1-2\lambda}\pi\,\Gamma(n + 2\lambda)}{(n + \lambda)\,[\Gamma(\lambda)]^2 n!}$$

15. By using Eq. (5.83) and the definition

$$T_n(x) = \frac{n}{2} \lim_{\lambda \to 0} \frac{C_n^\lambda(x)}{\lambda} \qquad n = 1, 2, 3, \ldots$$

   show that

$$T_n(x) = \frac{n}{2} \sum_{k=0}^{[n/2]} \frac{(-1)^k (n - k - 1)!}{k!(n - 2k)!} (2x)^{n-2k}$$

16. Using the recurrence formula (5.85), deduce the relations
   (a) $T_{n+1}(x) - 2xT_n(x) + T_{n-1}(x) = 0$
   (b) $U_{n+1}(x) - 2xU_n(x) + U_{n-1}(x) = 0$

In problems 17 to 22, derive the given relation for the Chebyshev polynomials.

17. $T_n(x) = U_n(x) - xU_{n-1}(x)$

18. $(1 - x^2)U_{n-1}(x) = xT_n(x) - T_{n+1}(x)$

---

* See problem 15 in Exercises 2.2 for definition of the symbol !!.

19. $T'_n(x) = nU_{n-1}(x)$

20. $2[T_n(x)]^2 = 1 + T_{2n}(x)$

21. $[T_n(x)]^2 - T_{n+1}(x)T_{n-1}(x) = 1 - x^2$

22. $[U_n(x)]^2 - U_{n+1}(x)U_{n-1}(x) = 1$

23. By making the substitution $x = \cos\phi$ in the orthogonality relation (5.101), show that

$$\int_0^\pi \cos n\phi \cos k\phi \, d\phi = 0 \qquad k \neq n$$

In problems 24 and 25, derive the generating-function relation.

24. $\dfrac{1 - t^2}{1 - 2xt + t^2} = T_0(x) + 2\sum_{n=1}^\infty T_n(x)t^n$

25. $\dfrac{1 - xt}{1 - 2xt + t^2} = \sum_{n=0}^\infty T_n(x)t^n$

26. Show that $T_n(1) = 1$, $n = 0, 1, 2, \ldots$, by using (a) problem 24 and (b) problem 25.

27. Verify the special values (for $n = 0, 1, 2, \ldots$)
    (a) $T_n(-1) = (-1)^n$
    (b) $T_{2n}(0) = (-1)^n$
    (c) $T_{2n+1}(0) = 0$

28. Verify the special values (for $n = 0, 1, 2, \ldots$)
    (a) $U_n(1) = n + 1$
    (b) $U_{2n}(0) = (-1)^n$
    (c) $U_{2n+1}(0) = 0$

29. Show that

$$\int_{-1}^1 (1 - x^2)^{-1/2}[T_n(x)]^2 \, dx = \begin{cases} \pi & n = 0 \\ \dfrac{\pi}{2} & n \geq 1 \end{cases}$$

30. Show that

$$\int_{-1}^1 (1 - x^2)^{1/2}[U_n(x)]^2 \, dx = \frac{\pi}{2}$$

In problems 31 to 38, verify the given relation for the Jacobi polynomials.

31. $P_n^{(a,b)}(-x) = (-1)^n P_n^{(b,a)}(x)$

32. $P_n^{(a,b)}(1) = \dbinom{a + n + 1}{n}$

33. $P_n^{(a,b)}(-1) = (-1)^n \dbinom{b + n + 1}{n}$

34. $P_n^{(a,b)}(x) = \dfrac{(-1)^n}{2^n n!}(1-x)^{-a}(1+x)^{-b}\dfrac{d^n}{dx^n}[(1-x)^{a+n}(1+x)^{b+n}]$

35. $\dfrac{d^k}{dx^k}P_n^{(a,b)}(x) = \dfrac{\Gamma(k+n+a+b+1)}{2^k\Gamma(n+a+b+1)}P_{n-k}^{(a+k,b+k)}(x)$

36. $P_n^{(a,b-1)}(x) - P_n^{(a-1,b)}(x) = P_{n-1}^{(a,b)}(x)$

37. $L_n^{(a)}(x) = \lim\limits_{b\to\infty}P_n^{(a,b)}\left(1-\dfrac{2x}{b}\right)$

38. $T_n(x) = \dfrac{2^{2n}(n!)^2}{(2n)!}P_n^{(-1/2,-1/2)}(x)$

6

Bessel Functions

6.1 Introduction

The German astronomer F. W. Bessel (1784–1846) first achieved fame by computing the orbit of Halley's comet. In addition to many other accomplishments in connection with his studies of planetary motion, he is credited with deriving the differential equation bearing his name and carrying out the first systematic study of the general properties of its solutions (now called *Bessel functions*) in his famous 1824 memoir. Nonetheless, Bessel functions were first discovered in 1732 by D. Bernoulli (1700–1782), who provided a series solution (representing a Bessel function) for the oscillatory displacements of a heavy hanging chain (see Sec. 6.7.1). Euler later developed a series similar to that of Bernoulli, which was also a Bessel function, and Bessel's equation appeared in a 1764 article by Euler dealing with the vibrations of a circular drumhead. J. Fourier (1768–1836) also used Bessel functions in his classical treatise on heat in 1822, but it was Bessel who first recognized their special properties.

Bessel functions are closely associated with problems possessing circular or cylindrical symmetry. For example, they arise in the study of free vibrations of a circular membrane and in finding the temperature distribution in a circular cylinder. They also occur in electromagnetic theory and numerous other areas of physics and engineering. In fact, Bessel functions occur so frequently in practice that they are undoubtedly the most important functions beyond the elementary ones.

Because of their close association with cylindrical domains, the solutions of Bessel's equation are also called *cylinder functions*. Bessel functions of the first and second kinds (studied in this chapter) are special cases of cylinder functions, as are modified

Bessel functions of the first and second kinds, Hankel functions, and spherical Bessel functions, among others (studied in Chap. 7).

6.2 Bessel Functions of the First Kind

Although Bessel functions frequently arise in practice as solutions of certain DEs, we wish to begin our discussion of them from the same point of view that we adopted in introducing the orthogonal polynomials in Chaps. 4 and 5, i.e., by a *generating function*.

6.2.1 The generating function

The function

$$w(x, t) = \exp\left[\frac{1}{2}x\left(t - \frac{1}{t}\right)\right] \qquad t \neq 0 \tag{6.1}$$

is called the **generating function** of the Bessel functions of integral order $J_n(x)$ since by expanding $w(x, t)$ in a particular series involving powers of t, it can be shown that

$$w(x, t) = \sum_{n=-\infty}^{\infty} J_n(x)t^n$$

To derive this relation, we begin by writing $w(x, t)$ as the product of two exponential functions and expand each in a Maclaurin series to get

$$w(x, t) = e^{xt/2} \cdot e^{-x/(2t)}$$

$$= \sum_{j=0}^{\infty} \frac{(xt/2)^j}{j!} \cdot \sum_{k=0}^{\infty} \frac{[-x/(2t)]^k}{k!}$$

$$= \sum_{j=0}^{\infty} \sum_{k=0}^{\infty} \frac{(-1)^k (x/2)^{j+k}}{j!\, k!} t^{j-k} \tag{6.2}$$

Our goal is to obtain a single series in powers of t. Thus, we make the change of index $n = j - k$, and because both j and k have an infinite range of values, it follows that $-\infty < n < \infty$. Consequently, Eq. (6.2) becomes

$$w(x, t) = \sum_{n=-\infty}^{\infty} \sum_{k=0}^{\infty} \frac{(-1)^k (x/2)^{2k+n}}{k!\, (k+n)!} t^n \tag{6.3}$$

The inside series in (6.3) is a function of x alone, which we define by the symbol

$$J_n(x) = \sum_{k=0}^{\infty} \frac{(-1)^k (x/2)^{2k+n}}{k!\,(k+n)!} \qquad -\infty < x < \infty \qquad (6.4)$$

called the **Bessel function of the first kind** of order n. It can readily be shown by the ratio test (Theorem 1.6) that the series (6.4) converges for all x. Finally, replacing the inside series in (6.3) with the symbol $J_n(x)$, we are led to the generating-function relation

$$w(x,t) = \exp\left[\frac{1}{2}x\left(t - \frac{1}{t}\right)\right] = \sum_{n=-\infty}^{\infty} J_n(x)t^n \qquad t \neq 0 \qquad (6.5)$$

Observe that (6.5) involves both positive and negative values of n. Therefore, we may wish to separately investigate the definition of $J_n(x)$ when $n < 0$. By first writing $(k+n)! = \Gamma(k+n+1)$, the formal replacement of n with $-n$ in (6.4) leads to

$$J_{-n}(x) = \sum_{k=0}^{\infty} \frac{(-1)^k (x/2)^{2k-n}}{k!\,\Gamma(k-n+1)} = \sum_{k=n}^{\infty} \frac{(-1)^k (x/2)^{2k-n}}{k!\,\Gamma(k-n+1)}$$

where we have used the fact that $1/\Gamma(k-n+1) = 0$ $(k = 0, 1, \ldots, n-1)$ by virtue of Theorem 2.1. To start the series at zero once again, we make the change of index $k = m + n$, which yields

$$J_{-n}(x) = \sum_{m=0}^{\infty} \frac{(-1)^{m+n}(x/2)^{2m+n}}{m!\,\Gamma(m+n+1)} = \sum_{m=0}^{\infty} \frac{(-1)^{m+n}(x/2)^{2m+n}}{m!\,(m+n)!}$$

This last series is a multiple of (6.4), from which we deduce

$$J_{-n}(x) = (-1)^n J_n(x) \qquad n = 0, 1, 2, \ldots \qquad (6.6)$$

Thus, $J_{-n}(x)$ equals either $J_n(x)$ (when n is even) or $-J_n(x)$ (when n is odd).

The Bessel functions that arise most frequently in practice are $J_0(x)$ and $J_1(x)$, whose series representations are

$$J_0(x) = \sum_{k=0}^{\infty} \frac{(-1)^k (x/2)^{2k}}{(k!)^2}$$

$$= 1 - \frac{x^2}{2^2} + \frac{x^4}{2^4(2!)^2} - \frac{x^6}{2^6(3!)^2} + \cdots \qquad (6.7a)$$

and

$$J_1(x) = \sum_{k=0}^{\infty} \frac{(-1)^k (x/2)^{2k+1}}{k!\,(k+1)!}$$

$$= \frac{x}{2} - \frac{x^3}{2^3 1!\,2!} + \frac{x^5}{2^5 2!\,3!} - \frac{x^7}{2^7 3!\,4!} + \cdots \qquad (6.7b)$$

At $x = 0$, we see that $J_0(0) = 1$ while $J_1(0) = 0$, and from (6.4) it follows that $J_n(0) = 0$, $n = 2, 3, 4, \ldots$. For x near zero, the $k = 0$ term of (6.4) yields the *asymptotic formulas**

$$J_0(x) \sim 1 \qquad x \to 0^+ \tag{6.8a}$$

$$J_n(x) \sim \frac{(x/2)^n}{n!} \qquad x \to 0^+, \qquad n = 1, 2, 3, \ldots \tag{6.8b}$$

The graphs of $J_0(x)$, $J_1(x)$, and $J_2(x)$ are shown in Fig. 6.1. Observe that these functions exhibit an oscillatory behavior somewhat like that of the sinusoidal functions, except that the amplitude (maximum departure from the x axis) of the Bessel functions diminishes as x increases and the (infinitely many) zeros of these functions are *not* evenly spaced. The location of these zeros is of great theoretical and practical importance, but the theory goes beyond the scope of this text.† For reference purposes, however, the first few zeros of some of the Bessel functions are listed in Table 6.1.

6.2.2 Bessel functions of nonintegral order

Thus far we have only discussed Bessel functions of integral order defined by the series (6.4). However, we can generalize this series definition of $J_n(x)$ to include nonintegral values of n by again writing $(k + n)! = \Gamma(k + n + 1)$ and then formally replacing n with p, where p

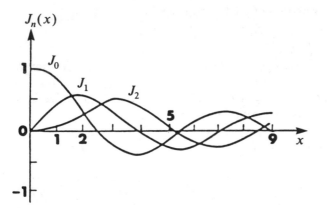

Figure 6.1 Graph of $J_n(x)$, $n = 0, 1, 2$.

---

* Asymptotic formulas for large x are developed in Sec. 7.6.

† For an introductory treatment of this topic, see chap. 5 in L. C. Andrews, *Introduction to Differential Equations with Boundary Value Problems*, Harper-Collins, New York, 1991.

TABLE 6.1 Zeros of Bessel Functions: $J_n(x_k) = 0$

$\overset{k}{\underset{n}{}}$	1	2	3	4	5
0	2.405	5.520	8.654	11.792	14.931
1	3.832	7.016	10.173	13.324	16.471
2	5.136	8.417	11.620	14.796	17.960
3	6.380	9.761	13.015	16.223	19.409
4	7.588	11.065	14.373	17.616	20.827

is any nonnegative real number. This action leads to

$$J_p(x) = \sum_{k=0}^{\infty} \frac{(-1)^k (x/2)^{2k+p}}{k!\,\Gamma(k+p+1)} \qquad p \geq 0 \qquad (6.9)$$

called the Bessel function of the first kind *of order p*. The replacement of p with $-p$ in (6.9) then yields

$$J_{-p}(x) = \sum_{k=0}^{\infty} \frac{(-1)^k (x/2)^{2k-p}}{k!\,\Gamma(k-p+1)} \qquad p > 0 \qquad (6.10)$$

Observe that $J_p(x)$ and $J_{-p}(x)$ have *asymptotic formulas*

$$J_p(x) \sim \frac{(x/2)^p}{\Gamma(p+1)} \qquad x \to 0^+ \qquad (6.11a)$$

$$J_{-p}(x) \sim \frac{(x/2)^{-p}}{\Gamma(1-p)} \qquad x \to 0^+ \qquad (6.11b)$$

For p *not* integral, it follows that the function $J_{-p}(x)$ becomes infinite as $x \to 0^+$ while $J_p(x)$ approaches zero. Hence, $J_{-p}(x)$ cannot be a multiple of $J_p(x)$, as in the case of $J_n(x)$ and $J_{-n}(x)$.

The cases where p is *half-integral* are of special interest because these Bessel functions are actually elementary functions. In particular, the case $p = \frac{1}{2}$ leads to the interesting results (see problem 13 in Exercises 6.2 and Sec. 7.4, too)

$$J_{1/2}(x) = \sqrt{\frac{2}{\pi x}} \sin x \qquad (6.12a)$$

$$J_{-1/2}(x) = \sqrt{\frac{2}{\pi x}} \cos x \qquad (6.12b)$$

6.2.3 Recurrence formulas

Because the generating-function relation (6.5) is restricted to Bessel functions of *integral order,* we will develop properties of Bessel functions of arbitrary order p without its use. Fortunately, many of the properties of $J_p(x)$ and $J_{-p}(x)$ can be developed directly from their series definition. For example, suppose we multiply the series for $J_p(x)$ by x^p and then differentiate the result with respect to x. This gives us

$$\frac{d}{dx}[x^p J_p(x)] = \frac{d}{dx}\sum_{k=0}^{\infty}\frac{(-1)^k x^{2k+2p}}{2^{2k+p}k!\,\Gamma(k+p+1)}$$

$$= \sum_{k=0}^{\infty}\frac{(-1)^j 2(k+p)x^{2k+2p-1}}{2^{2k+p}k!\,\Gamma(k+p+1)}$$

$$= x^p \sum_{k=0}^{\infty}\frac{(-1)^k x^{2k+(p-1)}}{k!\,\Gamma(k+p)}$$

or
$$\frac{d}{dx}[x^p J_p(x)] = x^p J_{p-1}(x) \tag{6.13}$$

Similarly, if we multiply $J_p(x)$ by x^{-p}, we find that (see problem 14 in Exercises 6.2)

$$\frac{d}{dx}[x^{-p}J_p(x)] = -x^{-p}J_{p+1}(x) \tag{6.14}$$

If we carry out the differentiation on the left-hand sides in (6.13) and (6.14) and divide the results by the factors x^p and x^{-p}, respectively, we deduce that

$$J_p'(x) + \frac{p}{x}J_p(x) = J_{p-1}(x) \tag{6.15}$$

and
$$J_p'(x) - \frac{p}{x}J_p(x) = -J_{p+1}(x) \tag{6.16}$$

Setting $p = 0$ in (6.16), we obtain the special case

$$J_0'(x) = -J_1(x) \tag{6.17}$$

Finally, the sum and difference of (6.15) and (6.16) yield, respectively,

$$2J_p'(x) = J_{p-1}(x) - J_{p+1}(x) \tag{6.18}$$

and
$$\frac{2p}{x}J_p(x) = J_{p-1}(x) + J_{p+1}(x) \tag{6.19}$$

This last relation is the three-term **recurrence formula** for the Bessel functions. Observe that it is *not* restricted to integral values of p.

6.2.4 Bessel's differential equation

By using the above recurrence formulas, we can find a derivative relation which involves only the Bessel function $J_p(x)$. To start, we rewrite Eqs. (6.15) and (6.16) as

$$xJ_p'(x) - xJ_{p-1}(x) + pJ_p(x) = 0 \qquad (6.20)$$

$$xJ_{p-1}'(x) = (p-1)J_{p-1}(x) - xJ_p(x) \qquad (6.21)$$

and differentiate (6.20) to find

$$xJ_p''(x) + (p+1)J_p'(x) - xJ_{p-1}'(x) - J_{p-1}(x) = 0 \qquad (6.22)$$

Next we multiply (6.20) by p and subtract it from (6.22), multiplied by x, which yields

$$x^2 J_p''(x) + xJ_p'(x) - p^2 J_p(x) + (p-1)xJ_{p-1}(x) - x^2 J_{p-1}'(x) = 0$$

Finally, using (6.21) to eliminate $J_{p-1}'(x)$ and $J_{p-1}(x)$ from this last expression, we obtain the desired relation

$$x^2 J_p''(x) + xJ_p'(x) + (x^2 - p^2)J_p(x) = 0 \qquad (6.23)$$

From (6.23), we deduce that $y_1 = J_p(x)$ is a solution of the second-order linear DE

$$x^2 y'' + xy' + (x^2 - p^2)y = 0 \qquad p \geq 0* \qquad (6.24)$$

called **Bessel's equation.** It can be shown that $y_2 = J_{-p}(x)$ is also a solution of Bessel's equation (see problem 17 in Exercises 6.2). Moreover, for p not an integer, we have shown that $J_p(x)$ and $J_{-p}(x)$ are not proportional and hence are *linearly independent*. Therefore, under these conditions a general solution of Bessel's equation (6.24) is given by

$$y = C_1 J_p(x) + C_2 J_{-p}(x) \qquad p \neq n \ (n = 0, 1, 2, \dots) \qquad (6.25)$$

where C_1 and C_2 are arbitrary constants.

Among other areas of application, Bessel's equation arises in the solution of various partial differential equations of mathematical physics, particularly in problems featuring either circular or cylindrical symmetry (see Sec. 8.4).

---

* Since only p^2 appears in (6.24), it is customary to assume $p \geq 0$.

Exercises 6.2

1. Show that the generating-function relation (6.5) can also be written in the form

$$\exp\left[\frac{1}{2}x\left(t-\frac{1}{t}\right)\right] = J_0(x) + \sum_{n=1}^{\infty} J_n(x)[t^n + (-1)^n t^{-n}] \qquad t \neq 0$$

2. Show that $J_n(-x) = (-1)^n J_n(x)$, $n = 0, 1, 2, \ldots$.

3. By using the series representation (6.4), show that
   (a) $J_1'(0) = \frac{1}{2}$ (b) $J_n'(0) = 0$, $n > 1$

4. It is given that $w(x, t) = \exp\left[\frac{1}{2}x(t - 1/t)\right]$.
   (a) Show that $w(x + y, t) = w(x, t)w(y, t)$.
   (b) From (a), deduce the *addition formula*

$$J_n(x + y) = \sum_{k=-\infty}^{\infty} J_k(x)J_{n-k}(y)$$

   (c) From (b), derive the result

$$J_0(2x) = [J_0(x)]^2 + 2\sum_{k=1}^{\infty} (-1)^k [J_k(x)]^2$$

5. The generating function (6.1) is given.
   (a) Show that it satisfies the identity

$$\frac{\partial w}{\partial t} - \frac{1}{2}x\left(1 + \frac{1}{t^2}\right)w = 0$$

   Use (a) to derive the recurrence relation

$$\frac{2n}{x}J_n(x) = J_{n-1}(x) + J_{n+1}(x) \qquad n = 1, 2, 3, \ldots$$

6. The generating function (6.1) is given.
   (a) Show that it satisfies the identity

$$\frac{\partial w}{\partial x} - \frac{1}{2}\left(t - \frac{1}{t}\right)w = 0$$

   (b) Use (a) to derive the relation
$$2J_n'(x) = J_{n-1}(x) - J_{n+1}(x) \qquad n = 1, 2, 3, \ldots$$

7. Show that $(k \neq 0, t \neq 0)$

$$\exp\left[-\frac{x}{2t}\left(k - \frac{1}{k}\right)\right] \sum_{n=-\infty}^{\infty} J_n(x)k^n t^n = \sum_{n=-\infty}^{\infty} J_n(kx)t^n$$

8. (a) From the product $w(x, t)w(-x, t)$, show that

$$1 = [J_0(x)]^2 + 2\sum_{n=1}^{\infty} [J_n(x)]^2$$

(b)   From (a), deduce for all x that

$$|J_0(x)| \leq 1 \quad \text{and} \quad |J_n(x)| \leq \frac{1}{\sqrt{2}} \quad n = 1, 2, 3, \ldots$$

9. Use the generating function to derive

(a) $e^{ix \sin \theta} = \sum\limits_{n=-\infty}^{\infty} J_n(x) e^{in\theta}$

(b) $e^{ix \cos \theta} = J_0(x) + 2 \sum\limits_{n=1}^{\infty} i^n J_n(x) \cos n\theta$

10. Use the result of problem 9a to deduce that

(a) $\cos(x \sin \theta) = J_0(x) + 2 \sum\limits_{n=1}^{\infty} J_{2n}(x) \cos 2n\theta$

(b) $\sin(x \sin \theta) = 2 \sum\limits_{n=1}^{\infty} J_{2n-1}(x) \sin(2n-1)\theta$

(c) $\cos x = J_0(x) + 2 \sum\limits_{n=1}^{\infty} (-1)^n J_{2n}(x)$

(d) $\sin x = 2 \sum\limits_{n=1}^{\infty} (-1)^{n-1} J_{2n-1}(x)$

11. Use the results of problem 10 to deduce that

(a) $x = 2 \sum\limits_{n=1}^{\infty} (2n-1)J_{2n-1}(x)$

   Hint: Differentiate the result of problem 10b.

(b) $x \sin x = 2 \sum\limits_{n=1}^{\infty} (2n)^2 J_{2n}(x)$

12. Set $t = e^\theta$ in the generating function and deduce that

(a) $\cosh(x \sinh \theta) = J_0(x) + 2 \sum\limits_{n=1}^{\infty} J_{2n}(x) \cosh 2n\theta$

(b) $\sinh(x \sinh \theta) = 2 \sum\limits_{n=1}^{\infty} J_{2n-1}(x) \sinh(2n-1)\theta$

13. Establish the following identities:

(a) $J_{1/2}(x) = \sqrt{\dfrac{2}{\pi x}} \sin x$

(b) $J_{-1/2}(x) = \sqrt{\dfrac{2}{\pi x}} \cos x$

(c) $J_{1/2}(x)J_{-1/2}(x) = \dfrac{\sin 2x}{\pi x}$

(d) $[J_{1/2}(x)]^2 + [J_{-1/2}(x)]^2 = \dfrac{2}{\pi x}$

14. Show that

$$\frac{d}{dx}[x^{-p}J_p(x)] = -x^{-p}J_{p+1}(x)$$

15. Show that

$$\frac{d}{dx}J_p(kx) = -kJ_{p+1}(kx) + \frac{p}{x}J_p(kx) \qquad k > 0$$

16. Show that

(a) $\dfrac{d}{dx}[xJ_p(x)J_{p+1}(x)] = x\{[J_p(x)]^2 - [J_{p+1}(x)]^2\}$

(b) $\dfrac{d}{dx}[x^2 J_{p-1}(x)J_{p+1}(x)] = 2x^2 J_p(x)J_p'(x)$

17. Use the series (6.10) to show directly that $y = J_{-p}(x)$ is a solution of

$$x^2 y'' + xy' + (x^2 - p^2)y = 0$$

18. If y_1 and y_2 are solutions of the second-order linear DE

$$y'' + a(x)y' + b(x)y = 0$$

their wronksian $W(y_1, y_2) = y_1 y_2' - y_1' y_2$ is given by *Abel's formula*

$$W(y_1, y_2)(x) = C \exp\left[-\int a(x)\,dx\right]$$

for some constant C. Use this result to deduce that the wronksian of the solutions of Bessel's equation is

$$W(y_1, y_2)(x) = \frac{C}{x}$$

19. From the result of problem 18, show that

$$W(J_p, J_{-p})(x) = -\frac{2\sin p\pi}{\pi x} \qquad p \neq n \qquad (n = 0, 1, 2, \dots)$$

Hint: Use $C = \lim_{x \to 0^+} xW(J_p, J_{-p})(x)$.

20. Use problem 19 to derive *Lommel's formula*

$$J_p(x)J_{1-p}(x) + J_{-p}(x)J_{p-1}(x) = \frac{2\sin p\pi}{\pi x}$$

21. Use the Cauchy product to deduce that

   (a) $[J_0(x)]^2 = \sum\limits_{n=0}^{\infty} \dfrac{(-1)^n (2n)!}{(n!)^4} \left(\dfrac{x}{2}\right)^{2n}$

   Hint: $\sum\limits_{k=0}^{n} \binom{n}{k}^2 = \binom{2n}{n}$

   (b) $J_0(x) \cos x = \dfrac{1}{\sqrt{\pi}} \sum\limits_{n=0}^{\infty} \dfrac{(-1)^n \Gamma(2n + \frac{1}{2})}{[(2n)!]^2} (2x)^{2n}$

   Hint: $\sum\limits_{k=0}^{n} (-1)^k \binom{2n}{2k}\binom{-\frac{1}{2}}{k} = \dfrac{2^{2n}\Gamma(2n + \frac{1}{2})}{\sqrt{\pi}\,(2n)!}$

In Problems 22 to 25, derive the given identity.

22. $J_0(\sqrt{x^2 - 2xt}) = \sum\limits_{n=0}^{\infty} J_n(x)\dfrac{t^n}{n!}$

23. $\left(\dfrac{x - 2t}{x}\right)^{-p/2} J_p(\sqrt{x^2 - 2xt}) = \sum\limits_{n=0}^{\infty} J_{p+n}(x)\dfrac{t^n}{n!}$

24. $e^{t \cos \phi} J_0(t \sin \phi) = \sum\limits_{n=0}^{\infty} P_n(\cos \phi)\dfrac{t^n}{m!}$, where $P_n(x)$ is the nth *Legendre polynomial*

25. $e^t (xt)^{-m/2} J_m(2\sqrt{xt}) = \sum\limits_{n=0}^{\infty} L_n^{(m)}(x)\dfrac{t^n}{(n+m)!}$, $m = 0, 1, 2, \ldots$,
   where $L_n^{(m)}(x)$ is the nth *associated Laguerre polynomial*

26. A waveform with phase modulation distortion may be represented by

   $$s(t) = R \cos [\omega_0 t + \epsilon(t)]$$

   where R is the amplitude and $\epsilon(t)$ represents the "distortion term." It often suffices to approximate $\epsilon(t)$ by the first term of its Fourier series, i.e.,

   $$\epsilon(t) \cong a \sin \omega_m t$$

   where a is the peak phase error and ω_m is the fundamental frequency of the phase error. Thus, the original waveform becomes

   $$s(t) \cong R \cos (\omega_0 t + a \sin \omega_m t)$$

   (a) Show that $s(t)$ can be decomposed into its harmonic components according to

   $$s(t) \cong RJ_0(a) \cos \omega_0 t + R \sum_{n=1}^{\infty} J_n(a)[\cos (\omega_0 t + n\omega_m t)$$

   $$+ (-1)^n \cos (\omega_0 t - n\omega_m t)]$$

   Hint: Use problem 10.

(b)   Whenever the peak phase error satisfies $a \leq 0.4$, we can use the approximations

$$J_0(a) \cong 1 \qquad J_1(a) \cong \frac{a}{2} \qquad J_n(a) \cong 0 \qquad n = 2, 3, 4, \ldots$$

Show that under these conditions the phase modulation error term produces only the effect of "paired sidebands" with a frequency displacement of $\pm \omega_m$ with respect to ω_0 and a relative amplitude of $a/2$.

6.3   Integral Representations

In many situations an *integral representation* of $J_p(x)$ is more convenient to use than its series representation. There are several such representations, but foremost is one involving Bessel functions of integral order. To derive it, we start with the generating-function relation

$$\exp\left[\frac{1}{2}x\left(t - \frac{1}{t}\right)\right] = \sum_{k=-\infty}^{\infty} J_k(x)t^k \qquad t \neq 0$$

and set $t = e^{-i\phi}$ to get

$$e^{-ix \sin \phi} = \sum_{k=-\infty}^{\infty} J_k(x)e^{-ik\phi} \tag{6.26}$$

where we have made the observation

$$t - \frac{1}{t} = e^{-i\phi} - e^{i\phi} = -2i \sin \phi \; .$$

Next we multiply both sides of (6.26) by $e^{in\phi}$ and integrate the result from 0 to π, which yields

$$\int_0^\pi e^{i(n\phi - x \sin \phi)}\, d\phi = \sum_{k=-\infty}^{\infty} J_k(x) \int_0^\pi e^{i(n-k)\phi}\, d\phi \tag{6.27}$$

assuming that termwise integration is permitted. By use of Euler's formula $e^{ix} = \cos x + i \sin x$, we can separate (6.27) into real and imaginary parts, i.e.,

$$\int_0^\pi \cos(n\phi - x \sin \phi)\, d\phi + i \int_0^\pi \sin(n\phi - x \sin \phi)\, d\phi$$

$$= \sum_{k=-\infty}^{\infty} J_k(x) \int_0^\pi \cos(n-k)\phi\, d\phi + i \sum_{k=-\infty}^{\infty} J_k(x) \int_0^\pi \sin(n-k)\phi\, d\phi$$

Equating the *real parts* of this last expression and using the result

$$\int_0^\pi \cos(n-k)\phi\,d\phi = \begin{cases} 0 & k \neq n \\ \pi & k = n \end{cases} \qquad (6.28)$$

we find that all terms of the infinite sum vanish except for the term corresponding to $k = n$; thus, we are left with the **integral representation**

$$J_n(x) = \frac{1}{\pi}\int_0^\pi \cos(n\phi - x\sin\phi)\,d\phi \qquad n = 0, 1, 2, \ldots \quad (6.29)$$

Using properites of the trigonometric functions, we can also write

$$J_n(x) = \frac{1}{2\pi}\int_0^{2\pi} \cos(n\phi - x\sin\phi)\,d\phi \qquad n = 0, 1, 2, \ldots \quad (6.30)$$

and the special case $n = 0$ leads to

$$J_0(x) = \frac{1}{\pi}\int_0^\pi \cos(x\sin\phi)\,d\phi = \frac{1}{2\pi}\int_0^{2\pi} \cos(x\sin\phi)\,d\phi \quad (6.31)$$

Another representation of Bessel functions, due to S. D. Poisson (1781–1840), is given by

$$J_p(x) = \frac{(x/2)^p}{\sqrt{\pi}\,\Gamma(p+\frac{1}{2})}\int_{-1}^1 (1-t^2)^{p-1/2}e^{ixt}\,dt \qquad p > -\tfrac{1}{2}, \qquad x > 0$$

$$(6.32)$$

where p is not restricted to integral values. To verify (6.32), we start with the relation

$$\int_{-1}^1 (1-t^2)^{p-1/2}e^{ixt}\,dt = 2\int_0^1 (1-t^2)^{p-1/2}\cos xt\,dt$$

$$= 2\sum_{k=0}^\infty \frac{(-1)^k x^{2k}}{(2k)!}\int_0^1 (1-t^2)^{p-1/2}t^{2k}\,dt \qquad (6.33)$$

where we are using properties of even and odd functions and have expressed $\cos xt$ in a power series. The residual integral in (6.33) can be evaluated in terms of the beta function by making the change of variable $u = t^2$, from which we get

$$\int_0^1 (1-t^2)^{p-1/2} t^{2k}\, dt = \tfrac{1}{2} \int_0^1 (1-u)^{p-1/2} u^{k-1/2}\, du$$

$$= \tfrac{1}{2} B(k + \tfrac{1}{2}, p + \tfrac{1}{2})$$

$$= \frac{\Gamma(k + \tfrac{1}{2})\Gamma(p + \tfrac{1}{2})}{2\Gamma(k + p + 1)}$$

$$= \frac{\sqrt{\pi}\,(2k)!\,\Gamma(p + \tfrac{1}{2})}{2^{2k+1} k!\,\Gamma(k + p + 1)} \qquad (6.34)$$

the last step of which follows from the identity

$$\Gamma\left(k + \frac{1}{2}\right) = \frac{\sqrt{\pi}\,(2k)!}{2^{2k} k!} \qquad k = 0, 1, 2, \ldots$$

Finally, by substituting the result of (6.34) into (6.33), we obtain

$$\int_{-1}^1 (1-t^2)^{p-1/2} e^{ixt}\, dt = \sqrt{\pi}\,\Gamma\left(p + \frac{1}{2}\right) \sum_{k=0}^{\infty} \frac{(-1)^k (x/2)^{2k}}{k!\,\Gamma(k + p + 1)}$$

$$= \sqrt{\pi}\,\Gamma\left(p + \frac{1}{2}\right)\left(\frac{x}{2}\right)^{-p} J_p(x)$$

from which we deduce (6.32).

By making the change of variable $t = \cos\theta$ in (6.32), we obtain still another representation

$$J_p(x) = \frac{(x/2)^p}{\sqrt{\pi}\,\Gamma(p + \tfrac{1}{2})} \int_0^{\pi} \cos(x\cos\theta)\sin^{2p}\theta\, d\theta \qquad p > -\tfrac{1}{2},\ x > 0$$

$$(6.35)$$

which was originally due to Bessel. The direct verification of (6.35) is left to the exercises (see problem 7 in Exercises 6.3).

6.3.1 Bessel's problem

During the early 1800s, the French physicist J. Fourier presented basic papers to the Academy of Sciences in Paris concerning the representation of functions by trigonometric series. This new concept intrigued many of the researchers of the era who subsequently tried to use Fourier's series representation in their own work. For example, in 1824 Bessel was working on a problem associated with elliptic planetary motion and found that an astronomical quantity ϕ, called the *eccentric anomaly*, could be represented by an infinite series of trigonometric functions.

The problem solved by Bessel was actually one proposed by Johann Kepler (1571–1630). In ideal planetary motion, a planet P moves in an elliptic orbit with the sun S situated at one of the foci (see Fig. 6.2), and the area swept out by the radius vector $\mathbf{r} = \overrightarrow{SP}$ during any interval of time is proportional to that interval. The ellipse $A'PA$ in Fig. 6.2 has major axis $A'A$, which is also the diameter of an auxiliary circle $A'QA$, the center of which is denoted by C. A line drawn through P perpendicular to $A'A$ defines the point Q on the circle. Let M define a point on the circle such that the radius CM rotates with constant angular speed and M coincides with P at A and A'. The angles ϕ, ψ, and θ are defined by the major axis $A'A$ and the line segments CQ, CM, and SP.

We assume that time t is measured from an instant when P is at A. In this case the **eccentric anomaly*** of P is angle ACQ, which we denote by ϕ, the **true anomaly** θ is angle CSP, and the **mean anomaly** ψ, which is proportional to t, is angle ACM. Kepler's problem was to express the variables r, θ, and ϕ in terms of time t, where $r = |\mathbf{r}| = SP$.

Bessel's approach to solving Kepler's problem was to use Fourier's method of trigonometric series. By using Kepler's principles of planetary motion, it can be shown that the mean anomaly ψ is related to the eccentric anomaly ϕ according to

$$\psi = \phi - e \sin \phi \qquad (6.36)$$

where e is the *eccentricity*. Clearly, the difference $\phi - \psi$ is an odd periodic function of ϕ, and so based on Fourier's work, Bessel was led

Figure 6.2 Planetary orbit.

---

* A polar angle is sometimes called the *anomaly* or *azimuth*.

to consider the sine series

$$\phi = \psi + \sum_{k=1}^{\infty} A_k \sin k\psi \qquad (6.37)$$

To determine the constants $A_1, A_2, A_3, \ldots$, we begin by forming the differential of each side of (6.37), which yields

$$d\phi = \left(1 + \sum_{k=1}^{\infty} kA_k \cos k\psi\right) d\psi$$

Next we multiply each side by $\cos n\psi$ and integrate over an interval of length π. (Notice that ψ varies from 0 to π as ϕ varies the same.) This action leads to

$$\int_0^\pi \cos n\psi \, d\phi = \int_0^\pi \cos n\psi \, d\psi + \sum_{k=1}^{\infty} kA_k \int_0^\pi \cos n\psi \cos k\psi \, d\psi$$

$$(6.38)$$

but since

$$\int_0^\pi \cos n\psi \, d\psi = 0 \qquad n = 1, 2, 3, \ldots \qquad (6.39)$$

$$\int_0^\pi \cos n\psi \cos k\psi \, d\psi = \begin{cases} 0 & k \neq n \\ \dfrac{\pi}{2} & k = n \end{cases} \qquad (6.40)$$

it follows that (6.38) reduces to

$$\int_0^\pi \cos n\psi \, d\phi = \frac{n\pi}{2} A_n \qquad (6.41)$$

Writing $\psi = \phi - e \sin \phi$ and solving for A_n, we find

$$A_n = \frac{2}{n\pi} \int_0^\pi \cos (n\phi - ne \sin \phi) \, d\phi \qquad n = 1, 2, 3, \ldots \quad (6.42)$$

The integral in (6.42) was Bessel's original definition of the function now bearing his name. By comparing (6.42) with Eq. (6.29), we can write

$$A_n = \frac{2}{n} J_n(ne) \qquad n = 1, 2, 3, \ldots \qquad (6.43)$$

Finally, by substituting this result into (6.37), we obtain the series representation

$$\phi = \psi + 2 \sum_{n=1}^{\infty} \frac{J_n(ne)}{n} \sin n\psi \qquad 0 < \psi < \pi \qquad (6.44)$$

6.3.2 Geometric problems

There are numerous problems similar in nature to Bessel's problem which relate Bessel functions to Fourier trigonometric series. Some of these, such as Example 1 following, are simple problems of a geometric nature. Additional problems of this type are taken up in the exercises.

Example 1: Find a Fourier series for the semicircle

$$y = \sqrt{\pi^2 - x^2} \qquad -\pi \le x \le \pi$$

Solution: Because the function $y = f(x) = \sqrt{\pi^2 - x^2}$ is an even function, we seek a cosine series, i.e.

$$f(x) = \tfrac{1}{2}a_0 + \sum_{n=1}^{\infty} a_n \cos nx \qquad -\pi < x < \pi$$

where $\quad a_n = \dfrac{2}{\pi} \displaystyle\int_0^{\pi} \sqrt{\pi^2 - x^2} \cos nx\, dx \qquad n = 0, 1, 2, \dots$

By making a change of variable $x = \pi \cos \theta$, we obtain

$$a_n = \frac{2}{\pi} \int_{\pi/2}^{0} (\pi \sin \theta) \cos(n\pi \cos \theta)(-\pi \sin \theta)\, d\theta$$

$$= 2\pi \int_0^{\pi/2} \cos(n\pi \cos \theta) \sin^2 \theta\, d\theta$$

$$= \pi \int_0^{\pi} \cos(n\pi \cos \theta) \sin^2 \theta\, d\theta$$

where we are using symmetry of the integrand to obtain the last result. By comparing this last expression with (6.35), we see that

$$a_n = \pi \frac{\sqrt{\pi}\, \Gamma(^3\!/_2)}{n\pi/2} J_1(n\pi)$$

$$= \frac{\pi}{n} J_1(n\pi) \qquad n = 1, 2, 3, \dots$$

whereas

$$a_0 = \pi \int_0^\pi \sin^2 d\theta = \frac{\pi^2}{2}$$

Hence, the series we seek takes the form

$$f(x) = \frac{\pi^2}{4} + \pi \sum_{n=1}^\infty \frac{J_1(n\pi)}{n} \cos nx \qquad -\pi < x < \pi$$

Exercises 6.3

In problems 1 to 4, use Eq. (6.29) to deduce the given integral representation for $n = 0, 1, 2, \ldots$.

1. $[1 + (-1)^n]J_n(x) = \dfrac{2}{\pi} \displaystyle\int_0^\pi \cos n\theta \cos (x \sin \theta)\, d\theta$

2. $[1 - (-1)^n]J_n(x) = \dfrac{2}{\pi} \displaystyle\int_0^\pi \sin n\theta \sin (x \sin \theta)\, d\theta$

3. $J_{2n}(x) = \dfrac{1}{\pi} \displaystyle\int_0^\pi \cos 2n\theta \cos (x \sin \theta)\, d\theta$

4. $J_{2n+1}(x) = \dfrac{1}{\pi} \displaystyle\int_0^\pi \sin (2n + 1)\theta \sin (x \sin \theta)\, d\theta$

5. Use problems 1 and 2 to deduce that

   (a) $\displaystyle\int_0^\pi \cos (2n + 1)\theta \cos (x \sin \theta)\, d\theta = 0,\ n = 0, 1, 2, \ldots$

   (b) $\displaystyle\int_0^\pi \sin 2n\theta \sin (x \sin \theta)\, d\theta = 0,\ n = 0, 1, 2, \ldots$

6. By writing $\cos xt$ in an infinite series and using termwise integration, deduce that

$$J_0(x) = \frac{2}{\pi} \int_0^1 \frac{\cos xt}{\sqrt{1 - t^2}}\, dt$$

7. By setting $t = \cos \theta$ in (6.32), show that

$$J_p(x) = \frac{(x/2)^p}{\sqrt{\pi}\,\Gamma(p + \frac{1}{2})} \int_0^\pi \cos (x \cos \theta) \sin^{2p} \theta\, d\theta \qquad p > -\tfrac{1}{2},\ x > 0$$

8. Replacing $J_m(xt)$ by its series representation and using termwise integration, deduce the integral relation

$$J_p(x) = \frac{2(x/2)^{p-m}}{\Gamma(p - m)} \int_0^1 (1 - t^2)^{p-m-1} t^{m+1} J_m(xt)\, dt$$

$$p > m > -1,\ x > 0$$

9. Show that

(a) $J_n'(x) = \dfrac{1}{\pi} \displaystyle\int_0^\pi \sin(n\theta - x \sin\theta) \sin\theta\, d\theta$

(b) $J_n''(x) = -\dfrac{1}{\pi} \displaystyle\int_0^\pi \cos(n\theta - x \sin\theta) \sin^2\theta\, d\theta$

(c) $J_n(x) = \dfrac{x}{2n\pi} \displaystyle\int_0^\pi \cos(n\theta - x \sin\theta) \cos\theta\, d\theta$

10. The amplitude of a diffracted wave through a circular aperture is given by

$$U = k \int_0^a \int_0^{2\pi} e^{\iota br \sin\theta} r\, d\theta\, dr$$

where k is a physical constant, a is the radius of the aperture, θ is the azimuthal angle in the plane of the aperture, and b is a constant inversely proportional to the wavelength of the incident wave. Show that the intensity of light in the diffraction pattern is given by

$$I = |U|^2 = \frac{4\pi^2 k^2 a^2}{b^2} [J_1(ab)]^2$$

11. Given that $r = a(1 - e \cos\phi)$, where $2a$ is the length of the major axis of the ellipse in Fig. 6.2, deduce that

$$\frac{r}{a} = 1 + \frac{1}{2}e^2 - 2e \sum_{n=1}^{\infty} J_n'(ne) \frac{\cos n\psi}{n}$$

12. If ϕ is the eccentric anomaly in Bessel's problem, show that

(a) $\dfrac{1}{1 - e \cos\phi} = 1 + 2 \displaystyle\sum_{n=1}^{\infty} J_n(ne) \cos n\psi$

(b) $(1 - e^2)^{-1/2} = 1 + 2 \displaystyle\sum_{n=1}^{\infty} [J_n(ne)]^2$

13. A light spot dancing around is described parametrically by

$$x = \pi \cos\omega t \qquad y = \alpha \sin 2\omega t$$

This is known as a *Lissajous figure* (i.e., a figure eight). Show that it has the Fourier series representation

$$y = 2\alpha \sum_{n=1}^{\infty} \frac{J_2(n\pi)}{n} \sin nx \qquad -\pi < x < \pi$$

14. A circle of unit radius rolls along a straight line, the original point of contact describing a curve called a *cycloid*. Given that the

parametric equations of a cycloid are

$$x = \theta - \sin \theta \qquad y = 1 - \cos \theta$$

show that the Fourier series representation of y is given by the Fourier cosine series

$$y = \frac{3}{2} - 2 \sum_{n=1}^{\infty} J_n'(n) \frac{\cos nx}{n}$$

Hint: See problem 9.

6.4   Integrals of Bessel Functions

Integrals whose integrands involve Bessel functions arise in a variety of applications. These may appear in the form of either *indefinite* or *definite integrals*. We discuss each case separately.

6.4.1   Indefinite integrals

Many of the indefinite integrals that arise in practice are simple products of some Bessel function and x raised to a power. In such cases the identities [previously (6.13) and (6.14)]

$$\frac{d}{dx}[x^p J_p(x)] = x^p J_{p-1}(x) \tag{6.45}$$

and

$$\frac{d}{dx}[x^{-p} J_p(x)] = -x^{-p} J_{p+1}(x) \tag{6.46}$$

may prove useful. For example, we find that direct integration of them leads to the integral formulas

$$\int x^p J_{p-1}(x)\, dx = x^p J_p(x) + C \tag{6.47}$$

and

$$\int x^{-p} J_{p+1}(x)\, dx = -x^{-p} J_p(x) + C \tag{6.48}$$

where C denotes a constant of integration. These integral formulas are valid for any $p \geq 0$.

Example 2:   Reduce $\int x^2 J_2(x)\, dx$ to an integral involving only $J_0(x)$.

Solution: The given integral does not exactly match either (6.47) or (6.48). However, by writing

$$\int x^2 J_2(x)\, dx = \int x^3 [x^{-1} J_2(x)]\, dx$$

we can use integration by parts with

$$u = x^3 \qquad dv = x^{-1}J_2(x)\,dx$$
$$du = 3x^2\,dx \qquad v = -x^{-1}J_1(x)$$

where we have used (6.48). Thus, we have

$$\int x^2 J_2(x)\,dx = -x^2 J_1(x) + 3\int x J_1(x)\,dx$$

and a second integration by parts finally leads to

$$\int x^3 J_2(x)\,dx = -x^3 J_1(x) - 3x J_0(x) + 3\int J_0(x)\,dx$$

The last integral involving $J_0(x)$ cannot be evaluated in closed form, and so our integration is complete.

As a general rule, if m and n are any integers such that $m + n > 0$, then every integral of the form

$$I = \int x^m J_n(x)\,dx$$

can be integrated to a closed-form expression when $m + n$ is *odd*, but will ultimately depend on the residual integral $\int J_0(x)\,dx$ when $m + n$ is *even*. Since it cannot be evaluated in closed form, the integral $\int_0^x J_0(t)\,dt$ has been tabulated.*

Other integral relations can be derived from the identities

$$x^p J_1(x)J_3(x) = x^{p+4}[x^{-1}J_1(x)][x^{-3}J_3(x)]$$
$$x^p J_2(x)J_4(x) = x^{p-6}[x^2 J_2(x)][x^4 J_4(x)]$$

the derivatives of which yield

$$\frac{d}{dx}[x^p J_1(x)J_3(x)] = (p+4)x^{p-1}J_1(x)J_3(x)$$
$$- x^p[J_2(x)J_3(x) + J_1(x)J_4(x)]$$

$$\frac{d}{dx}[x^p J_2(x)J_4(x)] = (p-6)x^{p-1}J_2(x)J_4(x)$$
$$+ x^p[J_2(x)J_3(x) + J_1(x)J_4(x)]$$

---

* See, for example, M. Abramowitz and I. Stegun (eds.), *Handbook of Mathematical Functions,* Dover, New York, 1965.

where we have used (6.45) and (6.46). Upon addition, we get

$$\frac{d}{dx}\{x^p[J_1(x)J_3(x)+J_2(x)J_4(x)]\}$$

$$=x^{p-1}[(p+4)J_1(x)J_3(x)+(p-6)J_2(x)J_4(x)]$$

which, for $p=6$, reduces to

$$\frac{d}{dx}\{x^6[J_1(x)J_3(x)+J_2(x)J_4(x)]\}=10x^5J_1(x)J_3(x)$$

Finally, direct integration of this last expression leads to

$$\int x^5J_1(x)J_3(x)\,dx = \frac{1}{10}x^6[J_1(x)J_3(x)+J_2(x)J_4(x)]+C \quad (6.49)$$

The choice $p=-4$ yields the similar result

$$\int x^{-5}J_2(x)J_4(x)\,dx = -\frac{1}{10}x^{-4}[J_1(x)J_3(x)+J_2(x)J_4(x)]+C \quad (6.50)$$

Integral formulas of the type given by (6.49) and (6.50) are called **Lommel integrals.** Some additional integral relations of this general type appear in the exercises.

6.4.2 Definite integrals

In practice we are often faced with the necessity of evaluating definite integrals involving Bessel functions in combination with various elementary functions or, in some instances, other special functions. The usual procedure is to replace the Bessel function by its series representation (or an integral representation) and then interchange the order in which the operations are carred out.

To illustrate the technique, let us consider the integral

$$I=\int_0^\infty e^{-st}t^{p/2}J_p(2\sqrt{t})\,dt \quad p>-\tfrac{1}{2}, \quad s>0 \quad (6.51)$$

which is important in the theory of Laplace transforms. To start, let us replace the product $t^{p/2}J_p(2\sqrt{t})$ with its series representation, i.e.,

$$t^{p/2}J_p(2\sqrt{t})=t^{p/2}\sum_{k=0}^\infty \frac{(-1)^k t^{k+p/2}}{k!\,\Gamma(k+p+1)}$$

$$=\sum_{k=0}^\infty \frac{(-1)^k t^{k+p}}{k!\,\Gamma(k+p+1)}$$

By substituting this series into the integrand in (6.51) and interchanging the order of integration and summation, we find

$$I = \sum_{k=0}^{\infty} \frac{(-1)^k}{k!\,\Gamma(k+p+1)} \int_0^{\infty} e^{-st} t^{k+p}\, dt$$

$$= \sum_{k=0}^{\infty} \frac{(-1)^k}{k!\,\Gamma(k+p+1)} \frac{\Gamma(k+p+1)}{s^{k+p+1}}$$

where we are using properties of the gamma function to evaluate the above integral. Finally, simplifying this last expression leads to

$$I = \frac{1}{s^{n+1}} \sum_{k=0}^{\infty} \frac{(-1)^k (1/s)^k}{k!} = \frac{1}{s^{p+1}} e^{-1/s}$$

and we deduce that

$$\int_0^{\infty} e^{-st} t^{p/2} J_p(2\sqrt{t})\, dt = \frac{1}{s^{p+1}} e^{-1/s} \qquad p > -\tfrac{1}{2}, \qquad s > 0 \quad (6.52)$$

Along similar lines, suppose we replace $J_p(bx)$ with its series representation in the integral

$$I = \int_0^{\infty} e^{-ax} x^p J_p(bx)\, dx \qquad p > -\tfrac{1}{2}, \qquad a, b > 0 \qquad (6.53)$$

and integrate the resulting series termwise. This action leads to

$$I = \sum_{k=0}^{\infty} \frac{(-1)^k (b/2)^{2k+p}}{k!\,\Gamma(k+p+1)} \int_0^{\infty} e^{-ax} x^{2k+2p}\, dx$$

$$= b^p \sum_{k=0}^{\infty} \frac{(-1)^k \Gamma(2k+2p+1)}{2^{2k+p} k!\,\Gamma(k+p+1)} (a^2)^{-(p+1/2)-k} (b^2)^k$$

where the integral has again been evaluated through properties of the gamma function. This time the resulting series is more difficult to identify, but it is actually a binomial series. Recalling the Legendre duplication formula and Eq. (1.27) in Sec. 1.2.4, we see that $(p > -\tfrac{1}{2})$

$$\frac{(-1)^k \Gamma(2k+2p+1)}{2^{2k+p} k!\,\Gamma(k+p+1)} = \frac{(-1)^k 2^p \Gamma(p+k+\tfrac{1}{2})}{\sqrt{\pi}\, k!}$$

$$= \frac{(-1)^k}{\sqrt{\pi}} 2^p \Gamma\!\left(p+\frac{1}{2}\right)\binom{p+k-\tfrac{1}{2}}{k}$$

$$= \frac{2^p \Gamma(p+\tfrac{1}{2})}{\sqrt{\pi}} \binom{-(p+1/2)}{k}$$

and thus

$$I = \frac{(2b)^p \Gamma(p + \frac{1}{2})}{\sqrt{\pi}} \sum_{k=0}^{\infty} \binom{-(p + \frac{1}{2})}{k} (a^2)^{-(p+1/2)-k} (b^2)^k \quad (6.54)$$

Finally, by summing this binomial series, we are led to*

$$\int_0^{\infty} e^{-ax} x^p J_p(bx)\, dx = \frac{(2b)^p \Gamma(p + \frac{1}{2})}{\sqrt{\pi}\,(a^2 + b^2)^{p+1/2}} \qquad p > -\frac{1}{2}, \qquad a, b > 0$$

$$(6.55)$$

By setting $p = 0$ in (6.55), we obtain the special result

$$\int_0^{\infty} e^{-ax} J_0(bx)\, dx = \frac{1}{\sqrt{a^2 + b^2}} \qquad a, b > 0 \qquad (6.56)$$

Strictly speaking, the validity of (6.56) rests on the condition that $a > 0$ (or at least the real part of a positive if a is complex). Yet it is possible to justify a limiting procedure whereby the real part of a approaches zero. Thus, if we formally replace a in (6.56) with the pure imaginary number ia, we get

$$\int_0^{\infty} e^{-iax} J_0(bx)\, dx = \frac{1}{\sqrt{b^2 - a^2}} \qquad a, b > 0$$

The separation of this expression into real and imaginary parts leads to

$$\int_0^{\infty} \cos ax\, J_0(bx)\, dx - i \int_0^{\infty} \sin ax\, J_0(bx)\, dx = \begin{cases} \dfrac{1}{\sqrt{b^2 - a^2}} & b > a \\[2ex] -\dfrac{i}{\sqrt{a^2 - b^2}} & b < a \end{cases}$$

and by equating the real and imaginary parts, we deduce the pair of integral formulas

$$\int_0^{\infty} \cos ax\, J_0(bx)\, dx = \begin{cases} \dfrac{1}{\sqrt{b^2 - a^2}} & b > a \\[2ex] 0 & b < a \end{cases} \qquad (6.57)$$

---

*Summing the series (6.54) requires that $a \neq b$, although the result (6.55) is valid even when $a = b$.

and

$$\int_0^\infty \sin ax \, J_0(bx) \, dx = \begin{cases} 0 & b > a \\ \dfrac{1}{\sqrt{a^2 - b^2}} & b < a \end{cases} \tag{6.58}$$

These last two integrals are important in the theory of Fourier integrals. Both (6.57) and (6.58) diverge when $b = a$.

Example 3: Derive *Weber's integral formula*

$$\int_0^\infty x^{2m-p-1} J_p(x) \, dx = \frac{2^{2m-p-1}\Gamma(m)}{\Gamma(p-m+1)} \qquad 0 < m < \tfrac{1}{2}, \qquad p > -\tfrac{1}{2}$$

Solution: Replacing $J_p(x)$ by its integral representation (6.35), we have

$$\int_0^\infty x^{2m-p-1} J_p(x) \, dx$$

$$= \frac{2^{-p}}{\sqrt{\pi}\,\Gamma(p+\tfrac{1}{2})} \int_0^\infty x^{2m-1} \int_0^\pi \cos(x \cos\theta) \sin^{2p}\theta \, d\theta \, dx$$

$$= \frac{2^{-p}}{\sqrt{\pi}\,\Gamma(p+\tfrac{1}{2})} \int_0^\pi \sin^{2p}\theta \int_0^\infty x^{2m-1} \cos(x \cos\theta) \, dx \, d\theta$$

where we have reversed the order of integration. By making the substitution $t = x \cos\theta$ in the inner integral and using the result of problem 37 in Exercises 2.2, we obtain

$$\int_0^\infty x^{2m-1} \cos(x \cos\theta) \, dx = \cos^{-2m}\theta \int_0^\infty t^{2m-1} \cos t \, dt$$

$$= \cos^{-2m}\theta\, \Gamma(2m) \cos m\pi$$

$$= \pi^{-1/2} 2^{2m-1} \Gamma(m) \Gamma(m + \tfrac{1}{2}) \cos m\pi \cos^{-2m}\theta$$

The last step follows from the Legendre duplication formula. The remaining integral above now leads to

$$\int_0^\pi \sin^{2p}\theta \cos^{-2m}\theta \, d\theta = 2 \int_0^{\pi/2} \sin^{2p}\theta \cos^{-2m}\theta \, d\theta$$

$$= \frac{\Gamma(p+\tfrac{1}{2})\Gamma(\tfrac{1}{2}-m)}{\Gamma(p-m+1)}$$

and hence, we deduce that

$$\int_0^\infty x^{2m-p-1} J_p(x)\, dx = \frac{2^{2m-p-1}\Gamma(m)\Gamma(m+\frac{1}{2})\Gamma(\frac{1}{2}-m)\cos m\pi}{\pi\Gamma(p-m+1)}$$

$$= \frac{2^{2m-p-1}\Gamma(m)}{\Gamma(p-m+1)}$$

where we are recalling the identity (problem 42b in Exercises 2.2)

$$\Gamma(\tfrac{1}{2}+m)\Gamma(\tfrac{1}{2}-m) = \pi \sec m\pi$$

(Although we won't show it, Weber's integral is valid for a much wider range of values on m and p than indicated above.)

Exercises 6.4

In problems 1 to 12, use recurrence relations, integration by parts, etc., to verify the given result.

1. $\displaystyle\int x J_0(x)\, dx = x J_1(x) + C$

2. $\displaystyle\int x^2 J_0(x)\, dx = x^2 J_1(x) + x J_1(x) - \int J_0(x)\, dx$

3. $\displaystyle\int x^3 J_0(x)\, dx = (x^3 - 4x) J_1(x) + 2x^2 J_0(x) + C$

4. $\displaystyle\int J_1(x)\, dx = -J_0(x) + C$

5. $\displaystyle\int x J_1(x)\, dx = -x J_0(x) + \int J_0(x)\, dx$

6. $\displaystyle\int x^2 J_1(x)\, dx = 2x J_1(x) - x^2 J_0(x) + C$

7. $\displaystyle\int x^3 J_1(x)\, dx = 3x^2 J_1(x) - (x^3 - 3x) J_0(x) - 3\int J_0(x)\, dx$

8. $\displaystyle\int J_3(x)\, dx = -J_2(x) - 2x^{-1} J_1(x) + C$

9. $\displaystyle\int x^{-1} J_1(x)\, dx = -J_1(x) + \int J_0(x)\, dx$

10. $\displaystyle\int x^{-2} J_2(x)\, dx = -\frac{2}{3x^2} J_1(x) - \frac{1}{3} J_1(x) + \frac{1}{3x} J_0(x) + \frac{1}{3}\int J_0(x)\, dx$

11. $\int J_0(x) \cos x \, dx = x J_0(x) \cos x + x J_1(x) \sin x + C$

12. $\int J_0(x) \sin x \, dx = x J_0(x) \sin x - x J_1(x) \cos x + C$

13. Verify the identity

$$\frac{d}{dx}\left(x^p\{[J_2(x)]^2 + [J_3(x)]^2\}\right) = x^{p-1}\{(p+4)[J_2(x)]^2$$
$$+ (p-6)[J_3(x)]^2$$

and use it to deduce that

(a) $\int x^5[J_2(x)]^2 \, dx - \frac{1}{10}x^6\{[J_2(x)]^2 + [J_3(x)]^2\} + C$

(b) $\int x^{-5}[J_3(x)]^2 \, dx = -\frac{1}{10}x^{-4}\{[J_2(x)]^2 + [J_3(x)]^2\} + C$

14. Show that

$$\int x\{[J_p(x)]^2 - [J_{p+1}(x)]^2\} \, dx = x^p J_p(x) J_{p+1}(x) + C$$

Hint: Use problem 16a in Exercises 6.2

15. Using repeated integration by parts, derive the recurrence formula $(n = 1, 2, 3, \ldots)$

$$\int J_0(x) \, dx = J_1(x) + \frac{J_2(x)}{x} + \frac{1 \cdot 3}{x^2} J_3(x) + \cdots$$

$$+ \frac{(2n-2)!}{2^{n-1}(n-1)! x^{n-1}} J_n(x) + \frac{(2n)!}{2^n n!} \int \frac{J_n(x)}{x^n} \, dx$$

In problems 16 to 34, derive the given integral relation.

16. $\int_0^\infty J_0(bx) \, dx = \frac{1}{b}, \qquad b > 0$

Hint: Let $a \to 0^+$ in Eq. (6.56).

17. $\int_0^\infty e^{-ax} x J_0(bx) \, dx = \frac{a}{(a^2 + b^2)^{3/2}}, \qquad a, b > 0$

18. $\int_0^\infty e^{-st} t J_1(t) \, dt = \frac{1}{(s^2 + 1)^{3/2}}, \quad s > 0$

19. $\int_0^\infty e^{-st} t^n J_n(at) \, dt = \frac{(2n)!}{2^n n!} (s^2 + a^2)^{-n-1/2}, \quad a, s > 0, \quad n = 0, 1, 2, \ldots$

20. $\int_0^\infty e^{-st} t^{n+1} J_n(at) \, dt = \frac{(2n+1)!}{2^n n!} s(s^2 + a^2)^{-n-3/2}, \qquad a, s > 0,$

$$n = 0, 1, 2, \ldots$$

Hint: Differentiate problem 19 with respect to s.

21. $\displaystyle\int_0^\infty e^{-ax}x^2 J_0(bx)\,dx = \frac{2a^2 - b^2}{(a^2 + b^2)^{5/2}}, \qquad a, b > 0$

Hint: Differentiate Eq. (6.56) with respect to a.

22. $\displaystyle\int_0^\infty e^{-ax^2}x^{p+1}J_p(bx)\,dx = \frac{b^p}{(2a)^{p+1}}e^{-b^2/(4a)}, \qquad p > -1, \qquad a, b > 0$

23. $\displaystyle\int_0^\infty e^{-ax^2}x^{p+3}J_p(bx)\,dx = \frac{b^p}{a(2a)^{p+1}}\left(p + 1 - \frac{b^2}{4a}\right)e^{-b^2/(4a)},$

$$p > -1, \qquad a, b > 0$$

Hint: Differentiate problem 22 with respect to a.

24. $\displaystyle\int_0^\infty \frac{\sin x}{x} J_0(bx)\,dx = \sin^{-1}\frac{1}{b}, \qquad b > 1$

Hint: Integrate Eq. (6.58) with respect to a.

25. $\displaystyle\int_0^\infty e^{-ax}J_p(bx)\,dx = \frac{(\sqrt{a^2 + b^2} - a)^p}{b^p\sqrt{a^2 + b^2}}, \qquad p > -1, \qquad a, b > 0$

26. $\displaystyle\int_0^{\pi/2} J_0(x \cos \phi) \cos \phi\,d\phi = \frac{\sin x}{x}$

27. $\displaystyle\int_0^{\pi/2} J_1(x \cos \phi)\,d\phi = \frac{1 - \cos x}{x}$

28. $\displaystyle\int_0^\pi J_0(2x \cos \phi)\,d\phi = \pi[J_0(x)]^2$

Hint: See problem 21 in Exercises 6.2.

29. $\displaystyle\int_0^{\pi/2} \frac{[J_1(x \sin \phi)]^2}{\sin \phi}\,d\phi = \frac{1}{2}\left[1 - \frac{J_1(2x)}{x}\right]$

30. $\displaystyle\int_0^a x(a^2 - x^2)^{-1/2}J_0(kx \sin \phi)\,dx = \frac{\sin (ka \sin \phi)}{k \sin \phi}$

31. $\displaystyle\int_0^\pi e^{t \cos \phi}J_0(t \sin \phi) \sin \phi\,d\phi = 2$

Hint: Use problem 24 in Exercises 6.2.

32. $\displaystyle\int_0^\infty e^{-t \cos \phi}J_0(t \sin \phi)t^n\,d\phi = n!\,P_n(\cos \phi), \quad 0 \le \phi < \pi,$ where $P_n(x)$ is the nth Legendre polynomial.

Hint: Use problem 24 in Exercises 6.2.

33. $\displaystyle\int_0^\infty e^{-t}t^{n+m/2}J_m(2\sqrt{xt})\,dt = n!\,e^{-x}x^{m/2}L_n^{(m)}(x),$ where $L_n^{(m)}(x)$ is the nth associated Laguerre polynomial.

Hint: Use problem 25 in Exercises 6.2.

34. $\displaystyle\int_0^\infty x(x^2+a^2)^{-1/2}J_0(bx)\,dx = \frac{1}{b}e^{-ab}, \qquad a \geq 0, \qquad b > 0$

Hint: Use the integral representation

$$(x^2+a^2)^{-1/2} = \frac{1}{\sqrt{\pi}}\int_0^\infty e^{-(x^2+a^2)t}t^{-1/2}\,dt$$

and then interchange the order of integration.

6.5 Series Involving Bessel Functions

There are a variety of applications that lead to series involving one or more Bessel functions. These usually appear in the form of *addition formulas* or *orthogonal expansions*.

6.5.1 Addition formulas

By recognizing that $w(x, t) = \exp\left[\frac{1}{2}x(t - 1/t)\right]$ satisfies the identity $w(x + y, t) = w(x, t)w(y, t)$, it can be shown that the Bessel function of order n satisfies the simple **addition formula** (see problem 4 in Exercises 6.2)

$$J_n(x + y) = \sum_{k=-\infty}^{\infty} J_k(x)J_{n-k}(y) \tag{6.59}$$

We wish now to develop the more general **addition formulas**

$$J_0(R) = \sum_{k=-\infty}^{\infty} J_k(a)J_k(b)e^{ik\theta}$$

$$= J_0(a)J_0(b) + 2\sum_{k=1}^{\infty} J_k(a)J_k(b)\cos k\theta \tag{6.60}$$

and

$$J_n(R)\begin{Bmatrix} \cos n\psi \\ \sin n\psi \end{Bmatrix} = \sum_{k=-\infty}^{\infty} J_k(a)J_{n+k}(b)\begin{Bmatrix} \cos k\theta \\ \sin k\theta \end{Bmatrix} \tag{6.61}$$

where $R = \sqrt{a^2+b^2-2ab\cos\theta}$, $\sin\psi = (a/R)\sin\theta$, and a and b are any positive constants (see Fig. 6.3).

A further generalization of these last two formulas is given by

$$\frac{J_n(R)}{R^n} = 2^n\Gamma(n)\sum_{k=0}^{\infty}(n+k)\frac{J_{n+k}(a)J_{n+k}(b)}{(ab)^n}C_k^n(\cos\theta) \tag{6.62}$$

where $n \neq 0, -1, -2, \ldots$ and where $C_k^n(x)$, $k = 0, 1, 2, \ldots$ are the *Gegenbauer polynomials* (see Sec. 5.4). We do derive (6.60) and

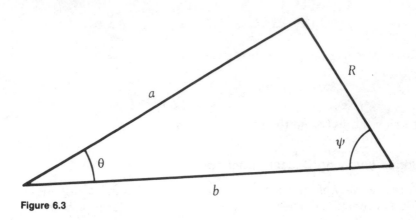

Figure 6.3

(6.61), but for the derivation of (6.62) we refer the reader to G. N. Watson, A *Treatise on the Theory of Bessel Functions*, 2d ed., Cambridge University Press, London, 1952. It can be shown that both (6.61) and (6.62) are also valid for nonintegral values of n, but in this case we must impose the additional condition $b > a$.

Addition theorems such as (6.60) to (6.62) play an important role in a number of applications in mathematical physics. To prove them, let us start with the identity (see problem 7 in Exercises 6.2)

$$\exp\left[-\frac{x}{2t}\left(k - \frac{1}{k}\right)\right] \sum_{n=-\infty}^{\infty} J_n(x)k^n t^n = \sum_{n=-\infty}^{\infty} J_n(kx)t^n \qquad (6.63)$$

and make the substitutions $x = a$, $k = e^{i\phi}$ and $x = b$, $k = e^{i\psi}$. This action leads to the equations

$$\exp\left(-\frac{ia}{t}\sin\phi\right) \sum_{n=-\infty}^{\infty} J_n(a)e^{in\phi}t^n = \sum_{n=-\infty}^{\infty} J_n(ae^{i\phi})t^n$$

$$\exp\left(-\frac{ib}{t}\sin\psi\right) \sum_{n=-\infty}^{\infty} J_n(b)e^{in\psi}t^n = \sum_{n=-\infty}^{\infty} J_n(be^{i\psi})t^n \qquad (6.64)$$

Forming the product of the expressions on the left-hand sides in (6.64) gives us

$$\exp\left[-\frac{i}{t}(a\sin\phi + b\sin\psi)\right] \sum_{n=-\infty}^{\infty} J_n(a)e^{in\phi}t^n \cdot \sum_{n=-\infty}^{\infty} J_n(b)e^{in\psi}t^n$$

$$= \exp\left[-\frac{i}{t}(a\sin\phi + b\sin\psi)\right] \sum_{n=-\infty}^{\infty} \sum_{m=-\infty}^{\infty} J_n(a)e^{in\phi}J_m(b)e^{im\psi}t^{n+m}$$

$$= \exp\left[-\frac{i}{t}(a\sin\phi + b\sin\psi)\right] \sum_{n=-\infty}^{\infty} \sum_{k=-\infty}^{\infty} J_k(a)J_{n-k}(b)e^{ik(\phi-\psi)}e^{in\psi}t^n$$

where the last step follows a change in index. Recognizing that the product of the series on the right-hand sides in (6.64) yields [based on the development of Eq. (6.59)]

$$\sum_{n=-\infty}^{\infty} J_n(ae^{i\phi})t^n \cdot \sum_{n=-\infty}^{\infty} J_n(be^{i\psi})t^n = \sum_{n=-\infty}^{\infty} J_n(ae^{i\phi} + be^{i\psi})t^n$$

we can equate like terms in these last two expressions to deduce that

$$J_n(ae^{i\phi} + be^{i\psi}) = \exp\left[-\frac{i}{t}(a\sin\phi + b\sin\psi)\right]e^{in\psi}$$

$$\times \sum_{k=-\infty}^{\infty} J_k(a)J_{n-k}(b)e^{ik(\phi-\psi)} \quad (6.65)$$

If we now require that $ae^{i\phi} + be^{i\psi}$ be a real number, say R, then we set

$$a\cos\phi + b\cos\psi = R$$

$$a\sin\phi + b\sin\psi = 0$$

and find that $R^2 = a^2 + b^2 + 2ab\cos(\phi - \psi)$. If we further set $\phi - \psi = \pi - \theta$, then $R^2 = a^2 + b^2 - 2ab\cos\theta$, and hence (6.65) takes the form

$$J_n(R)e^{-in\psi} = \sum_{k=-\infty}^{\infty} J_k(a)J_{n-k}(b)(-1)^k e^{-ik\theta} \quad (6.66)$$

Last, by replacing k with $-k$ and using $J_{-n}(x) = (-1)^n J_n(x)$, we finally obtain Eq. (6.61). The special case $n = 0$ reduces to (6.60).

6.5.2 Orthogonality of Bessel functions

Among other applications, the solution of certain partial DEs involving radial symmetry requires the expansion of a given function in a series of Bessel functions. Such series belong to the class of *generalized Fourier series.** Their theory closely parallels that of *Legendre series* and rests heavily on the **orthogonality property**

$$\int_0^b x J_p(k_m x)J_p(k_n x)\, dx = 0 \qquad m \neq n \quad (6.67)$$

---

* See the discussion in Sec. 4.4 on generalized Fourier series.

where k_m and k_n are distinct roots of

$$J_p(kb) = 0 \tag{6.68}$$

To prove (6.67), first we note that since $y = J_p(x)$ is a solution of Bessel's equation

$$x^2 y'' + xy' + (x^2 - p^2)y = 0$$

it follows that $y = J_p(kx)$ satisfies the more general equation (see problem 1 in Exercises 6.5)

$$x^2 y'' + xy' + (k^2 x^2 - p^2)y = 0 \tag{6.69}$$

For our purposes we rewrite (6.69) in the more useful form

$$x \frac{d}{dx}(xy') + (k^2 x^2 - p^2)y = 0$$

and hence $J_p(k_m x)$ and $J_p(k_n x)$ satisfy, respectively, the DEs

$$x \frac{d}{dx}\left[x \frac{d}{dx} J_p(k_m x) \right] + (k_m^2 x^2 - p^2)J_p(k_m x) = 0 \tag{6.70}$$

$$x \frac{d}{dx}\left[x \frac{d}{dx} J_p(k_n x) \right] + (k_n^2 x^2 - p^2)J_p(k_n x) = 0 \tag{6.71}$$

If we multiply (6.70) by $x^{-1}J_p(k_n x)$ and (6.71) by $x^{-1}J_p(k_m x)$, subtract the resulting equations, and integrate from 0 to b, we find upon rearranging the terms

$$(k_m^2 - k_n^2) \int_0^b x J_p(k_m x) J_p(k_n x)\, dx = \int_0^b J_p(k_m x) \frac{d}{dx}\left[x \frac{d}{dx} J_p(k_n x) \right] dx$$

$$- \int_0^b J_p(k_n x) \frac{d}{dx}\left[x \frac{d}{dx} J_p(k_m x) \right] dx$$

Performing integration by parts on the right-hand side and dividing by the factor $k_m^2 - k_n^2$ lead to

$$\int_0^b x J_p(k_m x) J_p(k_n x)\, dx$$

$$= \frac{x}{k_m^2 - k_n^2}\left[J_p(k_m x) \frac{d}{dx} J_p(k_n x) - J_p(k_n x) \frac{d}{dx} J_p(k_m x) \right]\Bigg|_{x=0}^{x=b} \tag{6.72}$$

By hypothesis, $k_m \neq k_n$ and $J_p(k_m b) = J_p(k_n b) = 0$, and thus the right-hand side of (6.72) vanishes, which proves the orthogonality property (6.67).

When $k_m = k_n$, the resulting integral

$$I = \int_0^b x [J_p(k_n x)]^2 \, dx$$

is also of interest to us. To deduce its value, we take the limit of (6.72) as $k_m \to k_n$. Because the right-hand side of (6.72) approaches the indeterminate form $0/0$ in the limit, we need to employ L'Hôpital's rule, which leads to (treating k_m as the variable and all other parameters constant)

$$I = \frac{x}{2k_n} \left[\frac{d}{dx} J_p(k_n x) \frac{d}{dk_n} J_p(k_n x) - J_p(k_n x) \frac{d}{dk_n} \frac{d}{dx} J_p(k_n x) \right] \Bigg|_{x=0}^{x=b} \quad (6.73)$$

Now, using the recurrence relations (see problem 15 in Exercises 6.2)

$$\frac{d}{dx} J_p(kx) = \frac{p}{x} J_p(kx) - k J_{p+1}(kx)$$

$$\frac{d}{dx} J_p(kx) = \frac{p}{k} J_p(kx) - x J_{p+1}(kx)$$

we find that (6.73) reduces to

$$I = \left\{ \frac{1}{2} \frac{p^2}{k_n^2} [J_p(k_n x)]^2 + \frac{1}{2} x^2 [J_{p+1}(k_n x)]^2 - \frac{p}{k_n} J_p(k_n x) J_{p+1}(k_n x) \right\} \Bigg|_{x=0}^{x=b}$$

or finally

$$\int_0^b x [J_p(k_n x)]^2 \, dx = \frac{1}{2} b^2 [J_{p+1}(k_n b)]^2 \quad (6.74)$$

6.5.3 Fourier–Bessel series

Having the orthogonality property (6.67), we now consider the representation of a given function f in a series of the form

$$f(x) = \sum_{n=1}^{\infty} c_n J_p(k_n x) \qquad 0 < x < b, \qquad p > -\frac{1}{2} \quad (6.75)$$

where $J_p(k_n b) = 0$ $(n = 1, 2, 3, \dots)$. Such a series is called a **Fourier–Bessel series** or simply a **Bessel series**. Let us assume that such a representation is valid and attempt to identify the constants c_n $(n = 1, 2, 3, \dots)$.

To begin, we multiply both sides of (6.75) by $xJ_p(k_mx)$, where $J_p(k_mb) = 0$, and we integrate the resulting series termwise from 0 to b. Doing so, we find that

$$\int_0^b xf(x)J_p(k_mx)\,dx = \sum_{n=1}^{\infty} c_n \int_0^b xJ_p(k_mx)J_p(k_nx)\,dx$$

$$= c_m \int_0^b x[J_p(k_mx)]^2\,dx$$

where all terms in the series vanish except for that which corresponds to $n = m$. This last integral is simply (6.74), and so solving for c_m and changing the index back to n, we obtain

$$c_n = \frac{2}{b^2[J_{p+1}(k_nb)]^2} \int_0^b xf(x)J_p(k_nx)\,dx \qquad n = 1, 2, 3, \ldots \quad (6.76)$$

Theorem 6.1. If f and f' are piecewise continuous functions on $0 \leq x \leq b$, then the Bessel series (6.75) with constants defined by (6.76) converges pointwise to $f(x)$ at points of continuity of f and to the average value $\frac{1}{2}[f(x^+) + f(x^-)]$ at points of discontinuity of f on the interval $0 < x < b$.*

Example 4: For the function

$$f(x) = \begin{cases} x & 0 \leq x < 1 \\ 0 & 1 < x < 2 \end{cases}$$

find a Bessel series of the form

$$f(x) = \sum_{n=1}^{\infty} c_n J_1(k_nx) \qquad 0 < x < 2$$

where $J_1(2k_n) = 0$ ($n = 1, 2, 3, \ldots$).

Solution: The series we seek is

$$f(x) = \sum_{n=1}^{\infty} c_n J_1(k_nx) \qquad 0 < x < 2$$

---

*The series always converges to zero for $x = b$, and it converges to zero at $x = 0$ if $p > 0$.

where
$$c_n = \frac{1}{2[J_2(2k_n)]^2} \int_0^2 xf(x)J_1(k_nx)\,dx$$

$$= \frac{1}{2[J_2(2k_n)]^2} \int_0^1 x^2J_1(k_nx)\,dx$$

In this last integral we make the change of variable $t = k_nx$, which yields

$$\int_0^1 x^2J_1(k_nx)\,dx = \frac{1}{k_n^3} \int_0^{k_n} t^2J_1(t)\,dt$$

$$= \frac{1}{k_n} J_2(k_n)$$

where we have made use of the integral formula (6.47). From this result we see that

$$c_n = \frac{J_2(k_n)}{2k_n[J_2(2k_n)]^2} \qquad n = 1, 2, 3, \ldots$$

and hence the desired series is given by

$$f(x) = \frac{1}{2} \sum_{n=1}^{\infty} \frac{J_2(k_n)}{k_n[J_2(2k_n)]^2} J_1(k_nx) \qquad 0 < x < 2$$

Generalizations of the Bessel series can be developed where the k_n satisfy the more general condition

$$hJ_p(k_nb) + k_nJ_p'(k_nb) = 0 \qquad (h \text{ constant}) \qquad (6.77)$$

The theory in such cases requires only a slight modification of that presented here and is taken up in the exercises.

Exercises 6.5

1. Show that $y = J_p(kx)$ is a solution of
$$x^2y'' + xy' + (k^2x^2 - p^2)y = 0$$

2. From Eq. (6.60) deduce the following:

(a) $J_0(\sqrt{x^2+y^2}) = J_0(x)J_0(y) + 2\sum_{k=1}^{\infty} (-1)^kJ_{2k}(x)J_{2k}(y)$

(b) $J_0\left(2x \cos\frac{\theta}{2}\right) = [J_0(x)]^2 + 2\sum_{k=1}^{\infty} (-1)^k[J_k(x)]^2 \cos k\theta$

(c) $J_0\left(2x \sin \dfrac{\theta}{2}\right) = [J_0(x)]^2 + 2 \sum\limits_{k=1}^{\infty} [J_k(x)]^2 \sin k\theta$

(d) $J_0(\sqrt{a^2 + b^2 + 2ab \cos \theta})$

$$= J_0(a)J_0(b) + 2 \sum\limits_{k=1}^{\infty} (-1)^k J_k(a)J_k(b) \cos k\theta$$

3. Show that the special case $\theta = \pi$ in Eq. (6.61) reduces that addition theorem to the simple result given by (6.59).

4. By setting $n = \frac{1}{2}$ in (6.62), deduce that

$$\frac{\sin R}{R} = \frac{\pi}{2\sqrt{ab}} \sum\limits_{k=0}^{\infty} (2k + 1)J_{k+1/2}(a)J_{k+1/2}(b)P_k(\cos \theta)$$

where $P_k(x)$ is the kth Legendre polynomial.

In problems 5 and 6, verify the series relation given that $J_0(k_n) = 0$ $(n = 1, 2, 3, \dots)$.

5. $1 - x^2 = 8 \sum\limits_{n=1}^{\infty} \dfrac{J_0(k_n x)}{k_n^3 J_1(k_n)}, \qquad 0 < x < 1$

6. $\ln x = -2 \sum\limits_{n=1}^{\infty} \dfrac{J_0(k_n x)}{[k_n J_1(k_n)]^2}, \qquad 0 < x < 1$

In problems 7 to 9, find the Bessel series for $f(x)$ in terms of the set $\{J_0(k_n x)\}$, given that $J_0(k_n) = 0$ $(n = 1, 2, 3, \dots)$.

7. $f(x) = 0.1J_0(k_3 x), \qquad 0 < x < 1$

8. $f(x) = 1, \qquad 0 < x < 1$

9. $f(x) = x^4, \qquad 0 < x < 1$

10. If $p \ge -\frac{1}{2}$ and $J_p(k_n) = 0$ $(n = 1, 2, 3, \dots)$, show that

$$x^p = 2 \sum\limits_{n=1}^{\infty} \frac{J_p(k_n x)}{k_n J_{p+1}(k_n)} \qquad 0 < x < 1$$

11. If $p \ge -\frac{1}{2}$ and $J_p(k_n) = 0$ $(n = 1, 2, 3, \dots)$, show that

(a) $x^{p+1} = 2^2(p + 1) \sum\limits_{n=1}^{\infty} \dfrac{J_{p+1}(k_n x)}{k_n^2 J_{p+1}(k_n)}, \qquad 0 < x < 1$

(b) $x^{p+2} = 2^3(p + 1)(p + 2) \sum\limits_{n=1}^{\infty} \dfrac{J_{p+2}(k_n x)}{k_n^3 J_{p+1}(k_n)}, \qquad 0 < x < 1$

12. Given that $J_p(k_n) = 0$ $(n = 1, 2, 3, \dots)$, $p \ge 0$, express $f(x) = x^{-p}$, $0 < x < 1$, in a series of the form

$$x^{-p} = \sum\limits_{n=1}^{\infty} c_n J_p(k_n x), \qquad 0 < x < 1$$

13. Given that $J_p'(k_n b) = 0$ $(n = 1, 2, 3, \dots)$ for $p > -\frac{1}{2}$, show that

   (a) $\displaystyle\int_0^b x J_p(k_m x) J_p(k_n x)\, dx = 0,\; m \neq n$

   (b) $\displaystyle\int_0^b x [J_p(k_n x)]^2\, dx = \frac{k_n^2 b^2 - p^2}{2k_n^2} [J_p(k_n b)]^2$

14. Given that $h J_p(k_n b) + k_n J_p'(k_n b) = 0$ $(n = 1, 2, 3, \dots)$ for constant h and $p > \frac{1}{2}$, show that

   (a) $\displaystyle\int_a^b x J_p(k_m x) J_p(k_n x)\, dx = 0,\; m \neq n$

   (b) $\displaystyle\int_0^b x [J_p(k_n x)]^2\, dx = \frac{(k_n^2 + h^2) b^2 - p^2}{2k_n^2} [J_p(k_n b)]^2$

15. It is given that $J_p'(k_n) = 0$ $(n = 1, 2, 3, \dots)$ for $p > 0$.

   (a) Use the result of problem 13 to derive the Bessel series

   $$x^p = 2 \sum_{n=1}^{\infty} \frac{k_n J_{p+1}(k_n)}{(k_n^2 - p^2)[J_p(k_n)]^2} J_p(k_n x), \qquad 0 < x < 1$$

   (b) Is the expansion valid when $p = 0$? Explain.

6.6 Bessel Functions of the Second Kind

We have previously shown that $y_1 = J_p(x)$ and $y_2 = J_{-p}(x)$ are both solutions of Bessel's equation

$$x^2 y'' + x y' + (x^2 - p^2) y = 0 \qquad p \geq 0 \qquad (6.78)$$

Moreover, for p not an integer, these are linearly independent solutions so that the general solution of (6.78) in this case can be expressed by

$$y = C_1 J_p(x) + C_2 J_{-p}(x) \qquad p \neq n \; (n = 0, 1, 2, \dots) \qquad (6.79)$$

where C_1 and C_2 are arbitrary constants.

For $p = n$ $(n = 0, 1, 2, \dots)$, the solutions $J_n(x)$ and $J_{-n}(x)$ are related by [see Eq. (6.6)]

$$J_{-n}(x) = (-1)^n J_{-n}(x) \qquad n = 0, 1, 2, \dots$$

and hence are *not* linearly independent. For purposes of constructing a general solution of (6.78), it is preferable to find a second solution y_2 whose independence is not restricted to certain values of p. Such a solution is defined by

$$Y_p(x) = \frac{(\cos p\pi) J_p(x) - J_{-p}(x)}{\sin p\pi} \qquad (6.80)$$

which is called the **Bessel function of the second kind** of order p. Because it is a linear combination of solutions of (6.78), the function $Y_p(x)$ is also a solution for all p. However, when $p = n$ $(n = 0, 1, 2, \ldots)$, it requires further investigation since the right-hand side of (6.80) assumes the indeterminate form $0/0$ in this case. Nonetheless, the limit as $p \to n$ does exist (see Sec. 6.6.1), and we define

$$Y_n(x) = \lim_{p \to n} Y_p(x) \qquad n = 0, 1, 2, \ldots \qquad (6.81)$$

The wronksian $W(J_p, Y_p)(x) = 2/(\pi x)$ for all p (see problem 7 in Exercises 6.6) and hence $J_p(x)$ and $Y_p(x)$ are *linearly independent* for all values of p. We conclude, therefore, that the general solution of Bessel's equation (6.78) for *all* values of p is given by

$$y = C_1 J_p(x) + C_2 Y_p(x) \qquad (6.82)$$

6.6.1   Series expansion for $Y_n(x)$

We wish to derive an expression for the Bessel function of the second kind when p takes on integral values. Because the limit (6.81) leads to the indeterminate form $0/0$, we must apply L'Hôpital's rule, from which we deduce

$$Y_n(x) = \lim_{p \to n} \frac{(\cos p\pi) J_p(x) - J_{-p}(x)}{\sin p\pi}$$

$$= \lim_{p \to n} \frac{(-\pi \sin p\pi) J_p(x) + (\cos p\pi) \dfrac{\partial}{\partial p} J_p(x) - \dfrac{\partial}{\partial p} J_{-p}(x)}{\pi \cos p\pi}$$

$$= \lim_{p \to n} \frac{1}{\pi} \left[\frac{\partial}{\partial p} J_p(x) - (-1)^n \frac{\partial}{\partial p} J_{-p}(x) \right] \qquad (6.83)$$

For $x > 0$, the derivative of the Bessel function with respect to order yields

$$\frac{\partial}{\partial p} J_p(x) = \sum_{k=0}^{\infty} \frac{(-1)^k}{k!} \left\{ \frac{(x/2)^{2k+p} \ln (x/2)}{\Gamma(k+p+1)} - \frac{(x/2)^{2k+p} \Gamma'(k+p+1)}{\Gamma(k+p+1)]^2} \right\}$$

$$= \sum_{k=0}^{\infty} \frac{(-1)^k (x/2)^{2k+p}}{k!\, \Gamma(k+p+1)} \left[\ln \frac{x}{2} - \psi(k+p+1) \right]$$

where $\psi(x)$ is the digamma function (see Sec. 2.5). We can further write this last expression as

$$\frac{\partial}{\partial p} J_p(x) = J_p(x) \ln\frac{x}{2} - \sum_{k=0}^{\infty} \frac{(-1)^k (x/2)^{2k+p}}{k!\,\Gamma(k+p+1)}\,\psi(k+p+1) \quad (6.84)$$

and by a similar analysis, it follows that

$$\frac{\partial}{\partial p} J_{-p}(x) = -J_{-p}(x) \ln\frac{x}{2} + \sum_{k=0}^{\infty} \frac{(-1)^k (x/2)^{2k-p}}{k!\,\Gamma(k-p+1)}\,\psi(k-p+1) \quad (6.85)$$

At this point, let us consider the special case $p \to 0$. From (6.83), we obtain

$$Y_0(x) = \lim_{p\to 0} \frac{1}{\pi}\left[\frac{\partial}{\partial p} J_p(x) - \frac{\partial}{\partial p} J_{-p}(x)\right]$$

and by using the results of (6.84) and (6.85), we get $(x > 0)$

$$Y_0(x) = \frac{2}{\pi} J_0(x) \ln\frac{x}{2} - \frac{2}{\pi} \sum_{k=0}^{\infty} \frac{(-1)^k (x/2)^{2k}}{(k!)^2}\,\psi(k+1) \quad (6.86)$$

Another form of (6.86) can be obtained by making the observation

$$\sum_{k=0}^{\infty} \frac{(-1)^k (x/2)^{2k}}{(k!)^2}\,\psi(k+1)$$

$$= -\gamma + \sum_{k=1}^{\infty} \frac{(-1)^k (x/2)^{2k}}{(k!)^2}\left(-\gamma + 1 + \frac{1}{2} + \cdots + \frac{1}{k}\right)$$

$$= -\gamma \sum_{k=0}^{\infty} \frac{(-1)^k (x/2)^{2k}}{(k!)^2} + \sum_{k=1}^{\infty} \frac{(-1)^k (x/2)^{2k}}{(k!)^2}\left(1 + \frac{1}{2} + \cdots + \frac{1}{k}\right)$$

where γ is Euler's constant. Hence, combining this last expression with (6.86), we have $(x > 0)$

$$Y_0(x) = \frac{2}{\pi} J_0(x)\left(\ln\frac{x}{2} + \gamma\right)$$

$$- \frac{2}{\pi} \sum_{k=1}^{\infty} \frac{(-1)^k (x/2)^{2k}}{(k!)^2}\left(1 + \frac{1}{2} + \cdots + \frac{1}{k}\right) \quad (6.87)$$

The derivation of the series for $Y_n(x)$ $(n = 1, 2, 3, \ldots)$ is a little more difficult to obtain. Proceeding as before and taking the limit in

(6.83), using (6.84) and (6.85), we find

$$Y_n(x) = \frac{1}{\pi}[J_n(x) + (-1)^n J_{-n}(x)] \ln \frac{x}{2}$$

$$- \frac{1}{\pi}\left[\sum_{k=0}^{\infty} \frac{(-1)^k (x/2)^{2k+n}}{k!\, \Gamma(k+n+1)} \psi(k+n+1) \right.$$

$$\left. + (-1)^n \sum_{k=0}^{\infty} \frac{(-1)^k (x/2)^{2k-n}}{k!\, \Gamma(k-n+1)} \psi(k-n+1) \right] \qquad (6.88)$$

Next, recalling that

$$|\Gamma(k-n+1)| \to \infty \qquad k = 0, 1, \ldots, n-1$$

and

$$|\psi(k-n+1)| \to \infty \qquad k = 0, 1, \ldots, n-1$$

we see that the first n terms in the last series in (6.88) become indeterminate. However, it can be shown that (see problem 9 in Exercises 6.6)

$$\lim_{p \to n} \frac{\psi(k-p+1)}{\Gamma(k-p+1)} = (-1)^{n-k}(n-k-1)! \qquad k = 0, 1, \ldots, n-1 \qquad (6.89)$$

and therefore we can write

$$(-1)^n \sum_{k=0}^{\infty} \frac{(-1)^k (x/2)^{2k-n}}{k!\, \Gamma(k-n+1)} \psi(k-n+1) = \sum_{k=0}^{n-1} \frac{(n-k-1)!}{k!} \left(\frac{x}{2}\right)^{2k-n}$$

$$+ (-1)^n \sum_{k=n}^{\infty} \frac{(-1)^k (x/2)^{2k-n}}{k!\, \Gamma(k-n+1)} \psi(k-n+1)$$

Finally, by making the change of index $m = k - n$ in the last summation in this last expression, (6.88) reduces to $(x > 0)$

$$Y_n(x) = \frac{2}{\pi} J_n(x) \ln \frac{x}{2} - \frac{1}{\pi} \sum_{k=0}^{n-1} \frac{(n-k-1)!}{k!} \left(\frac{x}{2}\right)^{2k-n}$$

$$- \frac{1}{\pi} \sum_{m=0}^{\infty} \frac{(-1)^m (x/2)^{2m+n}}{m!\, (m+n)!} [\psi(m+n+1) + \psi(m+1)] \qquad (6.90)$$

Graphs of $Y_0(x)$, $Y_1(x)$, and $Y_2(x)$ are shown in Fig. 6.4. Observe the *logarithmic behavior as $x \to 0^+$* (see Sec. 6.6.2). Also note that these functions have oscillatory characteristics similar to those of $J_n(x)$.

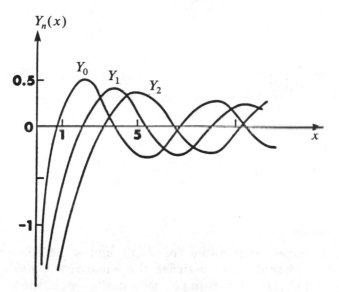

Figure 6.4 Graph of $Y_n(x)$, $n = 0, 1, 2$.

6.6.2 Asymptotic formulas for small arguments

To obtain asymptotic formulas of the Bessel functions $Y_p(x)$ for $x \to 0^+$, we start with the series representation (6.87) for $Y_0(x)$, and retaining only the most significant terms, we see that

$$Y_0(x) \sim \frac{2}{\pi} J_0(x) \left(\ln \frac{x}{2} + \gamma \right) \qquad x \to 0^+$$

However, since $J_0(x) \sim 1$ and $|\ln x| \gg \gamma - \ln 2$ for small x, we deduce that $Y_0(x)$ has the logarithmic behavior

$$Y_0(x) \sim \frac{2}{\pi} \ln x \qquad x \to 0^+ \qquad (6.91)$$

In the general case where $p > 0$, we start with

$$Y_p(x) = \frac{(\cos p\pi)J_p(x) - J_{-p}(x)}{\sin p\pi}$$

Here we make the observation that $J_p(x) \sim 0$ for $x \to 0^+$ [recall Eq.

(6.11a)], and thus

$$Y_p(x) \sim -\frac{J_{-p}(x)}{\sin p\pi}$$

$$\sim -\frac{(x/2)^{-p}}{\Gamma(1-p)\sin p\pi} \qquad p > 0, \qquad x \to 0^+$$

By use of the identity $\Gamma(x)\Gamma(1-x) = \pi/\sin \pi x$, we finally arrive at the asymptotic relation

$$Y_p(x) \sim -\frac{\Gamma(p)}{\pi}\left(\frac{2}{x}\right)^p \qquad p > 0, \qquad x \to 0^+ \tag{6.92}$$

6.6.3 Recurrence formulas

Because $Y_p(x)$ is a linear combination of $J_p(x)$ and $J_{-p}(x)$ for nonintegral p, it follows that $Y_p(x)$ satisfies the same recurrence formulas as $J_p(x)$ and $J_{-p}(x)$. For example, it is easily established that (see problems 10 and 11 in Exercises 6.6)

$$\frac{d}{dx}[x^p Y_p(x)] = x^p Y_{p-1}(x) \tag{6.93}$$

$$\frac{d}{dx}[x^{-p} Y_p(x)] = -x^{-p} Y_{p+1}(x) \tag{6.94}$$

and that

$$Y_{p-1}(x) + Y_{p+1}(x) = \frac{2p}{x} Y_p(x) \tag{6.95}$$

$$Y_{p-1}(x) - Y_{p+1}(x) = 2Y_p'(x) \tag{6.96}$$

For integral p the validity of these formulas can be deduced by considering the limit $p \to n$, noting that all functions are continuous with respect to the index p. Furthermore, it can be shown that (see problem 12 in Exercises 6.6)

$$Y_{-n}(x) = (-1)^n Y_n(x) \qquad n = 0, 1, 2, \ldots \tag{6.97}$$

Exercises 6.6

In problems 1 to 4, write the general solution of the DE in terms of Bessel functions.

1. $x^2 y'' + xy' + (x^2 - \frac{1}{4})y = 0$
2. $xy'' + y' + xy = 0$

3. $16x^2y'' + 16xy' + (16x^2 - 1)y = 0$

4. $x^2y'' + xy' + (4x^2 - 1)y = 0$

   Hint: Let $t = 2x$.

5. Show that the change of variable $y = u(x)/\sqrt{x}$ reduces Bessel's equation (6.78) to

$$u'' + \left(1 + \frac{1 - 4p^2}{4x^2}\right)u = 0$$

6. Use the result of problem 5 to find a general solution of Bessel's equation (6.78) when $p = \frac{1}{2}$ that does not involve Bessel functions.

7. From the results of problems 18 and 19 in Exercises 6.2, deduce that

$$W(J_p, Y_p)(x) = \frac{2}{\pi x}$$

8. Use problem 7 and appropriate recurrence relations to show that

$$J_p(x)Y_{p+1}(x) - J_{p+1}(x)Y_p(x) = -\frac{2}{\pi x}$$

9. Based on the identities $\Gamma(x)\Gamma(1-x) = \pi \csc \pi x$ and $\psi(1-x) - \psi(x) = \pi \cot \pi x$, show that

$$\lim_{p \to n} \frac{\psi(k - p + 1)}{\Gamma(k - p + 1)} = (-1)^{n-k}(n - k - 1)! \qquad k = 0, 1, \ldots, n - 1$$

10. Show that

    (a) $\dfrac{d}{dx}[x^p Y_p(x)] = x^p Y_{p-1}(x)$

    (b) $\dfrac{d}{dx}[x^{-p} Y_p(x)] = -x^{-p} Y_{p+1}(x)$

11. From the results of problem 10, deduce that

    (a) $Y_{p-1}(x) + Y_{p+1}(x) = \dfrac{2p}{x} Y_p(x)$

    (b) $Y_{p-1}(x) - Y_{p+1}(x) = 2Y_p'(x)$

12. Verify that

$$Y_{-n}(x) = (-1)^n Y_n(x) \qquad n = 0, 1, 2, \ldots$$

13. By making the change of variable $t = bx$, show that $(b > 0)$

$$y = C_1 J_p(bx) + C_2 Y_p(bx)$$

   is a general solution of

$$x^2y'' + xy' + (b^2x^2 - p^2)y = 0 \qquad p \geq 0$$

14. Show that the boundary-value problem $(p \geq 0)$

$$x^2 y'' + xy' + (k^2 x^2 - p^2)y = 0 \qquad |y(0)| < \infty, \qquad y(1) = 0$$

has only solutions which are multiples of $J_p(k_n x)$, where the k's are chosen to satisfy the relation

$$J_p(k_n) = 0 \qquad n = 1, 2, 3, \ldots$$

15. Assume a power series solution of the form (*Frobenius method**)

$$y = x^s \sum_{n=0}^{\infty} c_n x^n$$

   (a)   Show that Bessel's equation (6.78) has one solution corresponding to $s = p$ that leads to $y_1 = J_p(x)$.

   (b)   For $p = 0$, show that the method of Frobenius leads to the general solution

$$y = (A + B \ln x) \sum_{n=0}^{\infty} \frac{(-1)^n (x/2)^{2n}}{(n!)^2}$$

$$+ B \sum_{n=1}^{\infty} \frac{(-1)^{n-1}(x/2)^{2n}}{(n!)^2} \left(1 + \frac{1}{2} + \cdots + \frac{1}{n}\right)$$

   where A and B are arbitrary constants.

6.7   Differential Equations Related to Bessel's Equation

Elementary problems involving DEs are regarded as solved when their solutions can be expressed in terms of elementary functions, such as trigonometric and exponential functions. The same can be said of many problems of a more complicated nature when their solutions can be expressed in terms of Bessel functions.

A fairly large number of DEs occurring in physics and engineering problems are specializations of the form

$$x^2 y'' + (1 - 2a)xy' + [b^2 c^2 x^{2c} + (a^2 - c^2 p^2)]y = 0 \qquad p \geq 0, \qquad b > 0 \tag{6.98}$$

the general solution of which, expressed in terms of Bessel functions, is

$$y = x^a [C_1 J_p(bx^c) + C_2 Y_p(bx^c)] \tag{6.99}$$

where C_1 and C_2 are arbitrary constants.

---

* For an introductory discussion of the Frobenius method, see chap. 7 in L. C. Andrews, *Introduction to Differential Equations with Boundary Value Problems*, HarperCollins, New York, 1991.

To derive the solution formula (6.99) requires two transformations of variables. First, let us set

$$y = x^a z \tag{6.100}$$

from which we obtain

$$xy' = x^{a-1}z' + ax^a z$$

$$x^2 y'' = x^{a+2}z'' + 2ax^{a+1}z' + a(a-1)x^a z$$

Substituting these expressions into (6.98) and simplifying the results, we get

$$x^2 z'' + xz' + (b^2 c^2 x^{2c} - c^2 p^2)z = 0 \tag{6.101}$$

Next we make the change of independent variable

$$t = x^c \tag{6.102}$$

from which it follows, through application of the chain rule, that

$$xz' = cx^c \frac{dz}{dt}$$

$$x^2 z'' = c(c-1)x^c \frac{dz}{dt} + c^2 x^{2c} \frac{d^2 z}{dt^2}$$

Hence, Eq. (6.101) becomes

$$t^2 \frac{d^2 z}{dt^2} + t \frac{dz}{dt} + (b^2 t^2 - p^2)z = 0 \tag{6.103}$$

whose general solution is (see problem 13 in Exercises 6.6)

$$z(t) = C_1 J_p(bt) + C_2 Y_p(bt) \tag{6.104}$$

Transforming this result back to the original variables x and y leads us to the desired result (6.99). For those cases when p is not integral, we can express the general solution (6.99) in the alternative form

$$y = x^a [C_1 J_p(bx^c) + C_2 J_{-p}(bx^c)] \qquad p \neq n \qquad (n = 0, 1, 2, \dots) \tag{6.105}$$

Example 5: Find the general solution of

$$x^2 y'' + 5xy' + (4x^2 + 3)y = 0$$

Solution: This DE is of the form (6.98) with

$$1 - 2a = 5 \qquad 2x = 2 \qquad b^2 c^2 = 4 \qquad a^2 - c^2 p^2 = 3$$

These conditions are satisfied if $a = -2$, $b = 2$, $c = 1$, and $p = 1$. Thus, the general solution we seek is

$$y = \frac{1}{x^2}[C_1 J_1(2x) + C_2 Y_1(2x)]$$

Example 6: Find the general solution of *Airy's equation*

$$y'' + xy = 0$$

Solution: To compare this DE with (6.98), we must multiply through by x^2 ($x \neq 0$), putting it in the form

$$x^2 y'' + x^3 y = 0$$

Here we find that $a = \frac{1}{2}$, $b = \frac{2}{3}$, $c = \frac{3}{2}$, and $p = \frac{1}{3}$. Hence we obtain

$$y = \sqrt{x}\,[C_1 J_{1/3}(\tfrac{2}{3}x^{3/2}) + C_2 Y_{1/3}(\tfrac{2}{3}x^{3/2})]$$

or since p is nonintegral, we can also write

$$y = \sqrt{x}\,[C_1 J_{1/3}(\tfrac{2}{3}x^{3/2}) + C_2 J_{-1/3}(\tfrac{2}{3}x^{3/2})]$$

6.7.1 The oscillating chain

One of the classical problems of oscillation theory concerns the small oscillations of a flexible hanging chain. This problem was first discussed in 1732 by D. Bernoulli and later in 1781 by Euler, both many years prior to Bessel's legendary paper in 1824 on the properties of Bessel functions.

Consider a uniform heavy flexible chain of length L, fixed at the upper end and free at the lower end (see Fig. 6.5). When the chain is slightly disturbed from its position of equilibrium in a vertical plane, it undergoes "small" oscillations. Let ρ denote the constant mass per unit length and y the horizontal displacement of the chain at time t. Taking the origin at the bottom of the chain as shown in Fig. 6.5, the resultant force per unit length x is

$$\frac{\partial}{\partial x}\left(T\frac{\partial y}{\partial x}\right)$$

where T is the tension in the chain. Equating this resultant force to $\rho\, \partial^2 y/\partial t^2$, which represents the product of the mass (per unit length) and the acceleration, we obtain the equation of motion (*Newton's second law*)

$$\frac{\partial}{\partial x}\left(T\frac{\partial y}{\partial x}\right) = \rho\frac{\partial^2 y}{\partial t^2} \tag{6.106}$$

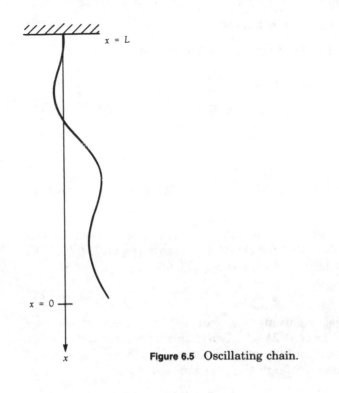

Figure 6.5 Oscillating chain.

Let us suppose the tension T is due entirely to the weight of the chain below a given point x. In this case $T = \rho g x$, where g is the gravitational constant, and (6.106) takes the form

$$g \frac{\partial}{\partial x}\left(x \frac{\partial y}{\partial x}\right) = \frac{\partial^2 y}{\partial t^2} \qquad (6.107)$$

Since the chain is fixed at the top where $x = L$ and free at the bottom where $x = 0$, the accompanying boundary conditions are given by

$$|y(0)| < \infty \qquad y(L) = 0 \qquad (6.108)$$

The first condition merely requires that the displacement remain bounded at $x = 0$.

Although (6.107) is a partial DE, we can reduce it to an ordinary DE by making an assumption regarding the variation of y with respect to time t. That is, if we assume the oscillations y are essentially sinusoidal with (angular) frequency ω, then we can find the *normal modes of vibration* by making the substitution

$$y = R \cos (\omega t - \phi) \qquad (6.109)$$

into (6.107), where the amplitude R is a function of x alone and ϕ is a phase angle. After doing so, and dividing the resulting equation by $\cos (\omega t - \phi)$, we obtain the ordinary DE

$$xR'' + R' + k^2 R = 0 \qquad (6.110)$$

where $k^2 = \omega^2/g$. We recognize (6.110) as a specialization of (6.98) for which $a = 0$, $b = 2k$, $c = \frac{1}{2}$, and $p = 0$. Hence, the general solution is

$$R(x) = C_1 J_0(2k \sqrt{x}) + C_2 Y_0(2k \sqrt{x}) \qquad (6.111)$$

where C_1 and C_2 are arbitrary constants.

To satisfy the boundary condition at $x = 0$, we must select $C_2 = 0$, since Y_0 becomes unbounded for small arguments [recall Eq. (6.91)]. The remaining boundary condition at $x = L$ leads to

$$J_0(2k \sqrt{L}) = 0 \qquad (6.112)$$

To satisfy this last relation, we can select values of k, say $k_1, k_2, k_3, \ldots, k_n, \ldots$ so that $2k \sqrt{L}$ always corresponds to a zero of J_0 (see Table 6.1). These values of k in turn determine the allowed frequencies of vibration through the relation

$$\omega_n = k_n \sqrt{g} \qquad n = 1, 2, 3, \ldots \qquad (6.113)$$

Hence, the corresponding normal modes of vibration are of the form

$$y_n(x, t) = A_n J_0(2k_n \sqrt{x}) \cos (\omega_n t - \phi_n) \qquad n = 1, 2, 3, \ldots \qquad (6.114)$$

where the A's and ϕ's are arbitrary constants.

For example, if the chain is 10 ft long, the lowest frequency of oscillation is determined by first solving

$$2k_1 \sqrt{10} = 2.405$$

where 2.405 is the first zero of J_0. Thus, $k_1 \cong 0.380$, and using $g = 32 \text{ ft/s}^2$, we find for the lowest frequency

$$\omega_1 = k_1 \sqrt{g} \cong 2.15 \text{ Hz}$$

Exercises 6.7

In problems 1 to 12, express the general solution in terms of Bessel functions.

1. $4xy'' + 4y' + y = 0$
2. $4x^2 y'' + 4xy' + (x^2 - n^2)y = 0$

3. $x^2y'' + xy' + 4(x^4 - k^2)y = 0$

4. $xy'' - y' + xy = 0$

5. $xy'' + (1 + 2n)y' + xy = 0$

6. $x^2y'' + (x^2 + \frac{1}{4})y = 0$

7. $x^2y'' - 7xy' + (36x^6 + \frac{175}{16})y = 0$

8. $y'' + y = 0$

9. $y'' + k^2x^2y = 0$

10. $y'' + k^2x^4y = 0$

11. $4x^2y'' + (1 + 4x)y = 0$

12. $x^2y'' + 5xy' + (9x^2 - 12)y = 0$

13. You are given the DE

$$y'' + ae^{mx}y = 0 \qquad m > 0$$

(a) Show that the substitution $t = e^{mx}$ transforms it to

$$t\frac{d^2y}{dt^2} + \frac{dy}{dt} + \frac{a}{m^2}y = 0$$

(b) Solve the DE in (a) in terms of Bessel functions, and use this to find the general solution of the original DE.

14. This DE is given:

$$x^2y'' + x(1 - 2x\tan x)y' - (x\tan x + n^2)y = 0$$

(a) Show that the transformation $y = u(x)\sec x$ leads to an equation in u solvable in terms of Bessel functions.

(b) Find the general solution of the original DE.

15. For the second mode $(n = 2)$ of the oscillating-chain problem, there is a *node* (stationary point) located on the interval $0 < x < L$. Show that this node always occurs at the point $x \cong 0.190L$.

16. The complete solution of the oscillating-chain problem consists of the superposition of all vibration modes, i.e.,

$$y(x, t) = \sum_{n=1}^{\infty} a_n J_0(2k_n\sqrt{x}) \cos(\omega_n t - \phi_n)$$

(a) If the initial velocity is $\partial y/\partial t(x, 0) = 0$, show that a sufficient condition is $\phi_n = 0$ $(n = 1, 2, 3, \dots)$.

(b) If also the initial displacement is $y(x, 0) = \epsilon(L - x)$, where ϵ is a small constant, show that

$$y(x, t) = \frac{\epsilon}{\sqrt{L}} \sum_{n=1}^{\infty} \frac{J_0(2k_n\sqrt{x})}{k_n^3 J_1(2k_n\sqrt{L})} \cos\omega_n t$$

Bessel Functions of Other Kinds

7.1 Introduction

The Bessel functions of the first and second kinds studied in Chap. 6 are often referred to as the *standard* Bessel or cylinder functions. In addition to these, there are a host of related functions also belonging to the general family of cylinder functions, the most notable of which are the *modified Bessel functions* of the first and second kinds. Although similar in definition to the standard Bessel functions, the modified Bessel functions are most clearly distinguished by their nonoscillatory behavior. For this reason, they often appear in applications that are different in nature from those for the standard functions.

The general family of cylinder functions also include *spherical Bessel functions, Hankel functions, Kelvin's functions, Lommel functions, Struve functions, Airy functions,* and *Anger* and *Weber functions.* Of these, Hankel functions have special significance in that they enable us to obtain asymptotic formulas for large arguments for all the other types of Bessel functions.

7.2 Modified Bessel functions

In Chap. 6 we found that the general solution of

$$x^2 y'' + xy' + (b^2 x^2 - p^2)y = 0 \qquad p \geq 0 \qquad (7.1)$$

is given by

$$y = C_1 J_p(bx) + C_2 Y_p(bx) \qquad (7.2)$$

where C_1 and C_2 are any constants. The related DE

$$x^2 y'' + xy' - (x^2 + p^2)y = 0 \qquad p \geq 0 \qquad (7.3)$$

which bears great resemblance to Bessel's equation, is **Bessel's modified equation**. It is of the form (7.1) with $b^2 = -1$, and so we formally write the solution of (7.3) as*

$$y = C_1 J_p(ix) + C_2 Y_p(ix) \qquad (7.4)$$

The disadvantage of the general solution (7.4) is that it is expressed in terms of functions with complex arguments, and in most situations we prefer real functions. The problem is similar to stating that

$$y = C_1 e^{ix} + C_2 e^{-ix}$$

is the general solution of $y'' + y = 0$. To avoid the imaginary arguments in (7.4), let us first formally replace x with ix in the series for $J_p(x)$ to obtain

$$J_p(ix) = \sum_{k=0}^{\infty} \frac{(-1)^k (ix/2)^{2k+p}}{k!\,\Gamma(k+p+1)}$$

$$= i^p \sum_{k=0}^{\infty} \frac{(x/2)^{2k+p}}{k!\,\Gamma(k+p+1)}$$

where we have used the fact that $i^{2k} = (-1)^k$. Next, observing that the above series itself is a real quantity multiplied by i^p, we are motivated to define the *real* function

$$I_p(x) = i^{-p} J_p(ix) = \sum_{k=0}^{\infty} \frac{(x/2)^{2k+p}}{k!\,\Gamma(k+p+1)} \qquad p > 0 \qquad (7.5)$$

This is called the **modified Bessel function of the first kind** of order p. Since $y = J_p(ix)$ is a solution of Eq. (7.3), it follows from (7.5) that $y = I_p(x)$ is also a solution.

Except that the alternating factor $(-1)^k$ is missing, the series (7.5) is identical to that of $J_p(x)$. Thus each term of the series (7.5) is positive for $x > 0$ and so continues to increase the sum with each additional term. We conclude, therefore, that $I_p(x)$ cannot have a positive zero and so cannot exhibit an oscillatory behavior like that of $J_p(x)$. In Fig. 7.1 we show the graphs of $I_0(x)$, $I_1(x)$, and $I_2(x)$, which

---

* Equation (7.3) also can be derived directly from Bessel's equation by replacing x with ix (see problem 1 in Exercises 7.2).

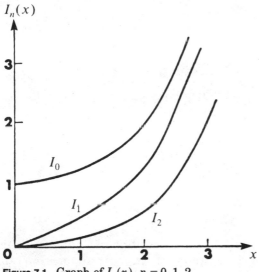

Figure 7.1 Graph of $I_n(x)$, $n = 0, 1, 2$.

are characteristic of all $I_p(x)$. Notice that $I_0(0) = 1$ while in general $I_p(0) = 0$ $(p > 0)$.

For negative p we define

$$I_{-p}(x) = \sum_{k=0}^{\infty} \frac{(x/2)^{2k-p}}{k!\Gamma(k-p+1)} \qquad p > 0 \qquad (7.6)$$

It is easy to verify that $y_2 = I_{-p}(x)$ is a solution of Bessel's modified equation (7.3) in addition to $y_1 = I_p(x)$. Moreover, by using an argument similar to that in Sec. 6.2.2 to show that $J_p(x)$ and $J_{-p}(x)$ are linearly independent for p not an integer, it follows that $I_p(x)$ and $I_{-p}(x)$ are *linearly independent functions* for nonintegral p. Hence for p not an integer, a general solution of (7.3) is

$$y = C_1 I_p(x) + C_2 I_{-p}(x) \qquad p \neq n \ (n = 0, 1, 2, \ldots) \qquad (7.7)$$

Recall that the Bessel functions $J_{1/2}(x)$ and $J_{-1/2}(x)$ are special in that they are actually elementary functions [see Eq. (6.12)]. In the same manner, we find that (see problem 10 in Exercises 7.2)

$$I_{1/2}(x) = \sqrt{\frac{2}{\pi x}} \sinh x \qquad (7.8a)$$

$$I_{-1/2}(x) = \sqrt{\frac{2}{\pi x}} \cosh x \qquad (7.8b)$$

Last, when $p = -n$ $(n = 0, 1, 2, \ldots)$, we find that negative-order functions lead to

$$I_{-n}(x) = i^n J_{-n}(ix)$$

$$= i^n(-1)^n J_n(ix)$$

$$= (-1)^{2n} I_n(x)$$

from which we deduce

$$I_{-n}(x) = I_n(x) \qquad n = 0, 1, 2, \ldots \qquad (7.9)$$

7.2.1 Modified Bessel functions of the second kind

Rather than use $Y_p(ix)$ to define a second linearly independent solution of (7.3) for general p, it is preferable in most applications to introduce the **modified Bessel function of the second kind** of order p (or **Macdonald's function**)

$$K_p(x) = \frac{\pi}{2} \frac{I_{-p}(x) - I_p(x)}{\sin p\pi} \qquad (7.10)$$

It follows directly from the definition that

$$K_{-p}(x) = K_p(x) \qquad (7.11)$$

for all values of p. Because it is a linear combination of solutions, the function $K_p(x)$ is also a solution of (7.3), which can be shown to be linearly independent of $I_p(x)$. Thus, for all values of p ($p \geq 0$), we write the *general solution* of Eq. (7.3) as

$$y = C_1 I_p(x) + C_2 K_p(x) \qquad (7.12)$$

For integral p we define

$$K_n(x) = \lim_{p \to n} K_p(x) \qquad n = 0, 1, 2, \ldots \qquad (7.13)$$

Then, following a procedure analogous to that in Sec. 6.6.1, we show that for $x > 0$ (see problem 15 in Exercises 7.2)

$$K_0(x) = -I_0(x)\left(\gamma + \ln\frac{x}{2}\right) + \sum_{k=1}^{\infty} \frac{(x/2)^{2k}}{(k!)^2}\left(1 + \frac{1}{2} + \cdots + \frac{1}{k}\right) \qquad (7.14a)$$

while for $x > 0$ and $n = 1, 2, 3, \ldots$ (see problem 16 in Exercises 7.2)

$$K_n(x) = (-1)^{n-1} I_n(x) \ln \frac{x}{2}$$

$$+ \frac{1}{2} \sum_{k=0}^{n-1} \frac{(-1)^k (n-k-1)!}{k!} \left(\frac{x}{2}\right)^{2k-n}$$

$$+ \frac{(-1)^n}{2} \sum_{m=0}^{\infty} \frac{(x/2)^{2m+n}}{m!(m+n)!} [\psi(m+n+1) + \psi(m+1)] \quad (7.14b)$$

Here γ is Euler's constant and $\psi(x)$ is the digamma function.

The graphs of $K_0(x)$, $K_1(x)$, and $K_2(x)$, characteristic of all $K_p(x)$, are shown in Fig. 7.2. Observe the exponential decay of $K_n(x)$ as $x \to \infty$, while the graphs of $I_n(x)$ appear to grow exponentially (see Fig. 7.1). Because of this behavior, the functions $I_p(x)$ and $K_p(x)$ (for general $p \geq 0$) are sometimes referred to as the *hyperbolic Bessel functions*.

7.2.2 Recurrence formulas

The recurrence formulas for the modified Bessel functions are very similar to those of the standard Bessel functions. By using techniques

$K_n(x)$

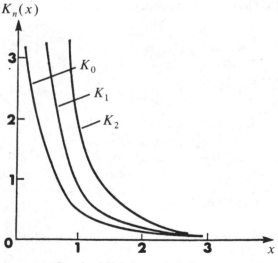

Figure 7.2 Graph of $K_n(x)$, $n = 0, 1, 2$.

analogous to those in Sec. 6.2, first it can be shown that

$$\frac{d}{dx}[x^p I_p(x)] = x^p I_{p-1}(x) \qquad (7.15)$$

$$\frac{d}{dx}[x^{-p} I_p(x)] = x^{-p} I_{p+1}(x) \qquad (7.16)$$

From these basic formulas, it readily follows that

$$I_p'(x) + \frac{p}{x} I_p(x) = I_{p-1}(x) \qquad (7.17)$$

$$I_p'(x) - \frac{p}{x} I_p(x) = I_{p+1}(x) \qquad (7.18)$$

$$I_{p-1}(x) + I_{p+1}(x) = 2I_p'(x) \qquad (7.19)$$

$$I_{p-1}(x) - I_{p+1}(x) = \frac{2p}{x} I_p(x) \qquad (7.20)$$

the details of which are left to the exercises.

Similarly, by using the relation (7.10) and the above recurrence formulas for $I_p(x)$, we find

$$\frac{d}{dx}[x^p K_p(x)] = -x^p K_{p-1}(x) \qquad (7.21)$$

$$\frac{d}{dx}[x^{-p} K_p(x)] = -x^{-p} K_{p+1}(x) \qquad (7.22)$$

$$K_p'(x) + \frac{p}{x} K_p(x) = -K_{p-1}(x) \qquad (7.23)$$

$$K_p'(x) - \frac{p}{x} K_p(x) = -K_{p+1}(x) \qquad (7.24)$$

$$K_{p-1}(x) + K_{p+1}(x) = -2K_p'(x) \qquad (7.25)$$

$$K_{p-1}(x) - K_{p+1}(x) = -\frac{2p}{x} K_p(x) \qquad (7.26)$$

7.2.3 Generating function and addition theorems

Replacing x with ix and t with $-it$ in the generating-function relation

$$\exp\left[\frac{1}{2}x\left(t - \frac{1}{t}\right)\right] = \sum_{n=-\infty}^{\infty} J_n(x)t^n \qquad t \neq 0$$

leads to

$$\exp\left[\frac{1}{2}x\left(t+\frac{1}{t}\right)\right] = \sum_{n=-\infty}^{\infty} J_n(ix)(-1)^n i^n t^n$$

$$= \sum_{n=-\infty}^{\infty} I_n(x)(-1)^n i^{2n} t^n$$

from which we deduce

$$\exp\left[\frac{1}{2}x\left(t+\frac{1}{t}\right)\right] = \sum_{n=-\infty}^{\infty} I_n(x)t^n \qquad t \neq 0 \qquad (7.27)$$

The function $w(x, t) = \exp[\frac{1}{2}x(t + 1/t)]$ is called the **generating function** for the modified Bessel functions.

By recognizing the identity $w(x + y, t) = w(x, t)w(y, t)$, we can use the above generating-function relation to obtain the result

$$\sum_{n=-\infty}^{\infty} I_n(x+y)t^n = \sum_{n=-\infty}^{\infty} I_n(x)t^n \cdot \sum_{n=-\infty}^{\infty} I_n(y)t^n$$

$$= \sum_{n=-\infty}^{\infty} \sum_{k=-\infty}^{\infty} I_k(x)I_{n-k}(y)t^n$$

the last step of which is a Cauchy product. Hence, comparing coefficients of t^n on both sides of the equation, we arrive at the simple **addition formula**

$$I_n(x+y) = \sum_{k=-\infty}^{\infty} I_k(x)I_{n-k}(y) \qquad (7.28)$$

More general forms of the addition formula are given by

$$I_n(R)\begin{Bmatrix} \cos n\psi \\ \sin n\psi \end{Bmatrix} = \sum_{k=-\infty}^{\infty} (-1)^k I_k(a)I_{n+k}(b)\begin{Bmatrix} \cos k\theta \\ \sin k\theta \end{Bmatrix} \qquad (7.29)$$

$$\frac{I_n(R)}{R^n} = 2^n \Gamma(n) \sum_{k=0}^{\infty} (-1)^k (n+k) \frac{I_{n+k}(a)I_{n+k}(b)}{(ab)^n} C_k^n(\cos\theta) \quad (7.30)$$

$$K_n(R)\begin{Bmatrix} \cos n\psi \\ \sin n\psi \end{Bmatrix} = \sum_{m=-\infty}^{\infty} I_m(a)K_{n+m}(b)\begin{Bmatrix} \cos m\theta \\ \sin m\theta \end{Bmatrix} \qquad (7.31)$$

and

$$\frac{K_n(R)}{R^n} = 2^n \Gamma(n) \sum_{m=0}^{\infty} (n+m) \frac{I_{m+k}(a)K_{m+k}(b)}{(ab)^m} C_m^n(\cos\theta) \quad (7.32)$$

where $R = \sqrt{a^2 + b^2 - 2ab \cos \theta}$, $\sin \psi = (a/R) \sin \theta$, and a and b are any positive constants (see Fig. 6.3 in Sec. 6.5). In Eqs. (7.30) and (7.32) we have the restriction $n \neq 0, -1, -2, \ldots,$ and $C_m^n(x)$, $m = 0, 1, 2, \ldots,$ are the *Gegenbauer polynomials* (see Sec. 5.4).

Exercises 7.2

1. Replace x by ix in Bessel's equation

$$x^2 y'' + xy' + (x^2 - p^2)y = 0$$

   and show that it leads to Bessel's modified equation

$$x^2 y'' + xy' - (x^2 + p^2)y = 0$$

In problems 2 through 5, write the general solution of the DE in terms of modified Bessel functions.

2. $x^2 y'' + xy' - (x^2 + 1)y = 0$

3. $xy'' + y' - xy = 0$

4. $4x^2 y'' + 4xy' - (4x^2 + 1)y = 0$

5. $x^2 y'' + xy' - (4x^2 + 1)y = 0$

6. Develop the asymptotic formulas.

   (a) $I_p(x) \sim \dfrac{1}{\Gamma(p+1)} \left(\dfrac{x}{2}\right)^p, \ p \neq -1, -2, -3, \ldots, x \to 0^+$

   (b) $K_0(x) \sim \ln x, \ x \to 0^+$

   (c) $K_p(x) \sim \dfrac{\Gamma(p)}{2} \left(\dfrac{2}{x}\right)^p, \ p > 0, \ x \to 0^+$

7. If y_1 and y_2 are any two solutions of Bessel's modified equation (7.3), show that for some constant C, the wronskian is

$$W(y_1, y_2)(x) = \frac{C}{x}$$

   Hint: See problem 18 in Sec. 6.2.

8. Use the result of problem 7 to deduce that

   (a) $W(I_p, I_{-p})(x) = -\dfrac{2 \sin p\pi}{\pi x}$

   (b) $W(I_p, K_p)(x) = -\dfrac{1}{x}$

9. Show that
   (a) $K_{-p}(x) = K_p(x)$ for all p

   (b) $I_p(x)K_{p+1}(x) + I_{p+1}(x)K_p(x) = \dfrac{1}{x}$

10. Show that

(a) $I_{1/2}(x) = \sqrt{\dfrac{2}{\pi x}} \sinh x$

(b) $I_{-1/2}(x) = \sqrt{\dfrac{2}{\pi x}} \cosh x$

(c) $K_{1/2}(x) = \sqrt{\dfrac{\pi}{2x}} e^{-x}$

(d) $[I_{-1/2}(x)]^2 - [I_{1/2}(x)]^2 = \dfrac{2}{\pi x}$

11. Show that

(a) $\dfrac{d}{dx}[x^p I_p(x)] = x^p I_{p-1}(x)$

(b) $\dfrac{d}{dx}[x^{-p} I_p(x)] = x^{-p} I_{p+1}(x)$

12. Using the results of problem 11, show that

(a) $I_p'(x) + \dfrac{p}{x} I_p(x) = I_{p-1}(x)$

(b) $I_p'(x) - \dfrac{p}{x} I_p(x) = I_{p+1}(x)$

(c) $I_{p-1}(x) + I_{p+1}(x) - 2I_p'(x)$

(d) $I_{p-1}(x) - I_{p+1}(x) = \dfrac{2p}{x} I_p(x)$

13. Verify that

(a) $\dfrac{d}{dx}[x^p K_p(x)] = -x^p K_{p-1}(x)$

(b) $\dfrac{d}{dx}[x^{-p} K_p(x)] = -x^{-p} K_{p+1}(x)$

14. Using the results of problem 13, show that

(a) $K_p'(x) + \dfrac{p}{x} K_p(x) = -K_{p-1}(x)$

(b) $K_p'(x) - \dfrac{p}{x} K_p(x) = -K_{p+1}(x)$

(c) $K_{p-1}(x) + K_{p+1}(x) = -2K_p'(x)$

(d) $K_{p-1}(x) - K_{p+1}(x) = -\dfrac{2p}{x} K_p(x)$

15. Show that

$$K_0(x) = -\lim_{p \to 0} \frac{\partial}{\partial p} I_p(x)$$

and use this result to deduce that for $x > 0$

$$K_0(x) = -I_0(x)\left(\gamma + \ln\frac{x}{2}\right)$$

$$+ \sum_{k=1}^{\infty} \frac{(x/2)^{2k}}{(k!)^2}\left(1 + \frac{1}{2} + \cdots + \frac{1}{k}\right)$$

16. For $n = 1, 2, 3, \ldots$, show that

$$K_n(x) = \frac{(-1)^n}{2} \lim_{p \to n}\left[\frac{\partial}{\partial p} I_{-p}(x) - \frac{\partial}{\partial p} I_p(x)\right]$$

and use this result to deduce that for $x > 0$

$$K_n(x) = (-1)^{n-1} I_n(x) \ln\frac{x}{2}$$

$$+ \frac{1}{2} \sum_{k=0}^{n-1} \frac{(-1)^k (n-k-1)!}{k!}\left(\frac{x}{2}\right)^{2k-n}$$

$$+ \frac{(-1)^n}{2} \sum_{m=0}^{\infty} \frac{(x/2)^{2m+n}}{m!(m+n)!}$$

$$\times [\psi(m+n+1) + \psi(m+1)]$$

17. Use the Cauchy product to show that

$$e^{-A^2/b} I_0\left(\frac{2A\sqrt{E}}{b}\right) = \sum_{n=0}^{\infty} \frac{(-1^n)(A^2/b)^n}{n!} L_n\left(\frac{E}{b}\right)$$

where $L_n(x)$ is the nth Laguerre polynomial.

18. Use the result of problem 7 in Exercises 6.2 to deduce that

$$I_n(x) = \sum_{m=0}^{\infty} \frac{x^m}{m!} J_{n+m}(x) \qquad n = 0, 1, 2, \ldots$$

19. Show that

$$xI_1(x) = 4\sum_{n=1}^{\infty} nI_{2n}(x)$$

Hint: Use problem 12d.

20. By expanding the function $w(x, t) = e^{xt/2} e^{x/(2t)}$ in a double series, deduce directly the generating-function relation

$$\exp\left[\frac{1}{2}x\left(t + \frac{1}{t}\right)\right] = \sum_{n=-\infty}^{\infty} I_n(x)t^n \qquad t \neq 0$$

21. Use problem 20 to show that

(a) $e^{x \cos \theta} = \sum\limits_{n=-\infty}^{\infty} I_n(x) \cos n\theta$

(b) $e^x = I_0(x) + 2\sum\limits_{n=1}^{\infty} I_n(x)$

(c) $e^{-x} = I_0(x) + 2\sum\limits_{n=1}^{\infty} (-1)^n I_n(x)$

22. Use problem 21 to verify the identities

(a) $1 = I_0(x) + 2\sum\limits_{n=1}^{\infty} (-1)^n I_{2n}(x)$

(b) $\cosh x = I_0(x) + 2\sum\limits_{n=1}^{\infty} I_{2n}(x)$

(c) $\sinh x = 2\sum\limits_{n=1}^{\infty} I_{2n-1}(x)$

23. If b in the DE

$$x^2 y'' + (1 - 2a)xy' + [b^2 c^2 x^{2c} + (a^2 - c^2 p^2)]y = 0 \qquad p \geq 0$$

is allowed to be pure imaginary, say, $b = i\beta$ ($\beta > 0$), show that the general solution can be expressed as

$$y = x^a [C_1 I_p(\beta x^c) + C_2 K_p(\beta x^c)]$$

In problems 24 through 28, use the result of problem 23 to express the general solution of each DE in terms of modified Bessel functions.

24. $y'' - y = 0$

25. $y'' - xy = 0$

26. $x^2 y'' + xy' - (4 + 36x^4)y = 0$

27. $xy'' - 3y' - 9x^5 y = 0$

28. $y'' - k^2 x^4 y = 0$

29. Replace a by ia and b by ib in Eq. (6.11) to deduce the addition formula (7.29).

30. Replace a by ia and b by ib in Eq. (6.12) to deduce the addition formula (7.30).

In problems 31 and 32, use addition theorems to derive the given result.

31. $I_0(\sqrt{x^2 + y^2}) = I_0(x)I_0(y) + 2\sum\limits_{k=1}^{\infty} I_{2k}(x)I_{2k}(y)$

32. $I_0(2\sqrt{x-y}) = \sum\limits_{k=-\infty}^{\infty} I_{2k}(2\sqrt{x})J_{2k}(2\sqrt{y})$

7.3   Integral Relations

Integral representations and integrals of modified Bessel functions are very similar to those involving the standard Bessel functions. In fact, results for $I_p(x)$ can often be formally obtained from corresponding results for $J_p(x)$.

7.3.1   Integral representations

The replacement of $t = e^{-i\phi}$ in the generating-function relation (7.27) leads to

$$e^{x \cos \phi} = \sum_{n=-\infty}^{\infty} I_n(x) e^{-in\phi} \qquad (7.33)$$

If we now follow the technique of Sec. 6.3, it can be readily shown in a similar manner that

$$I_n(x) = \frac{1}{\pi} \int_0^{\pi} e^{x \cos \phi} \cos n\phi \, d\phi \qquad n = 0, 1, 2, \ldots \qquad (7.34)$$

Also by starting with the integral representation [recall Eq. (6.32)]

$$J_p(x) = \frac{(x/2)^p}{\sqrt{\pi}\, \Gamma(p + \frac{1}{2})} \int_{-1}^1 (1 - t^2)^{p - 1/2} e^{ixt} \, dt \qquad p > -\frac{1}{2}, x > 0$$

we can replace x with ix and multiply both sides by i^{-p} to obtain

$$I_p(x) = \frac{(x/2)^p}{\sqrt{\pi}\, \Gamma(p + \frac{1}{2})} \int_{-1}^1 (1 - t^2)^{p - 1/2} e^{-xt} \, dt \qquad p > -\frac{1}{2}, x > 0 \quad (7.35)$$

An integral representation for $K_p(x)$ similar to (7.35), but more complicated to derive, is given by

$$K_p(x) = \frac{\sqrt{\pi}\, (x/2)^p}{\Gamma(p + \frac{1}{2})} \int_1^{\infty} (t^2 - 1)^{p - 1/2} e^{-xt} \, dt \qquad p > -\frac{1}{2}, x > 0 \quad (7.36)$$

while for arbitrary p, it can be shown that

$$K_p(x) = \frac{1}{2} \left(\frac{x}{2}\right)^p \int_0^{\infty} e^{-t - x^2/4t} t^{-(p+1)} \, dt \qquad x > 0 \qquad (7.37)$$

The derivations of (7.36) and (7.37) without the use of complex variable theory are quite involved and therefore are omitted.*

---

*The derivations of Eqs. (7.36) and (7.37) are discussed in N. N. Lebedev, *Special Functions and Their Applications*, Dover, New York, 1972, pp. 116–120.

7.3.2 Integrals of modified Bessel functions

Based on Eqs. (7.15) and (7.16), we obtain the indefinite integral relations

$$\int x^p I_{p-1}(x)\, dx = x^p I_p(x) + C \tag{7.38}$$

$$\int x^{-p} I_{p+1}(x)\, dx = x^{-p} I_p(x) + C \tag{7.39}$$

while (7.21) and (7.22) lead to the similar results

$$\int x^p K_{p-1}(x)\, dx = -x^p K_p(x) + C \tag{7.40}$$

$$\int x^{-p} K_{p+1}(x)\, dx = -x^{-p} K_p(x) + C \tag{7.41}$$

If we formally replace b with ib in the integral formula [recall Eq. (6.55)]

$$\int_0^\infty e^{-ax} x^p J_p(bx)\, dx = \frac{(2b)^p \Gamma(p + \tfrac{1}{2})}{\sqrt{\pi}\,(a^2 + b^2)^{p+1/2}} \qquad p > -\tfrac{1}{2},\ a > 0$$

we can deduce the result

$$\int_0^\infty e^{-ax} x^p I_p(bx)\, dx = \frac{(2b)^p \Gamma(p + \tfrac{1}{2})}{\sqrt{\pi}\,(a^2 - b^2)^{p+1/2}} \qquad p > -\tfrac{1}{2},\ a, b > 0 \tag{7.42}$$

Other integral formulas can be derived in a similar manner, some of which appear in the exercises.

Example 1: Derive the integral formula

$$\int_0^\infty x^\mu K_p(ax)\, dx = \frac{2^{\mu-1}}{a^{\mu+1}} \Gamma\!\left(\frac{1 + \mu + p}{2}\right)\Gamma\!\left(\frac{1 + \mu - p}{2}\right) \qquad \mu - p > -1,\ a > 0$$

Solution: In this case we replace $K_p(ax)$ with its integral representation [see Eq. (7.37)]

$$K_p(ax) = \frac{1}{2}\left(\frac{ax}{2}\right)^p \int_0^\infty e^{-t - a^2 x^2/4t}\, t^{-(p+1)}\, dt$$

This action leads to

$$I = \int_0^\infty x^\mu K_p(ax)\,dx$$

$$= \frac{1}{2}\left(\frac{a}{2}\right)^p \int_0^\infty x^\mu \int_0^\infty x^p e^{-t-a^2x^2/4t} t^{-(p+1)}\,dt\,dx$$

$$= \frac{1}{2}\left(\frac{a}{2}\right)^p \int_0^\infty e^{-t} t^{-(p+1)} \int_0^\infty x^{\mu+p} e^{-a^2x^2/4t}\,dx\,dt$$

where we have reversed the order of integration in the last step. Next, letting $a^2x^2/4t = y$ in the innermost integral and using properties of gamma functions, we obtain

$$I = \frac{1}{2}\left(\frac{a}{2}\right)^p \int_0^\infty e^{-t} t^{-(p+1)} \frac{1}{2}\left(\frac{4t}{a^2}\right)^{(1+\mu+p)/2} \Gamma\left(\frac{1+\mu+p}{2}\right)\,dt$$

$$= \frac{2^{\mu-1}}{a^{\mu+1}} \Gamma\left(\frac{1+\mu+p}{2}\right) \int_0^\infty e^{-t} t^{(1+\mu-p)/2-1}\,dt$$

but this last integral is simply $\Gamma[(1+\mu-p)/2]$, and hence we have our intended result.

Exercises 7.3

In problems 1 to 10, derive the given integral representation.

1. $I_0(x) = \dfrac{1}{\pi}\displaystyle\int_0^\pi e^{\pm x \cos\theta}\,d\theta = \dfrac{1}{2\pi}\displaystyle\int_{-\pi}^\pi e^{\pm x \cos\theta}\,d\theta$

2. $I_p(x) = \dfrac{(x/2)^p}{\sqrt{\pi}\,\Gamma(p+\frac{1}{2})}\displaystyle\int_{-1}^1 (1-t^2)^{p-1/2}\cosh xt\,dt,\ p > -\frac{1}{2},\ x > 0$

3. $I_p(x) = \dfrac{(x/2)^p}{\sqrt{\pi}\,\Gamma(p+\frac{1}{2})}\displaystyle\int_0^\pi e^{\pm x \cos\theta}\sin^{2p}\theta\,d\theta,\ p > -\frac{1}{2},\ x > 0$

4. $I_p(x) = \dfrac{(x/2)^p}{\sqrt{\pi}\,\Gamma(p+\frac{1}{2})}\displaystyle\int_{-\pi}^\pi \cosh(x\cos\theta)\sin^{2p}\theta\,d\theta,\ p > -\frac{1}{2},\ x > 0$

5. $K_p(x) = \sqrt{\dfrac{\pi}{2x}}\,\dfrac{e^{-x}}{\Gamma(p+\frac{1}{2})}\displaystyle\int_0^\infty e^{-s}s^{p-1/2}\left(1+\dfrac{s}{2x}\right)^{p-1/2}\,ds,\ p > -\frac{1}{2},\ x > 0$

   Hint: Use Eq. (7.36).

6. $K_p(x) = \frac{1}{2}\displaystyle\int_0^\infty e^{-x(t+1/t)/2} t^{-(p+1)}\,dt,\ x > 0$

7. $K_p(x) = \frac{1}{2} \displaystyle\int_0^\infty e^{-x(t+1/t)/2} t^{p-1} \, dt, \ x > 0$

8. $K_p(x) = \displaystyle\int_0^\infty e^{-x \cosh \theta} \cosh p\theta \, d\theta, \ x > 0$

9. $K_p(x) = \dfrac{\Gamma(p + \frac{1}{2})}{\sqrt{\pi}} (2x)^p \displaystyle\int_0^\infty \dfrac{\cos t}{(t^2 + x^2)^{p+1/2}} \, dt, \ p > -\frac{1}{2}, \ x > 0$

10. $K_p(x) = \dfrac{\sqrt{\pi} \, (x/2)^p}{\Gamma(p + \frac{1}{2})} \displaystyle\int_0^{\cdots} e^{-x \cosh \theta} \sin^{2p} \theta \, d\theta, \ p > -\frac{1}{2}, \ x > 0$

In problems 11 to 14, perform the integration, leaving at most the residual integral $\int I_0(x) \, dx$.

11. $\displaystyle\int x I_0(x) \, dx$

13. $\displaystyle\int x I_1(x) \, dx$

12. $\displaystyle\int x^2 I_0(x)$

14. $\displaystyle\int x^2 I_1(x) \, dx$

15. Show that if $p > -1$,

$$(\lambda^2 - \mu^2) \int x I_p(\lambda z) I_p(\mu x) \, dx = -x [\mu I_p(\lambda x) I_p'(\mu x) - \lambda I_p(\mu x) I_p'(\lambda x)]$$

16. From the result of problem 15, deduce that

$$\int x [I_p(\lambda x)]^2 \, dx = \frac{x^2}{2} \left\{ [I_p'(\lambda x)]^2 - \left(1 + \frac{p^2}{\lambda^2 x^2}\right) \right\}$$

In problems 17 to 20, verify the given integral relation.

17. $\displaystyle\int_0^\infty \dfrac{x^{p+1} J_p(bx)}{(x^2 + a^2)^{m+1}} \, dx = \dfrac{a^{p-m} b^m}{2^m \Gamma(m+1)} K_{p-m}(ab), \ a, \ b > 0,$

$$-1 < p < 2m + \tfrac{3}{2}$$

18. $\displaystyle\int_0^\infty \dfrac{x J_0(bx)}{\sqrt{x^2 + a^2}} \, dx = \dfrac{1}{b} e^{-ab}, \ a \geq 0, \ b > 0$

    Hint: Use problem 17.

19. $\displaystyle\int_0^\infty \dfrac{K_m(a\sqrt{x^2 + y^2})}{(x^2 + y^2)^{m/2}} J_p(bx) x^{p+1} \, dx = \dfrac{b^p}{a^m} \left(\dfrac{\sqrt{a^2 + b^2}}{y}\right)^{m-p-1}$

$$\times K_{m-p-1}(y\sqrt{a^2 + b^2}), \ a, b, y > 0, \ p > -1$$

20. $\displaystyle\int_0^\infty \dfrac{\exp(-a\sqrt{x^2 + y^2})}{\sqrt{x^2 + y^2}} J_0(bx) x \, dx = \dfrac{\exp(-y\sqrt{a^2 + b^2})}{\sqrt{a^2 + b^2}}, \ a, y > 0$

    Hint: Use problem 19.

7.4 Spherical Bessel Functions

Spherical Bessel functions are commonly associated with solving the *Helmholtz equation* in spherical coordinates.* In the solution of this DE we are often led to an ordinary DE in the radial variable which has the form

$$x^2y'' + 2xy' + [k^2x^2 - n(n+1)]y = 0 \qquad n = 0, 1, 2, \ldots \quad (7.43)$$

where the constant k enters directly from the Helmholtz equation and the integer n is a separation constant which often has the physical interpretation of angular momentum. We recognize (7.43) as a special case of (6.98) for which

$$a = -\tfrac{1}{2} \qquad b = k \qquad c = 1 \qquad \text{and} \qquad p = n + \tfrac{1}{2}$$

Hence the general solution of (7.43) can be expressed as

$$y = x^{-1/2}[C_1 J_{n+1/2}(kx) + C_2 Y_{n+1/2}(kx)] \qquad (7.44)$$

where C_1 and C_2 represent arbitrary constants.

Since n takes on integral values, all Bessel functions in (7.44) are of *half-integral order*. We previously found that the particular half-integral order Bessel function $J_{1/2}(x)$ is an elementary function given by [recall Eq. (6.12a)]

$$J_{1/2}(x) = \sqrt{\frac{2}{\pi x}} \sin x$$

Moreover, it turns out that *all* half-integral order Bessel functions reduce to elementary functions. To combine the multiplicative factor $x^{-1/2}$ appearing in front of (7.44) with the half-integral order Bessel functions, it has become customary to introduce functions defined by

$$j_n(x) = \sqrt{\frac{\pi}{2x}} J_{n+1/2}(x) \qquad n = 0, 1, 2, \ldots \quad (7.45)$$

and

$$y_n(x) = \sqrt{\frac{\pi}{2x}} Y_{n+1/2}(x) \qquad n = 0, 1, 2, \ldots \quad (7.46)$$

called, respectively, **spherical Bessel functions of the first and second kinds** of order n.

---

* See, for example, Sec. 8.4.3.

By using the series representation

$$J_{n+1/2}(x) = \sum_{m=0}^{\infty} \frac{(-1)^m (x/2)^{2m+n+1/2}}{m!\,\Gamma(m+n+3/2)}$$ (7.47)

together with the Legendre duplication formula

$$\sqrt{\pi}\,\Gamma(2x) = 2^{2x-1}\Gamma(x)\Gamma(x+1/2)$$

it can be shown that (see problem 1 in Exercises 7.4)

$$j_n(x) = 2^n x^n \sum_{m=0}^{\infty} \frac{(-1)^m (m+n)!\,x^{2m}}{m!\,(2m+2n+1)!} \qquad n = 0, 1, 2, \ldots$$ (7.48)

For instance, by setting $n = 0$ in (7.48), we find

$$j_0(x) = \sum_{m=0}^{\infty} \frac{(-1)^m x^{2m}}{(2m+1)!}$$

and thus deduce that [see also Eq. (7.45)]

$$j_0(x) = \frac{\sin x}{x}$$ (7.49)

Through repeated application of the recurrence formula

$$J_{p-1}(x) + J_{p+1}(x) = \frac{2p}{x} J_p(x)$$

and Eq. (6.12), it can be shown that

$$j_1(x) = \frac{\sin x}{x^2} - \frac{\cos x}{x}$$ (7.50)

$$j_2(x) = \left(\frac{3}{x^3} - \frac{1}{x}\right) \sin x - \frac{3}{x^2} \cos x$$ (7.51)

and so on. Similarly, recognizing that $Y_{1/2}(x) = -J_{-1/2}(x)$ and using the recurrence formula for $Y_p(x)$, we find

$$y_0(x) = -\frac{\cos x}{x}$$ (7.52)

$$y_1(x) = -\frac{\cos x}{x^2} - \frac{\sin x}{x}$$ (7.53)

$$y_2(x) = -\left(\frac{3}{x^3} - \frac{1}{x}\right) \cos x - \frac{3}{x^2} \sin x$$ (7.54)

The details of deriving Eqs. (7.50) to (7.54) are left to the exercises.

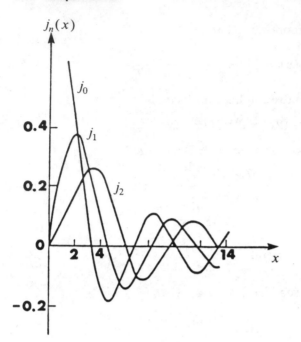

Figure 7.3   Graph of $j_n(x)$, $n = 0, 1, 2$.

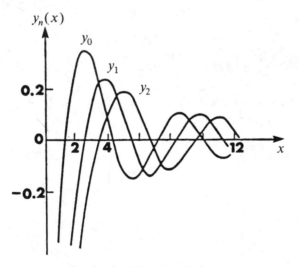

Figure 7.4   Graph of $y_n(x)$, $n = 0, 1, 2$.

The graphs of some of the spherical Bessel functions are shown in Figs. 7.3 and 7.4. Observe that the general behavior of these functions is that of the standard Bessel functions.

7.4.1   Recurrence formulas

Recurrence relations for the spherical Bessel functions, analogous to those of the standard Bessel functions, can be derived directly from the corresponding relations for $J_p(x)$ and $Y_p(x)$. This leads to (see the exercises)

$$\frac{d}{dx}[x^{n+1}j_n(x)] = x^{n+1}j_{n-1}(x) \tag{7.55}$$

$$\frac{d}{dx}[x^{-n}j_n(x)] = -x^{-n}j_{n+1}(x) \tag{7.56}$$

$$j_n'(x) = j_{n-1}(x) - \frac{n+1}{x}j_n(x) \tag{7.57}$$

$$j_n'(x) = \frac{n}{x}j_n(x) - j_{n+1}(x) \tag{7.58}$$

$$(2n+1)j_n'(x) = nj_{n-1}(x) - (n+1)j_{n+1}(x) \tag{7.59}$$

$$j_{n-1}(x) + j_{n+1}(x) = \frac{2n+1}{x}j_n(x) \tag{7.60}$$

The function $y_n(x)$ satisfies the same identities, so we do not list them separately.

7.4.2   Modified spherical Bessel functions

The modified Bessel functions of half-integral order also reduce to elementary functions. For example, when $p = {}^1\!/_2$, we obtain

$$I_{1/2}(x) = \sqrt{\frac{2}{\pi x}}\sinh x$$

$$K_{1/2}(x) = \sqrt{\frac{\pi}{2x}}e^{-x}$$

whereas in general it can be shown that

$$I_{n+1/2}(x) = \frac{1}{\sqrt{2\pi x}}\left[e^x \sum_{k=0}^{n} \frac{(-1)^k(n+k)!}{k!(n-k)!(2x)^k} \right.$$

$$\left. + (-1)^{n-1}e^{-x}\sum_{k=0}^{n}\frac{(n+k)!}{k!(n-k)!(2x)^k}\right] \tag{7.61}$$

and
$$K_{n+1/2}(x) = \sqrt{\frac{\pi}{2x}}e^{-x}\sum_{k=0}^{n}\frac{(n+k)!}{k!(n-k)!(2x)^k} \tag{7.62}$$

the details of which we leave to the exercises.

Based on the above results, we are motivated to introduce the **modified spherical Bessel functions of the first and second kinds,** respectively, by

$$i_n(x) = \sqrt{\frac{\pi}{2x}}I_{n+1/2}(x) \qquad n = 0, 1, 2, \ldots \tag{7.63}$$

and
$$k_n(x) = \sqrt{\frac{2}{\pi x}}K_{n+1/2}(x) \qquad n = 0, 1, 2, \ldots \tag{7.64}$$

In particular, when $n = 0$, we obtain

$$i_0(x) = \frac{\sinh x}{x} \tag{7.65}$$

$$k_0(x) = \frac{e^{-x}}{x} \tag{7.66}$$

while higher-order functions lead to other combinations of hyperbolic and exponential functions. Some of the properties associated with these modified spherical Bessel functions are taken up in the exercises.

Exercises 7.4

1. From the series representation (7.47), show that
$$j_n(x) = 2^n x^n \sum_{m=0}^{\infty}\frac{(-1)^m(m+n)!x^{2m}}{m!(2m+2n+1)!} \qquad n = 0, 1, 2, \ldots$$

2. Show that (for $n = 0, 1, 2, \ldots$)
$$j_n(x) = \frac{(x/2)^n}{2n!}\int_0^{\pi}\cos(x\cos\theta)\sin^{2n+1}\theta\,d\theta \qquad x > 0$$

In problems 3 to 8, verify the given recurrence relation.

3. $\frac{d}{dx}[x^{n+1}j_n(x)] = x^{n+1}j_{n-1}(x)$

4. $\frac{d}{dx}[x^{-n}j_n(x)] = -x^{-n}j_{n+1}(x)$

5. $j_n'(x) = j_{n-1}(x) - \frac{n+1}{x}j_n(x)$

6. $j_n'(x) = \dfrac{n}{x} j_n(x) - j_{n+1}(x)$

7. $(2n+1) j_n'(x) = n j_{n-1}(x) - (n+1) j_{n+1}(x)$

8. $j_{n-1}(x) + j_{n+1}(x) = \dfrac{2n+1}{x} j_n(x)$

9. By use of any of the recurrence relations, show that

(a) $j_1(x) = \dfrac{\sin x}{x^2} - \dfrac{\cos x}{x}$

(b) $j_2(x) = \left(\dfrac{3}{x^3} - \dfrac{1}{x}\right) \sin x - \dfrac{3}{x^2} \cos x$

10. Show that

(a) $y_0(x) = -\dfrac{\cos x}{x}$

(b) $y_1(x) = -\dfrac{\cos x}{x^2} - \dfrac{\sin x}{x}$

(c) $y_2(x) = -\left(\dfrac{3}{x^3} - \dfrac{1}{x}\right) \cos x - \dfrac{3}{x^2} \sin x$

11. Show that

(a) $j_n(x) y_{n-1}(x) - j_{n-1}(x) y_n(x) = \dfrac{1}{x^2}$

(b) $j_{n+1}(x) y_{n-1}(x) - j_{n-1}(x) y_{n+1}(x) = \dfrac{2n+1}{x^3}$

12. Develop the generating-function relations

(a) $\dfrac{1}{x} \cos \sqrt{x^2 - 2xt} = \displaystyle\sum_{n=0}^{\infty} \dfrac{t^n}{n!} j_{n-1}(x)$

(b) $\dfrac{1}{x} \sin \sqrt{x^2 + 2xt} = \displaystyle\sum_{n=0}^{\infty} \dfrac{(-1)^n t^n}{n!} y_{n-1}(x)$

13. Show that

(a) $i_1(x) = -\dfrac{\sinh x}{x^2} + \dfrac{\cosh x}{x}$

(b) $i_2(x) = \left(\dfrac{3}{x^3} + \dfrac{1}{x}\right) \sinh x - \dfrac{3}{x^2} \cosh x$

(c) $k_1(x) = \left(\dfrac{1}{x^2} + \dfrac{1}{x}\right) e^{-x}$

14. Verify Eq. (7.61).

15. Verify Eq. (7.62).

16. Verify the recurrence relations

$$(a) \quad i_{n-1}(x) - i_{n+1}(x) = \frac{2n+1}{x} i_n(x)$$

$$(b) \quad n i_{n-1}(x) + (n+1)i_{n+1}(x) = (2n+1)i_n'(x)$$

17. Show that the wronskian satisfies the relations

$$(a) \quad W(j_n, y_n)(x) = \frac{1}{x^2}$$

$$(b) \quad W(i_n, k_n)(x) = -\frac{1}{x^2}$$

18. Show that
    (a) $e^{x \cos \theta} = \sum_{n=0}^{\infty} (2n+1)i_n(x)P_n(\cos \theta)$, where $P_n(x)$ is the nth
    Legendre polynomial

$$(b) \quad J_0(x \sin \theta) = \sum_{n=0}^{\infty} (4n+1) \frac{(2n)!}{2^{2n}(n!)^2} j_{2n}(x) P_{2n}(\cos \theta)$$

7.5 Other Bessel Functions

In this section we briefly introduce some additional functions belonging to the general Bessel family, some occurring only in the exercises. For a more detailed discussion of these functions, consult G. N. Watson, A *Treatise on the Theory of Bessel Functions*, 2d ed., Cambridge University Press, London, 1952.

7.5.1 Hankel functions

The usefulness of the Euler formulas

$$e^{\pm ix} = \cos x \pm i \sin x$$

in a variety of applications suggests that linear combinations of $J_p(x)$ and $Y_p(x)$ of the form

$$H_p^{(1)}(x) = J_p(x) + iY_p(x) \tag{7.67}$$

$$H_p^{(2)}(x) = J_p(x) - iY_p(x) \tag{7.68}$$

may also be useful. We call these functions **Bessel functions of the third kind, or Hankel functions.** These functions are indeed useful in various applications, but perhaps their most useful property is that they enable us to readily develop asymptotic formulas for $J_p(x)$

and $Y_p(x)$ as $x \to \infty$ (see Sec. 7.6). Because $J_p(x)$ and $Y_p(x)$ each satisfy the same set of recurrence relations, it follows that $H_p^{(1)}(x)$ and $H_p^{(2)}(x)$ also satisfy these same recurrence relations.

It is common to likewise introduce the **spherical Hankel functions**

$$h_n^{(1)}(x) = j_n(x) + iy_n(x) \qquad (7.69)$$

$$h_n^{(2)}(x) - j_n(x) - iy_n(x) \qquad (7.70)$$

These functions, of course, satisfy the same recurrence relations as the spherical Bessel functions $j_n(x)$ and $y_n(x)$.

An important relation between the modified Bessel functions of the second kind and the Hankel functions can be derived through relations between $I_p(x)$ and $K_p(x)$ and the standard Bessel functions. We start with the relation

$$H_p^{(1)}(ix) = J_p(ix) + iY_p(ix)$$

$$= J_p(ix) + \frac{i}{\sin p\pi}[(\cos p\pi)J_p(ix) - J_{-p}(ix)]$$

$$= \frac{J_{-p}(ix) - e^{-ip\pi}J_p(ix)}{i \sin p\pi}$$

$$= \frac{e^{-ip\pi/2}I_{-p}(x) - e^{-ip\pi/2}I_p(x)}{i \sin p\pi}$$

$$= \frac{2}{i\pi}e^{-ip\pi/2}K_p(x)$$

from which we deduce

$$K_p(x) = \tfrac{1}{2}\pi i^{p+1}H_p^{(1)}(ix) \qquad (7.71)$$

Similarly, it can be shown that

$$K_p(x) = -\tfrac{1}{2}\pi i^{1-p}H_p^{(2)}(-ix) \qquad (7.72)$$

7.5.2 Struve functions

Struve functions are important in certain problems in optics and in loudspeaker design. Let us recall the integral representation [Eq. (6.32)]

$$J_p(x) = \frac{(x/2)^p}{\sqrt{\pi}\,\Gamma(p + \tfrac{1}{2})}\int_{-1}^{1}(1 - t^2)^{p-1/2}e^{ixt}\,dt$$

$$p > -\tfrac{1}{2}, \qquad x > 0 \quad (7.73)$$

If the above integral is multiplied by 2 and the integration performed over only the interval $0 \le t \le 1$, we are led to the result

$$\frac{2(x/2)^p}{\sqrt{\pi}\,\Gamma(p + \tfrac{1}{2})} \int_0^1 (1 - t^2)^{p-1/2} e^{ixt}\, dt$$

$$= \frac{2(x/2)^p}{\sqrt{\pi}\,\Gamma(p + \tfrac{1}{2})} \int_0^1 (1 - t^2)^{p-1/2} \cos xt\, dt$$

$$+ i\, \frac{2(x/2)^p}{\sqrt{\pi}\,\Gamma(p + \tfrac{1}{2})} \int_0^1 (1 - t^2)^{p-1/2} \sin xt\, dt \quad (7.74)$$

Using properties of even and odd functions, we see by comparison with (7.73) that the real part of (7.74) is once again $J_p(x)$. The imaginary part, however, can be used to define a new function

$$\mathbf{H}_p(x) = \frac{2(x/2)^p}{\sqrt{\pi}\,\Gamma(p + \tfrac{1}{2})} \int_0^1 (1 - t^2)^{p-1/2} \sin xt\, dt$$

$$p > -\tfrac{1}{2}, \qquad x > 0 \quad (7.75)$$

called the **Struve function** of order p. By a change of variable, it can also be shown that

$$\mathbf{H}_p(x) = \frac{2(x/2)^p}{\sqrt{\pi}\,\Gamma(p + \tfrac{1}{2})} \int_0^{\pi/2} \sin(x \cos\theta) \sin^{2p}\theta\, d\theta \quad (7.76)$$

The series representation of $\mathbf{H}_p(x)$, obtained from either (7.75) or (7.76), is readily found to be

$$\mathbf{H}_p(x) = \sum_{k=0}^{\infty} \frac{(-1)^k (x/2)^{2k+p+1}}{\Gamma(k + \tfrac{3}{2})\Gamma(k + p + \tfrac{3}{2})} \quad (7.77)$$

which is valid for all p. It can be shown that for $p > \tfrac{1}{2}$, the Struve function $\mathbf{H}_p(x)$ has no zeros in the range $x > 0$. Nonetheless, this function has an oscillatory behavior because it is an alternating series.

The related function

$$\mathbf{L}_p(x) = i^{-(p+1)} \mathbf{H}_p(ix) = \sum_{k=0}^{\infty} \frac{(x/2)^{2k+p+1}}{\Gamma(k + \tfrac{3}{2})\Gamma(k + p + \tfrac{3}{2})} \quad (7.78)$$

is called the **modified Struve function** of order p. Because its series is *not* an alternating series, this function is positive and hence has no zeros in the range $x > 0$. When p is half-integral, the function $\mathbf{H}_p(x)$ may be expressed in terms of circular functions and $\mathbf{L}_p(x)$ may be expressed in terms of hyperbolic functions.

7.5.3   Kelvin's functions

In obtaining the current density in a wire carrying an alternating current, it is sometimes necessary to solve DEs of the form

$$xy'' + y' - i\kappa^2 xy = 0 \tag{7.79}$$

in which κ is real. We recognize this equation as similar in form to that appearing in problem 23 in Exercises 7.2. Thus formally we can write the general solution as

$$y = C_1 I_0(\kappa x i^{1/2}) + C_2 K_0(\kappa x i^{1/2}) \tag{7.80}$$

The particular modified Bessel functions appearing in (7.80) with complex arguments are used to define four real functions known as *Kelvin's functions* (named in honor of Lord Kelvin). For example, directly from the series definition of $I_0(x)$, we find that

$$I_0(x i^{1/2}) = \sum_{m=0}^{\infty} \frac{i^m (x/2)^{2m}}{(m!)^2}$$

and by splitting this series into real and imaginary parts, we obtain

$$I_0(x i^{1/2}) = \text{ber}\,(x) + i\,\text{bei}\,(x) \tag{7.81}$$

where

$$\text{ber}\,(x) = \sum_{k=0}^{\infty} \frac{(-1)^k (x/2)^{4k}}{[(2k)!]^2} \tag{7.82}$$

and

$$\text{bei}\,(x) = \sum_{k=0}^{\infty} \frac{(-1)^k (x/2)^{4k+2}}{[(2k+1)!]^2} \tag{7.83}$$

The notation for these functions is based on Kelvin's notation; i.e., ber is "Bessel real" while bei is "Bessel imaginary."* Using the series (7.13a) for $K_0(x)$, we can define two additional functions for which

$$K_0(x i^{1/2}) = \text{ker}\,(x) + i\,\text{kei}\,(x) \tag{7.84}$$

The functions ber (x) and bei (x) oscillate about the x axis with increasing amplitude and therefore have real zeros on the interval $x > 0$. In fact, because they are even functions, their graphs are symmetric about the vertical axis.

---

*Although we used the modified Bessel function $I_0(x i^{1/2})$ to define ber (x) and bei (x), they can also be obtained from $J_0(x i^{3/2})$.

Generalizations of these functions to functions of order p are also possible by using Bessel functions $I_p(xi^{1/2})$ and $K_p(xi^{1/2})$, although we do not pursue it here.

7.5.4 Airy functions

Solutions of the second-order DE

$$y'' - xy = 0 \qquad (7.85)$$

are important in a variety of applications in mathematical physics, including the diffraction of radio waves around the earth's surface and the construction of asymptotic results near a two-dimensional caustic in the study of optical fields. By comparing (7.85) with the general equation form given in problem 23 of Exercises 7.2, we find that $a = \frac{1}{2}$, $\beta = \frac{2}{3}$, $c = \frac{3}{2}$, and $p = \frac{1}{3}$. Hence the general solution of (7.85) can be expressed in the form

$$y = \sqrt{x}\,[C_1 I_{1/3}(\tfrac{2}{3}x^{3/2}) + C_2 I_{-1/3}(\tfrac{2}{3}x^{3/2})] \qquad (7.86)$$

It is customary, however, to introduce the solutions

$$\mathrm{Ai}\,(x) = \tfrac{1}{3}\sqrt{x}\,[I_{-1/3}(\tfrac{2}{3}x^{3/2}) - I_{1/3}(\tfrac{2}{3}x^{3/2})]$$

$$= \frac{1}{\pi}\sqrt{\frac{x}{3}}\,K_{1/3}(\tfrac{2}{3}x^{3/2}) \qquad (7.87)$$

and

$$\mathrm{Bi}\,(x) = \sqrt{\frac{x}{3}}\,[I_{-1/3}(\tfrac{2}{3}x^{3/2}) + I_{1/3}(\tfrac{2}{3}x^{3/2})] \qquad (7.88)$$

called, respectively, **Airy functions of the first and second kinds.** In terms of these functions, the general solution of (7.85) can be expressed in the equivalent form

$$y = C_1\,\mathrm{Ai}\,(x) + C_2\,\mathrm{Bi}\,(x) \qquad (7.89)$$

Some basic properties of the Airy functions are taken up in the exercises.

Exercises 7.5

1. Show that both $H_p^{(1)}(x)$ and $H_p^{(2)}(x)$ satisfy the identities [where $Z_p(x)$ represents either function]

   (a) $\dfrac{d}{dx}\,[x^p Z_p(x)] = x^p Z_{p-1}(x)$

(b) $\dfrac{d}{dx}[x^{-p}Z_p(x)] = -x^{-p}Z_{p+1}(x)$

(c) $Z_{p-1}(x) + Z_{p+1}(x) = \dfrac{2p}{x}Z_p(x)$

(d) $Z_{p-1}(x) - Z_{p+1}(x) = 2Z_p'(x)$

2. Show that

$$K_p(x) = -\tfrac{1}{2}\pi i^{1-p}H_p^{(2)}(-ix)$$

3. By substituting $t = \cos\theta$ into (7.75), show that

$$\mathbf{H}_p(x) - \dfrac{2(x/2)^p}{\sqrt{\pi}\,\Gamma(p+\tfrac{1}{2})}\int_0^{\pi/2}\sin(x\cos\theta)\sin^{2p}\theta\,d\theta$$

4. Use Eq. (7.75) to develop the series representation

$$\mathbf{H}_p(x) = \sum_{k=0}^{\infty}\dfrac{(-1)^k(x/2)^{2k+p+1}}{\Gamma(k+\tfrac{3}{2})\Gamma(k+p+\tfrac{3}{2})}$$

5. Show that

$$\int_{-\pi/2}^{\pi/2}e^{-ix\cos\theta}\cos d\theta = 2 - \pi[\mathbf{H}_1(x) + iJ_1(x)]$$

6. Verify that $\mathbf{H}_p(x)$ is a particular solution of the DE

$$x^2y'' + xy' + (x^2 - p^2)y = \dfrac{(x/2)^{p+1}}{\sqrt{\pi}\,\Gamma(p+\tfrac{1}{2})}\qquad p > -\tfrac{1}{2}$$

7. Show that

(a) $\mathbf{H}_{p-1}(x) + \mathbf{H}_{p+1}(x) = \dfrac{2p}{x}\mathbf{H}_p(x) + \dfrac{(x/2)^p}{\sqrt{\pi}\,\Gamma(p+\tfrac{3}{2})}$

(b) $\mathbf{H}_{p-1}(x) - \mathbf{H}_{p+1}(x) = 2\mathbf{H}_p'(x) - \dfrac{(x/2)^p}{\sqrt{\pi}\,\Gamma(p+\tfrac{3}{2})}$

(c) $\mathbf{H}_0(x) = \dfrac{2}{\pi} - \mathbf{H}_1(x)$

8. Show that

(a) $\mathbf{H}_{-1/2}(x) = \sqrt{\dfrac{2}{\pi x}}\sin x$

(b) $\mathbf{H}_{1/2}(x) = \sqrt{\dfrac{2}{\pi x}}(1 - \cos x)$

(c) $\mathbf{L}_{-1/2}(x) = \sqrt{\dfrac{2}{\pi x}}\sinh x$

9. Show that

(a) $H_{-(n+1/2)}(x) = (-1)^n J_{n+1/2}(x)$, $n = 0, 1, 2, \ldots$

(b) $L_{-(n+1/2)}(x) = I_{n+1/2}(x)$, $n = 0, 1, 2, \ldots$

10. Show that

$$L_p(x) = \frac{2(x/2)^p}{\sqrt{\pi}\,\Gamma(p + 1/2)} \int_0^{\pi/2} \sinh(x \cos \theta) \sin^{2p} \theta \, d\theta$$

11. Show that

(a) $L_{p-1}(x) - L_{p+1}(x) = \dfrac{2p}{x} L_p(x) + \dfrac{(x/2)^p}{\sqrt{\pi}\,\Gamma(p + 3/2)}$

(b) $L_{p-1}(x) + L_{p+1}(x) = 2L_p'(x) - \dfrac{(x/2)^p}{\sqrt{\pi}\,\Gamma(p + 3/2)}$

(c) $L_0'(x) = \dfrac{2}{\pi} + L_1(x)$

12. Verify that

$$y = C_1[\text{ber}(x) + \text{ker}(x)] + C_2[\text{bei}(x) + \text{kei}(x)]$$

is a solution of Eq. (7.79).

13. Show that

$$\text{ber}^2(x) + \text{bei}^2(x) = \sum_{k=0}^{\infty} \frac{(x/2)^{4k}}{(k!)^2(2k+1)!}$$

14. *Kelvin's functions of order p* are defined by

$$I_p(xi^{1/2}) = \text{ber}_p(x) + i\,\text{bei}_p(x)$$

Show that

(a) $\text{ber}_1(x) = -\dfrac{x}{2\sqrt{2}}\left[1 + \dfrac{(x/2)^2}{1!2!} - \dfrac{(x/2)^4}{2!3!} - \dfrac{(x/2)^6}{3!4!} + \cdots\right]$

(b) $\text{bei}_1(x) = \dfrac{x}{2\sqrt{2}}\left[1 - \dfrac{(x/2)^2}{1!2!} - \dfrac{(x/2)^4}{2!3!} + \dfrac{(x/2)^6}{3!4!} + \cdots\right]$

15. The *integral Bessel function* of order p is defined by

$$\text{Ji}_p(x) = \int_\infty^x \frac{J_p(t)}{t}\, dt$$

Verify that

(a) $p\,\text{Ji}_p(x) = p\displaystyle\int_0^x \frac{J_p(t)}{t}\, dt - 1$

(b) $p\,\text{Ji}_p(x) = \displaystyle\int_0^x J_{p-1}(t)\, dt - J_p(x) - 1$

16. Referring to problem 15, show that

(a) $\mathrm{Ji_0}(x) = \ln\dfrac{x}{2} + \gamma + \displaystyle\sum_{k=1}^{\infty} \dfrac{(-1)^k (x/2)^{2k}}{2k\,(k!)^2}$

(b) $\mathrm{Ji_n}(x) = -\dfrac{1}{n} + \displaystyle\sum_{k=0}^{\infty} \dfrac{(-1)^k (x/2)^{2k+n}}{(2k+n)k!(n+k)!}$, $\quad n = 1, 2, 3, \ldots$

17. The *Anger function* is defined by

$$\mathbf{J}_p(x) = \frac{1}{\pi}\int_0^\pi \cos\,(p\theta - x\,\sin\,\theta)\,d\theta \qquad x \geq 0$$

Show that
(a) $\mathbf{J}_{p-1}(x) - \mathbf{J}_{p+1}(x) = 2\mathbf{J}_p'(x)$

(b) $\mathbf{J}_{p-1}(x) + \mathbf{J}_{p+1}(x) = \dfrac{2p}{x}\,\mathbf{J}_p(x) - \dfrac{2}{\pi x}\,\sin p\pi$

18. The *Weber function* is defined by

$$\mathbf{E}_p(x) = \frac{1}{p}\int_0^\pi \sin\,(p\theta - x\,\sin\,\theta)\,d\theta \qquad x \geq 0$$

Show that
(a) $\mathbf{E}_{p-1}(x) - \mathbf{E}_{p+1}(x) = 2\mathbf{E}_p'(x)$

(b) $\mathbf{E}_{p-1}(x) + \mathbf{E}_{p+1}(x) = \dfrac{2p}{x}\,\mathbf{E}_p(x) - \dfrac{2}{\pi x}\,(1 - \cos p\pi)$

19. Show that

(a) $\mathrm{Ai}'(x) = \dfrac{1}{\pi\sqrt{3}} K_{2/3}(\tfrac{2}{3}x^{3/2})$

(b) $\mathrm{Bi}'(x) = \dfrac{x}{\sqrt{3}}\,[I_{-2/3}(\tfrac{2}{3}x^{3/2}) + I_{2/3}(\tfrac{2}{3}x^{3/2})]$

20. Show that the Airy functions satisfy the initial conditions
    (a) $\mathrm{Ai}\,(0) = 3^{-2/3}/\Gamma(\tfrac{2}{3})$
    (b) $\mathrm{Bi}\,(0) = 3^{-1/6}/\Gamma(\tfrac{2}{3})$
    (c) $\mathrm{Ai}'\,(0) = 3^{-11/3}/\Gamma(\tfrac{4}{3})$
    (d) $\mathrm{Bi}'\,(0) = 3^{-5/6}/\Gamma(\tfrac{4}{3})$

21. Develop the wronskian relation

$$W[\mathrm{Ai}\,(x),\,\mathrm{Bi}\,(x)] = \frac{1}{\pi}$$

Hint: Use problem 40 in Exercises 2.2.

22. Show that a general solution of $y'' + xy = 0$ is given by
$$y = C_1\,\mathrm{Ai}\,(-x) + C_2\,\mathrm{Bi}\,(-x)$$

where

$$\text{Ai} (-x) = \tfrac{1}{3}\sqrt{x}\ [J_{-1/3}(\tfrac{2}{3}x^{3/2}) + J_{1/3}(\tfrac{2}{3}x^{3/2})]$$

$$\text{Bi} (-x) = \sqrt{\frac{x}{3}}\ [J_{-1/3}(\tfrac{2}{3}x^{3/2}) - J_{1/3}(\tfrac{2}{3}x^{3/2})]$$

23. Referring to problem 22, show that

(a) $\text{Ai}' (-x) = -\dfrac{x}{\sqrt{3}}\ [J_{-2/3}(\tfrac{2}{3}x^{3/2}) - J_{2/3}(\tfrac{2}{3}x^{3/2})]$

(b) $\text{Bi}' (-x) = \dfrac{x}{\sqrt{3}}\ [J_{-2/3}(\tfrac{2}{3}x^{3/2}) + J_{2/3}(\tfrac{2}{3}x^{3/2})]$

24. Verify the integral representations

(a) $\text{Ai} (x) = \dfrac{2\sqrt{x}}{3\pi} \displaystyle\int_0^\infty \cos\left(\dfrac{2}{3}x^{3/2}\sinh t\right)\cosh\dfrac{t}{3}\,dt,\ x > 0$

(b) $\text{Ai} (x) = \dfrac{1}{\pi} \displaystyle\int_0^\infty \cos\left(\dfrac{1}{3}t^3 + xt\right)dt,\ x > 0$

7.6 Asymptotic Formulas

For numerical computations it is usually convenient to use simplified asymptotic formulas when the argument of the Bessel function is either very small or very large. In fact, it can be shown that almost all computations involving Bessel functions can be performed with the use of these asymptotic formulas.*

7.6.1 Small arguments

For small arguments $(x \to 0^+)$, we simply utilize the first term or so of the series representation for the given function. Doing so, we have previously shown that the standard Bessel functions have the asymptotic formulas

$$J_p(x) \sim \frac{(x/2)^p}{\Gamma(p+1)} \qquad p \neq -1, -2, -3, \ldots, \qquad x \to 0^+ \qquad (7.90)$$

$$Y_0(x) \sim \frac{2}{\pi}\ln x \qquad x \to 0^+ \qquad (7.91)$$

$$Y_p(x) \sim -\frac{\Gamma(p)}{\pi}\left(\frac{2}{x}\right)^p \qquad p > 0, \qquad x \to 0^+ \qquad (7.92)$$

---

*For example, see chap. 11 in G. Arfken, *Mathematical Methods for Physicists,* 3d ed., Academic, New York, 1985.

Similarly, the modified Bessel functions have the asymptotic formulas

$$I_p(x) \sim \frac{(x/2)^p}{\Gamma(p+1)} \qquad p \neq -1, -2, -3, \ldots, \qquad x \to 0^+ \qquad (7.93)$$

$$K_0(x) \sim -\ln x \qquad x \to 0^+ \qquad (7.94)$$

$$K_p(x) \sim \frac{\Gamma(p)}{2} \left(\frac{2}{x}\right)^p \qquad p > 0, \qquad x \to 0^+ \qquad (7.95)$$

We leave it to the exercises to obtain the following results:

$$j_n(x) \sim \frac{n!(2x)^n}{(2n+1)!} \qquad n = 0, 1, 2, \ldots, \qquad x \to 0^+ \qquad (7.96)$$

$$y_n(x) \sim -\frac{2(2n)!}{n!(2x)^{n+1}} \qquad n = 0, 1, 2, \ldots, \qquad x \to 0^+ \qquad (7.97)$$

$$i_n(x) \sim \frac{n!(2x)^n}{(2n+1)!} \qquad n = 0, 1, 2, \ldots, \qquad x \to 0^+ \qquad (7.98)$$

$$k_n(x) \sim \frac{2(2n)!}{n!(2x)^{n+1}} \qquad n = 0, 1, 2, \ldots, \qquad x \to 0^+ \qquad (7.99)$$

7.6.2 Large arguments

To derive asymptotic formulas for large arguments, we start with the integral representation [recall Eq. (7.36)]

$$K_p(x) = \frac{\sqrt{\pi}\,(x/2)^p}{\Gamma(p+\tfrac{1}{2})} \int_1^\infty e^{-xt}(t^2-1)^{p-1/2}\, dt \qquad p > -\tfrac{1}{2}, x > 0$$

and make the substitution $t = 1 + u/x$, which leads to

$$K_p(x) = \frac{\sqrt{\pi}\,(x/2)^p}{\Gamma(p+\tfrac{1}{2})} \left(\frac{2}{x}\right)^{p-1/2} \frac{e^{-x}}{x}$$

$$\times \int_0^\infty e^{-u}\left(1 + \frac{u}{2x}\right)^{p-1/2} u^{p-1/2}\, du \qquad (7.100)$$

Now, for large x, we use the approximation

$$\left(1 + \frac{u}{2x}\right)^{p-1/2} \sim 1 \qquad x \gg u$$

and (7.100) reduces to

$$K_p(x) = \sqrt{\frac{\pi}{2x}} \frac{e^{-x}}{\Gamma(p + \frac{1}{2})} \int_0^\infty e^{-u} u^{p-1/2}\, du$$

from which we deduce

$$K_p(x) \sim \sqrt{\frac{\pi}{2x}} e^{-x} \qquad p \geq 0, \qquad x \to \infty \qquad (7.101)$$

Based on the asymptotic formula (7.101), we can develop asymptotic formulas for the other Bessel functions through interconnecting relations. For example, by starting with the relation [recall Eq. (7.71)]

$$K_p(x) = \frac{1}{2}\pi i^{p+1} H_p^{(1)}(ix)$$

it follows through the replacement of x with $-ix$ that

$$H_p^{(1)}(x) = \frac{2}{\pi} i^{-(p+1)} K_p(-ix) \qquad (7.102)$$

Under the assumption that (7.101) is valid also for complex arguments, we are led to

$$H_p^{(1)}(x) \sim \frac{2}{\pi} i^{-(p+1)} \sqrt{\frac{\pi}{-2ix}} e^{ix}$$

$$\sim \sqrt{\frac{2}{\pi x}} i^{-(p+1/2)} e^{ix} \qquad x \to \infty$$

and by writing $i = e^{i\pi/2}$, we obtain

$$H_p^{(1)}(x) \sim \sqrt{\frac{2}{\pi x}} \exp i\left[x - \frac{(p + \frac{1}{2})\pi}{2}\right]$$

$$\sim \sqrt{\frac{2}{\pi x}} \left\{\cos\left[x - \frac{(p + \frac{1}{2})\pi}{2}\right]\right.$$

$$\left. + i \sin\left[x - \frac{(p + \frac{1}{2})\pi}{2}\right]\right\} \qquad x \to \infty$$

Finally, recalling the relation

$$H_p^{(1)}(x) = J_p(x) + i Y_p(x)$$

and equating real and imaginary parts of these last two expressions, we obtain the set of asymptotic formulas

$$J_p(x) \sim \sqrt{\frac{2}{\pi x}} \cos\left[x - \frac{(p + \frac{1}{2})\pi}{2}\right] \qquad p \geq 0, \qquad x \to \infty \quad (7.103)$$

and

$$Y_p(x) \sim \sqrt{\frac{2}{\pi x}} \sin\left[x - \frac{(p + \frac{1}{2})\pi}{2}\right] \qquad p \geq 0, \qquad x \to \infty \quad (7.104)$$

By using the relation

$$I_p(x) = i^{-p} J_p(ix)$$

we also find that

$$I_p(x) \sim \frac{e^x}{\sqrt{2\pi x}} \qquad p \geq 0, \qquad x \to \infty \qquad (7.105)$$

Similarly, it can be shown that (see the exercises)

$$j_n(x) \sim \frac{1}{x} \sin\left(x - \frac{n\pi}{2}\right) \qquad n = 0, 1, 2, \ldots, x \to \infty \qquad (7.106)$$

$$y_n(x) \sim -\frac{1}{x} \cos\left(x - \frac{n\pi}{2}\right) \qquad n = 0, 1, 2, \ldots, x \to \infty \qquad (7.107)$$

$$i_n(x) \sim \frac{e^x}{2x} \qquad n = 0, 1, 2, \ldots, x \to \infty \qquad (7.108)$$

$$k_n(x) \sim \frac{e^{-x}}{x} \qquad n = 0, 1, 2, \ldots, x \to \infty \qquad (7.109)$$

Exercises 7.6

In problems 1 to 8, derive the given asymptotic formula for small arguments.

1. (a) $K_0(x) \sim -\ln x, \ x \to 0^+$

   (b) $K_p(x) \sim \dfrac{\Gamma(p)}{2} \left(\dfrac{2}{x}\right)^p, \ p > 0, \ x \to 0^+$

2. (a) $j_n(x) \sim \dfrac{n!(2x)^n}{(2n + 1)!}, \ n = 0, 1, 2, \ldots, x \to 0^+$

   (b) $y_n(x) \sim -\dfrac{2(2n)!}{n!(2x)^{n+1}}, \ n = 0, 1, 2, \ldots, x \to 0^+$

3. (a) $i_n(x) \sim \dfrac{n!(2x)^n}{(2n+1)!}$, $n = 0, 1, 2, \ldots, x \to 0^+$

   (b) $k_n(x) \sim \dfrac{2(2n)!}{n!(2x)^{n+1}}$, $n = 0, 1, 2, \ldots, x \to 0^+$

4. (a) $H_0^{(1)}(x) \sim \dfrac{2}{\pi} \ln x$, $x \to 0^+$

   (b) $H_p^{(1)}(x) \sim -\dfrac{\Gamma(p)}{\pi} \left(\dfrac{2}{x}\right)^p$, $x \to 0^+$

5. (a) $\mathbf{H}_0(x) \sim \dfrac{2}{\pi} x$, $x \to 0^+$

   (b) $\mathbf{H}_p(x) \sim \dfrac{2(x/2)^{p+1}}{\sqrt{\pi}\,\Gamma(p + \frac{3}{2})}$, $x \to 0^+$

   (c) $\mathbf{L}_p(x) \sim \dfrac{2(x/2)^{p+1}}{\sqrt{\pi}\,\Gamma(p + \frac{3}{2})}$, $x \to 0^+$

6. (a) $\text{ber}\,(x) \sim 1$, $x \to 0^+$
   (b) $\text{bei}\,(x) \sim \frac{1}{4}x^2$, $x \to 0^+$

7. (a) $\text{Ji}_0\,(x) \sim \ln \dfrac{x}{2}$, $x \to 0^+$

   (b) $\text{Ji}_n\,(x) \sim -\dfrac{1}{n}$, $n = 1, 2, 3, \ldots, x \to 0^+$

8. (a) $\text{Ai}\,(x) \sim \dfrac{\Gamma(\frac{1}{3})}{\pi\sqrt{6}} \left(\dfrac{x}{2}\right)^{1/6}$, $x \to 0^+$

   (b) $\text{Bi}\,(x) \sim \dfrac{\sqrt{2}}{\Gamma(\frac{2}{3})} \left(\dfrac{x}{2}\right)^{1/6}$, $x \to 0^+$

In problems 9 to 14, derive the given asymptotic formula for large arguments.

9. (a) $j_n(x) \sim \dfrac{1}{x} \sin\left(x - \dfrac{n\pi}{2}\right)$, $n = 0, 1, 2, \ldots, x \to \infty$

   (b) $y_n(x) \sim -\dfrac{1}{x} \cos\left(x - \dfrac{n\pi}{2}\right)$, $n = 0, 1, 2, \ldots, x \to \infty$

10. (a) $i_n(x) \sim \dfrac{e^x}{2x}$, $n = 0, 1, 2, \ldots, x \to \infty$

    (b) $k_n(x) \sim \dfrac{e^{-x}}{x}$, $n = 0, 1, 2, \ldots, x \to \infty$

11. (a) $\mathbf{H}_0(x) - Y_0(x) \sim \dfrac{2}{\pi x}$, $x \to \infty$

   (b) $\mathbf{H}_1(x) - Y_1(x) \sim \dfrac{2}{\pi}$, $x \to \infty$

   (c) $\mathbf{H}_p(x) - Y_p(x) \sim \dfrac{(2/x)^{1-p}}{\sqrt{\pi}\,\Gamma(p + \frac{1}{2})}$, $x \to \infty$

12. (a) $\mathbf{L}_0(x) - I_0(x) \sim -\dfrac{2}{\pi x}$, $x \to \infty$

   (b) $\mathbf{L}_p(x) - I_p(x) \sim -\dfrac{(2/x)^{1-p}}{\sqrt{\pi}\,\Gamma(p + \frac{1}{2})}$, $x \to \infty$

13. (a) $\text{ber}\,(x) \sim \dfrac{1}{\sqrt{2\pi x}} e^{x/\sqrt{2}} \cos\left(\dfrac{x}{\sqrt{2}} - \dfrac{\pi}{8}\right)$, $x \to \infty$

   (b) $\text{bei}\,(x) \sim \dfrac{1}{\sqrt{2\pi x}} e^{x/\sqrt{2}} \sin\left(\dfrac{x}{\sqrt{2}} - \dfrac{\pi}{8}\right)$, $x \to \infty$

14. (a) $\text{Ai}\,(x) \sim \frac{1}{2}\pi^{-1/2} x^{-1/4} \exp\left(-\frac{2}{3} x^{3/2}\right)$, $x \to \infty$
   (b) $\text{Bi}\,(x) \sim \pi^{-1/2} x^{-1/4} \exp\left(\frac{2}{3} x^{3/2}\right)$, $x \to \infty$

15. By expressing the factor $[1 + u/(2x)]^{p-1/2}$ in a binomial series in Eq. (7.100), show that

$$K_p(x) \sim \sqrt{\dfrac{\pi}{2x}} \dfrac{e^{-x}}{\Gamma(p + \frac{1}{2})} \sum_{n=0}^{\infty} \binom{p - \frac{1}{2}}{n} \dfrac{\Gamma(p + n + \frac{1}{2})}{(2x)^n}$$

$$p > -\frac{1}{2},\ x \to \infty$$

Applications Involving Bessel Functions

8.1 Introduction

Bessel functions are prominent in a variety of applications, some of which were discussed in Chaps. 6 and 7. Now we wish to consider some additional examples typical of those occurring in more than one field of application. To provide some variety in our discussions, we have chosen examples from the fields of mechanics, wave propagation and scattering, fiber optics, heat conduction in solids, and vibration phenomena. (A working knowledge of each subject is generally sufficient to follow the discussion.)

8.2 Problems in Mechanics

We begin this chapter on applications with some examples chosen from the fields of particle dynamics and the static displacements of beams and columns. Additional problems of a similar nature are taken up in the exercises.

8.2.1 The lengthening pendulum

In the absence of frictional forces, the angle of oscillation ϕ of a swinging pendulum of *fixed length* and mass μ is governed by the equation (see Sec. 3.5.2)

$$r\frac{d^2\phi}{dt^2} + g \sin \phi = 0 \qquad (8.1)$$

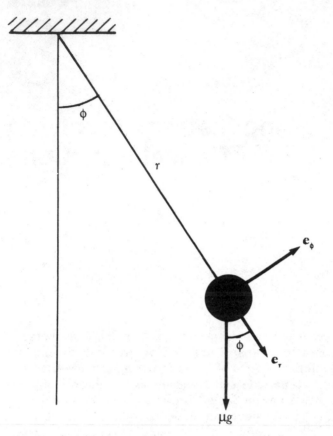

Figure 8.1 A swinging pendulum.

where r is the length of the pendulum rod and g is the gravitational constant (see Fig. 8.1). Suppose now we consider a variation of that problem in which the length r is not constant, but increases linearly with time from an initial length b at time $t = 0$. Thus the length of the rod for all $t \geq 0$ is described by

$$r(t) = at + b \tag{8.2}$$

where a is a constant equal to the rate of increase.

Let $\mathbf{e}_r$ and $\mathbf{e}_\phi$ denote *orthogonal* unit vectors, where $\mathbf{e}_r$ always points in the direction of the position vector $\mathbf{r}(t)$ and $\mathbf{e}_\phi$ is in the transverse direction, as shown in Fig. 8.1. These base vectors are related by

$$\frac{d\mathbf{e}_r}{dt} = \frac{d\phi}{dt}\,\mathbf{e}_\phi \qquad \frac{d\mathbf{e}_\phi}{dt} = -\frac{d\phi}{dt}\,\mathbf{e}_r$$

The position vector of the mass is therefore simply

$$\mathbf{r}(t) = r(t)\mathbf{e}_r \quad (at+b)\mathbf{e}_r \tag{0.3}$$

the derivative of which yields the velocity vector

$$\mathbf{v}(t) = \frac{d\mathbf{r}}{dt} = a\mathbf{e}_r + (at+b)\frac{d\mathbf{e}_r}{dt}$$

$$= a\mathbf{e}_r + (at+b)\frac{d\phi}{dt}\mathbf{e}_\phi \tag{8.4}$$

Differentiating once more, we obtain the acceleration vector

$$\mathbf{a}(t) = \frac{d^2\mathbf{r}}{dt^2}$$

$$= -(at+b)\left(\frac{d\phi}{dt}\right)^2\mathbf{e}_r + \left[(at+b)\frac{d^2\phi}{dt^2} + 2a\frac{d\phi}{dt}\right]\mathbf{e}_\phi \tag{8.5}$$

The transverse acceleration component of the mass is the coefficient of $\mathbf{e}_\phi$ in (8.5), and the particular term $2a\,d\phi/dt$ is called the *coriolis* acceleration component. By equating the mass times this transverse acceleration component to the gravitational force component in the transverse direction, that is, $-\mu g \sin \phi$, we obtain the *nonlinear* equation of motion*

$$(at+b)\frac{d^2\phi}{dt^2} + 2a\frac{d\phi}{dt} + g \sin \phi = 0 \tag{8.6}$$

Here we consider only the case of "small oscillations" for which we may replace $\sin \phi$ by ϕ. If we also make the change of variables $at + b = ax$, then Eq. (8.6) reduces to the linear equation

$$x\phi'' + 2\phi' + k^2\phi = 0 \tag{8.7}$$

where $k^2 = g/a$ and the primes denote differentiation with respect to x. Multiplying (8.7) by x and comparing the resulting equation with the general form given by (6.98), we see that

$$1 - 2a = 2 \qquad b^2c^2 = k^2 \qquad 2c = 1 \qquad a^2 - c^2p^2 = 0$$

---

*Since the mass μ of the pendulum bob is constant, it drops out in the final equation. Hence, it is sufficient to simply equate acceleration components.

from which we deduce $a = -\frac{1}{2}$, $b = 2k$, $c = \frac{1}{2}$, and $p = 1$. Therefore, the general solution of (8.7) is given by

$$\phi(x) = \frac{1}{\sqrt{x}}[C_1 J_1(2k\sqrt{x}) + C_2 Y_1(2k\sqrt{x})] \tag{8.8}$$

To determine the arbitrary constants C_1 and C_2, we must impose initial conditions on the solution. For instance, suppose that at time $t = 0$ the pendulum has angular displacement ϕ_0 with zero angular velocity. We then have the initial conditions

$$\phi(0) = \phi_0 \qquad \frac{d\phi}{dt}(0) = 0 \tag{8.9}$$

which, in terms of variable x, become

$$\phi\left(\frac{b}{a}\right) = \phi_0 \qquad \phi'\left(\frac{b}{a}\right) = 0 \tag{8.10}$$

The derivative of $\phi(x)$ leads to (see problem 4 in Exercises 8.2)

$$\phi'(x) = -\frac{k}{x}[C_1 J_2(2k\sqrt{x}) + C_2 Y_2(2k\sqrt{x})] \tag{8.11}$$

so that, upon imposing the initial conditions (8.10), we obtain

$$\phi\left(\frac{b}{a}\right) = \sqrt{\frac{a}{b}}[C_1 J_1(\lambda) + C_2 Y_1(\lambda)] = \phi_0$$

$$\phi'\left(\frac{b}{a}\right) = -\frac{ak}{b}[C_1 J_2(\lambda) + C_2 Y_2(\lambda)] = 0 \tag{8.12}$$

where $\lambda = 2k\sqrt{b/a}$. The simultaneous solution of (8.12) leads to

$$C_1 = \frac{\phi_0\sqrt{b/a}\, Y_2(\lambda)}{J_1(\lambda)Y_2(\lambda) - J_2(\lambda)Y_1(\lambda)}$$

$$C_2 = -\frac{\phi_0\sqrt{b/a}\, J_2(\lambda)}{J_1(\lambda)Y_2(\lambda) - J_2(\lambda)Y_1(\lambda)} \tag{8.13}$$

However, by virtue of the relationship (see problem 8 in Exercises 6.6)

$$J_1(\lambda)Y_2(\lambda) - J_2(\lambda)Y_1(\lambda) = -\frac{2}{\pi\lambda} \tag{8.14}$$

the constants C_1 and C_2 take the simpler form

$$C_1 = -\frac{1}{2}\pi\lambda\phi_0\sqrt{\frac{b}{a}}\,Y_2(\lambda) = -\frac{b}{a}\sqrt{\frac{g}{a}}\,\pi\phi_0 Y_2(\lambda)$$

$$C_2 = \frac{1}{2}\pi\lambda\phi_0\sqrt{\frac{b}{a}}\,J_2(\lambda) = \frac{b}{a}\sqrt{\frac{g}{a}}\,\pi\phi_0 J_2(\lambda) \qquad (8.15)$$

Thus, the angular displacement of the pendulum at any time t is given by

$$\phi(t) = \frac{\pi b\phi_0\sqrt{g}}{a\sqrt{at+b}}\,[J_2(\lambda)Y_1(s) - Y_2(\lambda)J_1(s)] \qquad (8.16)$$

where $\lambda = 2\sqrt{bg}/a$ and $s = 2\sqrt{g(at+b)}/a$.

8.2.2 Buckling of a long column

Vertical columns have been used extensively in Greek and Roman structures throughout the centuries. One of the oldest engineering problems concerns the buckling of such columns under a compressive load. Euler developed the first truly mathematical model that can rather accurately predict the *critical compressive load* that a column can withstand before deformation or *buckling* takes place.

Consider a long column or rod of length b that is simply supported at each end and is subject to an axial compressive load P applied at the top, as shown in Fig. 8.2. By "long" we mean that the length of

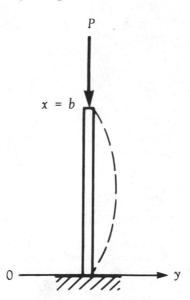

Figure 8.2 Buckling column.

the column is much greater than the largest dimension in its cross section. From the elementary theory of small deflections of beams and columns, the departure y from the vertical x axis for such a beam or column is governed by the equation

$$EIy'' = M \qquad (8.17)$$

where M is the *bending moment*, E is the *modulus of elasticity*, and I is the *moment of inertia* of the cross section of the column. When the column is deflected a small amount y from the vertical due to the loading P, the bending moment is $M = -Py$. Since the column is assumed to be simply supported, there can be no displacement at the ends. Thus, Eq. (8.17) together with the boundary conditions takes the form

$$EIy'' + Py = 0 \qquad y(0) = 0 \qquad y(b) = 0 \qquad (8.18)$$

Clearly, the problem described by (8.18) has the trivial solution $y = 0$, corresponding to the column not bending away from the x axis. However, if P is large enough, there may exist nontrivial solutions of (8.18). That is, if P is sufficiently large, the column may suddenly bow out of its equilibrium state, called a state of **buckling**. The smallest value of P that leads to buckling is called the **Euler critical load**, and the corresponding deflection mode is called the **fundamental buckling mode.**

The classic example of a buckling column involves the case when E, I, and P are all constant, and then it is easy to show that the Euler critical load P_1 and corresponding fundamental deflection mode are given, respectively, by[*]

$$P_1 = \frac{\pi^2 EI}{b^2} \qquad y = C \sin \frac{\pi x}{b}$$

where C is an arbitrary constant. The constant C remains undetermined in this analysis (because we chose a *linear* model). It can be determined only from a more exact *nonlinear* analysis, but the value of C is often of little concern in practice, i.e., usually only the critical load is of interest.

Suppose now we consider the case where the column is tapered so the moment of inertia is not constant, but is given by $I(x) = \alpha x$,

---

[*] For example, see chap. 2 of L. C. Andrews, *Elementary Partial Differential Equations with Boundary Value Problems*, Academic, New York, 1986.

where $\alpha > 0$. In this case Eq. (8.18) becomes

$$E\alpha xy'' + Py = 0 \qquad y(0) = 0 \qquad y(b) = 0 \tag{8.19}$$

In solving the DE, first we rewrite it in the form

$$x^2 y'' + k^2 xy = 0 \tag{8.20}$$

where $k^2 = P/E\alpha$. Then, by comparing this DE with the general form given by (6.98), we see that $a = \frac{1}{2}$, $b = 2k$, $c = \frac{1}{2}$, and $p = 1$, from which we deduce the solution

$$y = \sqrt{x}\,[C_1 J_1(2k\sqrt{x}) + C_2 Y_1(2k\sqrt{x})] \tag{8.21}$$

where C_1 and C_2 are arbitrary constants.

To apply the first boundary condition $y(0) = 0$, we use the asymptotic forms for both J_1 and Y_1; hence

$$y(0) = \lim_{x \to 0} \sqrt{x}\,[C_1 J_1(2k\sqrt{x}) + C_2 Y_1(2k\sqrt{x})]$$

$$= \lim_{x \to 0} \sqrt{x}\left[C_1 k\sqrt{x} - C_2 \frac{1}{\pi k\sqrt{x}}\right]$$

$$= -C_2 \frac{1}{\pi k} \tag{8.22}$$

which can vanish only if $C_2 = 0$. The second boundary condition then requires that

$$y(b) = C_1 \sqrt{b}\, J_1(2k\sqrt{b}) = 0 \tag{8.23}$$

There are infinitely many solutions of $J_1(2k\sqrt{b}) = 0$, the first of which yields the Euler critical load P_1. That is, if we let k_1 denote the smallest value of k for which (8.23) is satisfied, the corresponding Euler critical load is

$$P_1 = E\alpha k_1^2 \tag{8.24}$$

leading to the fundamental deflection mode

$$y = C_1 \sqrt{x}\, J_1(2k_1\sqrt{x}) \tag{8.25}$$

where C_1 remains undetermined.

Example 1: Given that $b = 400$ cm, $\alpha = 10^{-4}$ cm^3, and $E = 2 \times 10^9$ g/cm^2, calculate the Euler critical load given by Eq. (8.23) and determine the point along the vertical x axis where the maximum displacement occurs.

Solution: With $b = 400$, we first determine k_1 by solving

$$J_1(40k_1) = 0$$

From Table 6.1, the first zero of J_1 occurs when $40k_1 = 3.832$, or $k = 0.0958$. Hence, the Euler critical load is

$$P_1 = Eak_1^2 = (2 \times 10^9)(10^{-4})(0.0958)^2 \cong 1836 \text{ kg}$$

The maximum displacement of the column takes place at the point where $y' = 0$, which from (8.25) leads to (see problem 5 in Exercises 8.2)

$$y' = C_1 k_1 J_0(2k_1\sqrt{x}) = 0$$

The first admissible zero of J_0 occurs when $2k_1\sqrt{x} = 2.405$. Thus, using $k_1 = 0.0958$, we find that $x \cong 158$ cm, which is closer to the bottom (more tapered end) of the column.

Exercises 8.2

1. Assuming that $a \ll 1$, use asymptotic formulas to show that the solution of the lengthening pendulum (7.16) assumes the approximate form

$$\phi(t) \cong \phi_0 \left(\frac{b}{at + b}\right)^{3/4} \cos(s - \lambda) \qquad a \ll 1$$

2. Use the solution in problem 1 to show that the pendulum is in the vertical position at times

$$t \cong \frac{b}{a}\left\{\left[\frac{(2n + 1)\pi}{2\lambda} + 1\right]^2 - 1\right\}$$

where n assumes integer values.

3. By writing

$$s - \lambda = \frac{2\sqrt{bg}}{a}\left[\left(1 + \frac{at}{b}\right)^{1/2} - 1\right]$$

expand the factor $(1 + at/b)^{1/2}$ in a binomial series and deduce that in the limit $a \to 0$, the solution in problem 1 reduces to the classic result

$$\phi(t) \cong \phi_0 \cos\sqrt{\frac{g}{b}}\, t$$

4. Use the identity

$$\frac{d}{du}[u^{-1}J_1(u)] = -u^{-1}J_2(u)$$

(a) Show that

$$\frac{d}{dx}\left[\frac{1}{\sqrt{x}}J_1(2k\sqrt{x})\right] = -\frac{k}{x}J_2(2k\sqrt{x}) \qquad k>0$$

(b) Show that the identity in (a) also applies to the Bessel function of the second kind.

5. Using the identity

$$\frac{d}{du}[uJ_1(u)] = uJ_0(u)$$

show that

$$\frac{d}{dx}[\sqrt{x}\,J_1(2k\sqrt{x})] = kJ_0(2k\sqrt{x}) \qquad k>0$$

6. A particle of variable mass $m = (a + bt)^{-1}$, where a and b are constants, starts from rest at a distance r_0 from the origin O and is attracted by a force always directed toward O of magnitude k^2mr (k constant). Given that the equation of motion is given by

$$\frac{d}{dt}\left(m\,\frac{dr}{dt}\right) = -k^2mr$$

find a solution in terms of J_1 and Y_1 that satisfies the prescribed initial conditions.
Hint: Make the change of variable $bx = a + bt$.

7. A particle of variable mass $m = (a + t)/a$, where a is a constant, is initially ejected from the origin O with velocity v. If it is repelled from the origin with a force per unit mass proportional to the distance r from O, the equation of motion is given by

$$\frac{d}{dt}\left(m\,\frac{dr}{dt}\right) = k^2mr$$

Find a solution in terms of modified Bessel functions of order zero that satisfies the prescribed initial conditions.

8. Prove that, regardless of the values of α and E, the ratio of the point of maximum displacement x to the length of the column b is always $x/b \cong 0.395$ for the particular column discussed in Sec. 8.2.2 (i.e., it is independent of the material used to make the column).

9. When $I(x) = \alpha\sqrt{x}$, show that the solution of the buckling problem in Sec. 7.2.2 is

$$y = C_1\sqrt{x}\, J_{2/3}(2kx^{3/2})$$

10. In a problem on the stability of a tapered strut, the displacement y satisfies the boundary-value problem

$$4xy'' + k^2 y = 0 \qquad y'(a) = 0 \qquad y'(b) = 0 \qquad (a < b)$$

Show that the determination of the positive constant k for which nontrivial solutions exist leads to the relation

$$J_0(k\sqrt{a})Y_0(k\sqrt{b}) = J_0(k\sqrt{b})Y_0(k\sqrt{a})$$

11. The small deflections of a uniform column of length b bending under its own weight are governed by

$$\theta'' + k^2 x\theta = 0 \qquad \theta'(0) = 0 \qquad \theta(b) = 0$$

where θ is the angle of deflection from the vertical and k is a positive constant.

(a) Show that the solution of the DE satisfying the first boundary condition $\theta'(0) = 0$ is

$$\theta = C_1\sqrt{x}\, J_{-1/3}(\tfrac{2}{3}kx^{3/2})$$

where C_1 is any constant.

(b) Show that the shortest column length for which buckling may occur (denoted by b_0) is $b_0 \cong 1.986k^{-2/3}$.
    Hint: The first zero of $J_{-1/3}(u)$ is $u \cong 1.866$.

12. An axial load P is applied to a column whose circular cross section is tapered so that the moment of inertia is $I(x) = (x/a)^4$. The buckling modes are described by solutions of

$$x^4 y'' + k^2 y = 0 \qquad y(1) = 0 \qquad y(a) = 0 \qquad (a > 1)$$

where $k^2 = Pa^4/E$.

(a) Show that the Euler critical load is

$$P_1 = \frac{\pi^2 E}{a^2(a-1)^2}$$

(b) Show that the fundamental buckling mode is

$$y = x\sin\left[\frac{\pi a}{a-1}\left(1 - \frac{1}{x}\right)\right]$$

8.3   Statistical Communication Theory

Statistical communication theory is basically the application of probability and statistics to problems in the communication process.

This includes such areas as the transmission and reception of messages, the measurement and processing of data, design of decision systems, and so on. Our primary interest here, however, concerns only the propagation and detection of electromagnetic radiation.

Theoretical studies prior to 1960 concerning the *propagation of electromagnetic radiation* involved the propagation of starlight through the atmosphere, propagation of sound waves through the atmosphere and ocean, and propagation of radio waves through the ionosphere. With the introduction of the laser in 1960 new theoretical investigations were begun that included related phenomena such as microwave sea scatter and speckle patterns from optical waves scattered off rough surfaces.

8.3.1 Narrowband noise and envelope detection

In the general study of communication systems, it is important to understand the statistical properties of random noise at the input and output of various linear and nonlinear devices. Let us consider noise at the output of a narrowband linear filter with center frequency $f_0 = \omega_0/2\pi$ Hz and bandwidth $B \ll f_0$. In such cases the output noise $\mathbf{n}(t)$ is called **narrowband noise** and is expected to appear almost as a sine wave of frequency f_0, but with slowly varying and randomly modulated phase and amplitude. In general, therefore, we write

$$\mathbf{n}(t) = \mathbf{x}(t)\cos\omega_0 t - \mathbf{y}(t)\sin\omega_0 t \qquad (8.26)$$

where $\mathbf{x}(t)$ and $\mathbf{y}(t)$ are independent *gaussian* (or *normal*) random processes with zero means and mean-squared values given by the expected values

$$\langle \mathbf{x}^2(t)\rangle = \langle \mathbf{y}^2(t)\rangle = N = \langle \mathbf{n}^2(t)\rangle \qquad (8.27)$$

where $\langle\ \rangle$ denotes an ensemble average. Hence $\mathbf{x}(t)$ and $\mathbf{y}(t)$ have the probability distributions

$$p_{\mathbf{x}}(x) = \frac{1}{\sqrt{2\pi N}}e^{-x^2/2N} \qquad (8.28a)$$

$$p_{\mathbf{y}}(y) = \frac{1}{\sqrt{2\pi N}}e^{-y^2/2N} \qquad (8.28b)$$

and because $x(t)$ and $\mathbf{y}(t)$ are *independent,* their joint distribution is given by

$$p_{\mathbf{xy}}(x, y) = p_{\mathbf{x}}(x)p_{\mathbf{y}}(y)$$

$$= \frac{1}{2\pi N} e^{-(x^2+y^2)/2N} \tag{8.29}$$

In high-frequency communications we usually also have a sinusoidal signal at or near frequency f_0. For example, this signal could be a carrier in either AM (amplitude modulation) or FM (frequency modulation) transmission, or one of two transmitting states of an FSK (frequency-shift keying) binary transmission system. In most cases the signal is taken as *additive* to the noise, and it is the statistical properties of the envelope of the composite signal plus noise that we wish to investigate.

Let us assume the signal is an unmodulated carrier of amplitude A and frequency $f_0 = \omega_0/2\pi$. Because the phase of the signal is purely arbitrary and does not affect our results, we assume it is zero. The composite signal plus narrowband noise at the output of a *linear device* then has the representation

$$\mathbf{v}(t) = A \cos \omega_0 t + \mathbf{n}(t)$$

$$= [\mathbf{x}(t) + A] \cos \omega_0 t - \mathbf{y}(t) \sin \omega_0 t$$

$$= \mathbf{r}(t) \cos [\omega_0 t + \mathbf{\theta}(t)] \tag{8.30}$$

where $\mathbf{r}(t)$ and $\mathbf{\theta}(t)$ represent the random **envelope** (or **amplitude**) and **phase**, respectively, of the composite signal and noise. Clearly, $\mathbf{r}(t)$ and $\mathbf{\theta}(t)$ are related to the quadrature components $\mathbf{x}(t)$ and $\mathbf{y}(t)$ by the transformation

$$\mathbf{x}(t) = \mathbf{r}(t) \cos \mathbf{\theta} - A$$

$$\mathbf{y}(t) = \mathbf{r}(t) \sin \mathbf{\theta} \tag{8.31}$$

The joint density function of $\mathbf{r}(t)$ and $\mathbf{\theta}(t)$ is obtained from (8.29) by solving

$$p_{\mathbf{r\theta}}(r, \theta) \, dr \, d\theta = p_{\mathbf{xy}}(x, y) \, dx \, dy$$

Thus, by substituting Eqs. (8.31) into (8.29), we get

$$p_{\mathbf{r\theta}}(r, \theta) \, dr \, d\theta = \frac{r}{2\pi N} \exp\left\{ -\frac{[(r \cos \theta - A)^2 + (r \sin \theta)^2]}{2N} \right\} dr \, d\theta$$

where we have used $dx\,dy = r\,dr\,d\theta$. Simplifying the terms in the above exponential function, we obtain the joint distribution

$$p_{r\theta}(r,\theta) = \frac{r}{2\pi N}\exp\left[-\frac{(r^2+A^2-2Ar\cos\theta)}{2N}\right] \qquad (8.32)$$

The marginal distribution of the envelope $r(t)$ is found by integrating (8.32) over θ (modulo 2π) whereas that of the phase $\theta(t)$ is found by integrating over $0 \le r < \infty$. Performing integration over the interval $-\pi < \theta \le \pi$ (or over $0 \le \theta < 2\pi$), we have

$$p_r(r) = \int_{-\pi}^{\pi} p_{r\theta}(r,\theta)\,d\theta$$

which reduces to

$$p_r(r) = \frac{r}{N}e^{-(r^2+A^2)/2N}I_0\left(\frac{Ar}{N}\right) \qquad r>0 \qquad (8.33)$$

where we have used the Bessel function relation (see problem 1 in Exercises 7.3)

$$I_0(x) = \frac{1}{2\pi}\int_{-\pi}^{\pi} e^{x\cos\theta}\,d\theta$$

The distribution (8.33) is known in the literature as the **rician distribution**, or sometimes the **Rice-Nakagami distribution**.

The marginal phase distribution is likewise determined by evaluating the integral

$$p_\theta(\theta) = \int_0^\infty p_{r\theta}(r,\theta)\,dr$$

$$= \frac{1}{2\pi N}e^{-A^2/2N}\int_0^\infty e^{-(r^2-2Ar\cos\theta)/2N}\,dr$$

To perform the integration, we complete the square in the argument of the exponential to find

$$p_\theta(\theta) = \frac{1}{2\pi N}e^{-s}e^{s\cos^2\theta}\int_0^\infty re^{-(r/\sqrt{2N}-\sqrt{s}\cos\theta)^2}\,dr \qquad (8.34)$$

where $s = A^2/2N$ is the **signal-to-noise ratio** (SNR) of the power of the sinusoidal signal to that of the noise. Next we make the change of variable $t = r/\sqrt{2N} - \sqrt{s}\cos\theta$ and integrate the resulting expression to obtain (see problem 8 in Exercises 8.3)

$$p_{\theta}(\theta) = \frac{1}{2\pi} e^{-s}\{1 + \sqrt{\pi s}\, e^{s\,\cos^2\theta}\cos\theta[1 + \text{erf}(\sqrt{s}\cos\theta)]\}$$

$$-\pi < \theta \le \pi \quad (8.35)$$

where $\text{erf}\,x$ is the *error function* (see Sec. 3.2).

8.3.2 Non-Rayleigh radar sea clutter

The performance of *microwave sea echo* over the open sea, which is often limited by the presence of *sea clutter* (unwanted returns) in the return signal, has been the subject of a large number of investigations since World War II. When large areas of the sea are illuminated by radar, the return signal behaves as narrowband gaussian noise. Hence, the envelope $\mathbf{r}(t)$ of the return signal can usually be well approximated by the **Rayleigh distribution** [obtained from (8.33) by setting $A = 0$]

$$p_{\mathbf{r}}(r) = \frac{r}{\sigma^2} e^{-r^2/2\sigma^2} \quad r > 0 \quad (8.36)$$

where $\sigma^2 = N = \langle \mathbf{r}^2(t)\rangle/2$. However, when the sea is viewed at low grazing angles with a high-resolution radar, the envelope $\mathbf{r}(t)$ shows significant deviations from Rayleigh statistics.

Experimental evidence in the case of low grazing angles suggests that the envelope $\mathbf{r}(t)$ of the return radar signal can be represented by the product of two independent random processes with differing correlation times, i.e.,*

$$\mathbf{r}(t) = \mathbf{x}(t)\mathbf{y}(t) \quad (8.37)$$

where $\mathbf{x}(t)$ has a long correlation time compared with $\mathbf{y}(t)$. The probability density function of $\mathbf{r}(t)$ can then be determined by evaluating the integral

$$p_{\mathbf{r}}(r) = \int_0^\infty p_{\mathbf{x}}(x) p_{\mathbf{r}}(r\,|\,x)\,dx \quad (8.38)$$

where $p_{\mathbf{r}}(r\,|\,x)$ is the *conditional* density function of $\mathbf{r}(t)$ given $\mathbf{x}(t)$. From basic probability theory it is known that $p_{\mathbf{r}}(r\,|\,x)$ is related to

---

*See K. D. Ward, "Compound Representation of High Resolution Sea Clutter," *Electronics Lett.*, **17:** 561–563, 1981.

the distribution $p_y(y)$ according to

$$p_r(r \mid x) = |x|^{-1} p_y\left(\frac{r}{x} \middle| x\right) = |x|^{-1} p_y\left(\frac{r}{x}\right) \qquad (8.39)$$

where we are using the fact that $\mathbf{x}(t)$ and $\mathbf{y}(t)$ are independent. For modeling purposes, we assume that the density function for $\mathbf{x}(t)$ is the **chi-squared distribution**

$$p_x(x) = \frac{2}{\Gamma(v)} x^{2v-1} e^{-x^2} \qquad x > 0 \qquad (8.40)$$

where v is a positive parameter dependent upon range, height, aspect angle, and sea state. We further assume that the distribution of $\mathbf{y}(t)$ is Rayleigh, i.e.,

$$p_y(y) = \frac{y}{\sigma^2} e^{-y^2/2\sigma^2} \qquad y > 0 \qquad (8.41)$$

and thus (8.39) becomes

$$p_r(r \mid x) = \frac{r}{\sigma^2 x^2} e^{-r^2/2\sigma^2 x^2} \qquad r > 0 \qquad (8.42)$$

Finally, the substitution of (8.40) and (8.42) into the integral in (8.38) yields

$$p_r(r) = \int_0^\infty p_x(x) p_r(r \mid x) \, dx$$

$$= \frac{2r}{\Gamma(v)\sigma^2} \int_0^\infty x^{2v-3} e^{-x^2 - r^2/2\sigma^2 x^2} \, dx$$

and by making the change of variable $t = x^2$, this becomes

$$p_r(r) = \frac{r}{\Gamma(v)\sigma^2} \int_0^\infty t^{v-2} e^{-t - r^2/2\sigma^2 t} \, dt \qquad (8.43)$$

We recognize this last integral as that given by (7.37), and hence we obtain

$$p_r(r) = \frac{2^{2-v} r}{\Gamma(v)\sigma^2} \left(\frac{\sqrt{2}\, r}{\sigma}\right)^{v-1} K_{v-1}\left(\frac{\sqrt{2}\, r}{\sigma}\right) \qquad r > 0* \qquad (8.44)$$

which belongs to the general family of **K distributions**.

---

*We are using the fact that $K_{-p}(x) = K_p(x)$.

Exercises 8.3

1. (a) Show that in the absence of a signal (that is, $A = 0$) Eq. (8.33) reduces to the Rayleigh distribution

$$p_{\mathbf{r}}(r) = \frac{r}{N} e^{-r^2/2N} \qquad r > 0$$

   (b) Show that the statistical moments of the Rayleigh distribution are given by

$$\langle \mathbf{r}^v \rangle = \int_0^\infty r^v p_{\mathbf{r}}(r)\, dr$$

$$= (2N)^{v/2} \Gamma\left(1 + \frac{v}{2}\right) \qquad v > -\tfrac{1}{2}$$

2. The joint distribution of the envelopes and phases of two correlated narrowband processes is

$$p_2(r_1, r_2; \theta_1, \theta_2) = \frac{r_1 r_2}{(2\pi N)^2(1 - \rho^2)}$$

$$\times \exp\left\{ -\frac{[r_1^2 + r_2^2 - 2\rho r_1 r_2 \cos(\theta_2 - \theta_1)]}{2N(1 - \rho^2)} \right\}$$

$$-\pi < \theta_1, \theta_2 \le \pi, \qquad 0 \le r_1, r_2 < \infty$$

   where ρ is a normalized correlation coefficient. Show that

$$p_2(r_1, r_2) = \int_{-\pi}^{\pi} \int_{-\pi}^{\pi} p_2(r_1, r_2; \theta_1, \theta_2)\, d\theta_1\, d\theta_2$$

$$= \frac{r_1 r_2}{N^2(1 - \rho)^2} \exp\left[-\frac{(r_1^2 + r_2^2)}{2N(1 - \rho^2)} \right] I_0\left[\frac{\rho r_1 r_2}{N(1 - \rho^2)} \right]$$

$$0 \le r_1, r_2 < \infty$$

3. Show that when $v = \tfrac{3}{2}$, the K distribution (8.44) reduces to

$$p_{\mathbf{r}}(r) = \frac{2r}{\sigma^2} e^{-\sqrt{2}\, r/\sigma} \qquad r > 0$$

4. The intensity of a radar signal is defined by $\mathbf{E} = \mathbf{r}^2$, where $\mathbf{r}$ is the amplitude or envelope. If $\mathbf{r}$ is governed by the K distribution (8.44), show that $\mathbf{E}$ satisfies

$$p_{\mathbf{E}}(E) = \frac{2^{1-v}}{\Gamma(v)\sigma^2} \left(\frac{\sqrt{2E}}{\sigma} \right)^{v-1} K_{v-1}\left(\frac{\sqrt{2E}}{\sigma} \right) \qquad E > 0$$

5. If **E** is governed by the distribution in problem 4, find the *normalized* moments of **E** defined by $\langle \mathbf{E}^n \rangle / \langle \mathbf{E} \rangle^n$, where

$$\langle \mathbf{E}^n \rangle = \int_0^\infty E^n p_E(E)\, dE \qquad n = 1, 2, 3, \ldots$$

Hint: See Example 1 in Sec. 7.3.

6. If both **x** and **y** are zero mean gaussian variates with unit variances, show that the distribution of $\mathbf{r} = \mathbf{xy}$ is given by

$$p_r(r) = \frac{1}{\pi} K_0(|r|)$$

7. If the intensity **E** of a radar signal is governed by the modified rician distribution

$$p_E(E) = \frac{1}{b} e^{-(A^2+E)/b} I_0\left(\frac{2A}{b}\sqrt{E}\right) \qquad E > 0$$

show that the statistical moments are given by

$$\langle \mathbf{E}^n \rangle = b^n n!\, L_n\left(-\frac{A^2}{b}\right) \qquad n = 1, 2, 3, \ldots$$

where $L_n(x)$ is the nth Laguerre polynomial.

8. Derive Eq. (8.35) from (8.34).

8.4 Heat Conduction and Vibration Phenomena

The fundamental problem in the mathematical theory of **heat conduction** is to determine the temperature in a homogeneous solid when the distribution of temperature throughout the solid is known at time $t = 0$ and the temperature (or temperature gradient) at every boundary point is known. In **vibration phenomena** the corresponding problem is to describe the wave motion of a natural or mechanical system, such as a vibrating membrane, subject to certain initial and boundary conditions.

The governing equations for a wide variety of applications involving *vibration phenomena, diffusion processes,* and *potential theory* are primarily the **wave equation**

$$\nabla^2 u = c^{-2} \frac{\partial^2 u}{\partial t^2} \tag{8.45}$$

heat equation

$$\nabla^2 u = a^{-2} \frac{\partial u}{\partial t} \tag{8.46}$$

and **Laplace's equation**

$$\nabla^2 u = 0 \qquad (8.47)$$

where the *laplacian* $\nabla^2 u$ is defined in rectangular coordinates by

$$\nabla^2 u = \frac{\partial^2 u}{\partial x^2} + \frac{\partial^2 u}{\partial y^2} + \frac{\partial^2 u}{\partial z^2} \qquad (8.48)$$

8.4.1 Radial symmetric problems involving circles

We wish to begin by determining the small displacements u of a thin circular membrane (such as a drumhead) of unit radius whose edge is rigidly fixed (see Fig. 8.3). The governing equation for this problem is the wave equation (8.45), where c is a physical constant having the dimensions of velocity.

If the displacements u depend only upon the radial distance r from the center of the membrane and on time t, then (8.45) reduces to the *radial symmetric* form of the wave equation given by (see problem 1 in Exercises 8.4)

$$\frac{\partial^2 u}{\partial r^2} + \frac{1}{r}\frac{\partial u}{\partial r} = c^{-2}\frac{\partial^2 u}{\partial t^2} \qquad 0 < r < 1, \qquad t > 0 \qquad (8.49)$$

where $u = u(r, t)$. Since we have assumed the membrane is rigidly fixed on the boundary, we impose the boundary condition

$$u(1, t) = 0 \qquad t > 0 \qquad (8.50)$$

If the membrane is initially deflected to the shape $f(r)$ with velocity $g(r)$, we also prescribe the initial conditions

$$u(r, 0) = f(r) \qquad \frac{\partial u}{\partial t}(r, 0) = g(r) \qquad 0 < r < 1 \qquad (8.51)$$

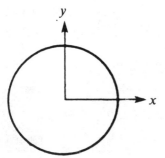

Figure 8.3 A circular membrane.

By expressing the unknown function u in (8.49) in the product form $u(r, t) = R(r)T(t)$, the separation of variables technique leads to

$$\frac{R'' + (1/r)R'}{R} = \frac{T''}{c^2 T} = -\lambda \tag{8.52}$$

where $-\lambda$ is the separation constant. (The choice of the negative sign is conventional, not necessary.) Hence (8.52) is equivalent to the system of ordinary DEs

$$rR'' + R' + \lambda r R = 0 \qquad 0 < r < 1 \tag{8.53}$$

$$T'' + \lambda c^2 T = 0 \qquad t > 0 \tag{8.54}$$

Under the assumption $u(r, t) = R(t)T(t)$, the boundary condition (8.50) becomes

$$u(1, t) = R(1)T(t) = 0$$

from which we deduce

$$R(1) = 0 \tag{8.55}$$

Equation (8.53) is recognized as a generalized form of Bessel's equation of order zero. By writing $\lambda = k^2 > 0$,* the general solution is

$$R(r) = C_1 J_0(kr) + C_2 Y_0(kr) \tag{8.56}$$

To maintain finite displacements of the membrane at $r = 0$, we must set $C_2 = 0$ since Y_0 becomes unbounded when the argument is zero. The remaining solution $R(r) = C_1 J_0(kr)$ must then satisfy the boundary condition (8.55), i.e.,

$$R(1) = C_1 J_0(k) = 0 \tag{8.57}$$

The Bessel function J_0 has infinitely many zeros (see Table 6.1) on the positive axis, but they are not evenly spaced. Thus, for $C_1 \neq 0$, we can satisfy (8.57) by selecting k as one of the zeros of J_0, which we denote by k_n ($n = 1, 2, 3, \ldots$). With k so restricted, we set $\lambda = k_n^2$ ($n = 1, 2, 3, \ldots$) in (8.54) to obtain

$$T'' + k_n^2 c^2 T = 0$$

---

*For $\lambda \leq 0$, Eq. (8.53) has no bounded nontrivial solutions satisfying (8.55) (see problem 15 in Exercises 8.4).

which has the general solution

$$T_n(t) = a_n \cos k_n ct + b_n \sin k_n ct \qquad n = 1, 2, 3, \ldots \qquad (8.58)$$

where the a's and b's are arbitrary constants.

Combining our results, we have the family of solutions

$$u_n(r, t) = (a_n \cos k_n ct + b_n \sin k_n ct)J_0(k_n r) \qquad n = 1, 2, 3, \ldots \qquad (8.59)$$

These solutions are called *standing waves*, since each can be viewed as having fixed shape $J_0(k_n r)$ with varying amplitude $T_n(t)$. The zeros of a standing wave, i.e., curves for which $J_0(k_n r) = 0$, are referred to as *nodal lines*. Clearly the number of nodal lines depends on the value of n. For example, when $n = 1$, there is no nodal line for $0 < r < 1$. There is one nodal line when $n = 2$; there are two nodal lines when $n = 3$; and so forth (see Fig. 8.4).

By forming a linear combination of the solutions (8.59) through the superposition principle, we obtain

$$u(r, t) = \sum_{n=1}^{\infty} (a_n \cos k_n ct + b_n \sin k_n ct)J_0(k_n r) \qquad (8.60)$$

The constants a_n and b_n are selected in such a way that the initial conditions (8.51) are satisfied. Therefore, we find

$$u(r, 0) = f(r) = \sum_{n=1}^{\infty} a_n J_0(k_n r) \qquad 0 < r < 1 \qquad (8.61)$$

and

$$\frac{\partial u}{\partial t}(r, 0) = g(r) = \sum_{n=1}^{\infty} k_n c b_n J_0(k_n r) \qquad 0 < r < 1 \qquad (8.62)$$

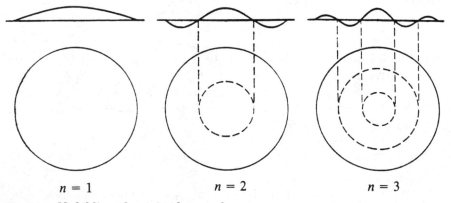

$$n = 1 \qquad\qquad n = 2 \qquad\qquad n = 3$$

Figure 8.4 Nodal lines for a circular membrane.

which are recognized as *Bessel series* for $f(r)$ and $g(r)$, respectively, where

$$a_n = \frac{2}{[J_1(k_n)]^2} \int_0^1 rf(r)J_0(k_n r)\, dr \qquad n = 1, 2, 3, \ldots \qquad (8.63)$$

and

$$k_n c b_n = \frac{2}{[J_1(k_n)]^2} \int_0^1 rg(r)J_0(k_n r)\, dr \qquad n = 1, 2, 3, \ldots \qquad (8.64)$$

8.4.2 Radial symmetric problems involving cylinders

Let us now consider a solid homogeneous cylinder with unit radius and height π units (see Fig. 8.5). If the temperatures on the surfaces of the cylinder are prescribed in such a way that they are a function of only the radial distance $r = \sqrt{x^2 + y^2}$ and height z, the temperature inside the cylinder will approach a *steady-state* distribution that depends only on r and z. In this case the governing DE is *Laplace's equation* which, in cylindrical coordinates, takes the form (see problem 1 in Exercises 8.4)

$$\frac{\partial^2 u}{\partial r^2} + \frac{1}{r}\frac{\partial u}{\partial r} + \frac{\partial^2 u}{\partial z^2} = 0 \qquad 0 < r < 1, \qquad 0 < z < \pi \qquad (8.65)$$

Let us assume for the sake of illustration that the boundary temperatures are prescribed by

$$u(r, 0) = 0 \qquad u(r, \pi) = 0 \qquad u(1, z) = f(z) \qquad (8.66)$$

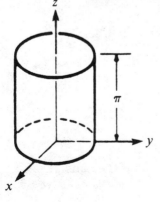

Figure 8.5 A solid cylinder.

If we set $u(r, z) = R(r)Z(z)$, then separation of variables leads to

$$rR'' + R' - \lambda rR = 0 \qquad 0 < r < 1 \qquad (8.67)$$

and

$$Z'' + \lambda Z = 0 \qquad 0 < z < \pi \qquad Z(0) = 0 \qquad Z(\pi) = 0 \qquad (8.68)$$

where λ is again the separation constant and where we have utilized the first two boundary conditions in (8.66). By assuming $\lambda = k^2 > 0$, the general solution of (8.68) is

$$Z(z) = C_1 \cos kz + C_2 \sin kz \qquad (8.69)$$

Imposing the first boundary condition in (8.68) requires that $C_1 = 0$, whereas the second boundary condition leads to

$$Z(\pi) = C_2 \sin k\pi = 0 \qquad (8.70)$$

For $C_2 \neq 0$, we can satisfy this last relation by choosing $k = n$ ($n = 1, 2, 3, \dots$). Hence, we find that

$$\lambda = k^2 = n^2 \qquad n = 1, 2, 3, \dots \qquad (8.71)$$

and consequently

$$Z_n(z) = \sin nz \qquad n = 1, 2, 3, \dots \qquad (8.72)$$

where we set $C_2 = 1$ for convenience. For $\lambda \leq 0$ there are no further (nontrivial) solutions of (8.68), the proof of which we leave to the exercises (see problem 16 in Exercises 8.4).

For values of λ given by (8.71), we see that (8.67) becomes

$$rR'' + R' - n^2 rR = 0$$

which we recognize as *Bessel's modified equation* of order zero, with general solution

$$R_n(r) = a_n I_0(nr) + b_n K_0(nr) \qquad n = 1, 2, 3, \dots \qquad (8.73)$$

However, because K_0 is unbounded at $r = 0$, we must choose $b_n = 0$ for all n. Then, combining solutions (8.72) and (8.73) through use of the superposition principle, we obtain

$$u(r, z) = \sum_{n=1}^{\infty} a_n I_0(nr) \sin nz \qquad (8.74)$$

The remaining task at this point is to determine the constants a_n, and this we do by imposing the last boundary condition in (8.66).

Thus we have

$$u(1, z) = f(z) = \sum_{n=1}^{\infty} a_n I_0(n) \sin nz \qquad 0 < z < \pi \qquad (8.75)$$

from which we deduce (see Sec. 1.4)

$$a_n I_0(n) = \frac{2}{\pi} \int_0^{\pi} f(z) \sin nz \, dz \qquad n = 1, 2, 3, \dots \qquad (8.76)$$

8.4.3 The Helmholtz equation

Bessel functions, in combination with the Legendre functions, are also prominent in the solution of the **Helmholtz equation**

$$\nabla^2 \psi + k^2 \psi = 0 \qquad (8.77)$$

in spherical coordinates (r, θ, ϕ). This partial DE is of particular importance in various problems in mathematical physics featuring both the heat equation and the wave equation (see problems 18 and 19 in Exercises 8.4).

By assuming that (8.77) has solutions of the form

$$\psi(r, \theta, \phi) = R(r)\Theta(\theta)\Phi(\phi) \qquad (8.78)$$

the separation of variables technique leads to the system of ordinary DEs (see problem 21 in Exercises 8.4)

$$\Theta'' + \lambda \Theta = 0 \qquad -\pi < \theta < \pi \qquad (8.79)$$

$$\frac{1}{\sin \phi} \frac{d}{d\phi} \left(\sin \phi \frac{d\Phi}{d\phi} \right) + \left(\mu - \frac{\lambda}{\sin^2 \phi} \right) \Phi = 0 \qquad 0 < \phi < \pi \quad (8.80)$$

and

$$\frac{d}{dr} \left(r^2 \frac{dR}{dr} \right) + (k^2 r^2 - \mu)R = 0 \qquad 0 < r < 1 \qquad (8.81)$$

where λ and μ are both separation constants. If we require that the solutions be bounded and periodic with period 2π, we arrive at the conclusion that the separation constants must assume the restricted values

$$\lambda = m^2 \qquad m = 0, 1, 2, \dots \qquad (8.82)$$

and

$$\mu = n(n+1) \qquad n = 0, 1, 2, \dots \qquad (8.83)$$

With the restrictions given by (8.82) and (8.83), we see that (8.79) has the solutions

$$\Theta_m(\theta) = \begin{cases} a_0 & m = 0 \\ a_m \cos m\theta + b_m \sin m\theta & m = 1, 2, 3, \ldots \end{cases} \quad (8.84)$$

where the a's and b's are arbitrary constants, and (8.80) is the *associated Legendre's equation* with bounded solutions

$$\phi_{mn}(\phi) = P_n^m(\cos \phi) \qquad m, n = 0, 1, 2, \ldots \quad (8.85)$$

Finally, with $\mu = n(n+1)$, we recognize Eq. (8.81) as a generalized form of Bessel's equation for which the bounded solutions are given by the *spherical Bessel functions* of the first kind (recall Sec. 7.4)

$$R_n(r) = j_n(kr) \qquad n = 0, 1, 2, \ldots \quad (8.86)$$

We conclude, therefore, that all bounded periodic solutions of the Helmholtz equation (8.77) in spherical coordinates are various linear combinations of the family of solutions

$$\psi_{mn}(r, \theta, \phi)$$
$$= \begin{cases} A_{0n} j_n(kr) P_n(\cos \phi) & m = 0 \\ (A_{mn} \cos m\theta + B_{mn} \sin m\theta) j_n(kr) P_n^m(\cos \phi) & m = 1, 2, 3, \ldots \end{cases}$$
$$(8.87)$$

where $n = 0, 1, 2, \ldots$ and the A's and B's are arbitrary constants.

Exercises 8.4

1. Polar coordinates are defined by
$$x = r \cos \theta \qquad y = r \sin \theta$$

   (a) Use the chain rule to show that
   $$\frac{\partial^2 u}{\partial x^2} + \frac{\partial^2 u}{\partial y^2} = \frac{\partial^2 u}{\partial r^2} + \frac{1}{r}\frac{\partial u}{\partial r} + \frac{1}{r^2}\frac{\partial^2 u}{\partial \theta^2}$$

   (b) If u does not depend upon θ, deduce that
   $$\frac{\partial^2 u}{\partial x^2} + \frac{\partial^2 u}{\partial y^2} = \frac{\partial^2 u}{\partial r^2} + \frac{1}{r}\frac{\partial u}{\partial r}$$

2. If the initial conditions (8.51) are given by
$$u(r, 0) = A J_0(k_3 r) \qquad (A \text{ constant})$$
$$\frac{\partial u}{\partial r}(r, 0) = 0$$

where $J_0(k_3) = 0$, show that the subsequent displacements of the membrane are described by

$$u(r, t) = AJ_0(k_3 r) \cos k_3 ct$$

3. If the initial conditions (8.51) are given by $f(r) = 0$ and $g(r) = 1$, show that the subsequent displacements of the membrane are described by

$$u(r, t) = \frac{2}{c} \sum_{n=1}^{\infty} \frac{\sin k_n ct}{k_n^2 J_1(k_n)} J_0(k_n r)$$

where $J_0(k_n) = 0$ $(n = 1, 2, 3, \dots)$.

4. Solve the problem described in Sec. 8.4.1 for a circular membrane of radius ρ.

5. The temperature distribution in a circular plate, independent of the polar angle θ, is described by solutions of

$$\frac{\partial^2 u}{\partial r^2} + \frac{1}{r} \frac{\partial u}{\partial r} = a^{-2} \frac{\partial u}{\partial t} \qquad 0 < r < 1, \qquad t > 0$$

$$\frac{\partial u}{\partial r}(1, t) = 0 \qquad u(r, 0) = f(r)$$

where a^2 is a physical constant. Show that solutions are of the form

$$u(r, t) = c_0 + \sum_{n=1}^{\infty} c_n J_0(k_n r) e^{-a^2 k_n^2 t}$$

where $J_0'(k_n) = 0$ $(n = 1, 2, 3, \dots)$.

6. Solve explicitly for the c's in problem 5 when the initial temperature distribution is prescribed by (see problem 13 in Exercises 6.5)
   (a) $f(r) = J_0(k_1 r)$, where $J_0'(k_1) = 0$
   (b) $f(r) = T_0$ (constant)
   (c) $f(r) = 1 - r^2$

7. Over a long, solid cylinder of unit radius at uniform temperature T_1 is fitted a long, hollow cylinder $(1 \leq r \leq 2)$ of the same material at uniform temperature T_2. Show that the temperature distribution through the two cylinders is given by

$$u(r, t) = T_2 + \frac{1}{2}(T_1 - T_2) \sum_{n=1}^{\infty} \frac{J_1(k_n)}{k_n [J_1(2k_n)]^2} J_0(k_n r) e^{-a^2 k_n^2 t}$$

where $J_0(2k_n) = 0$ $(n = 1, 2, 3, \dots)$. What temperature is approached in the limit as $t \to \infty$?

8. The temperature distribution $u(r, t)$ in a thin circular plate with heat exchanges from its faces into the surrounding medium at 0°C satisfies the boundary-value problem

$$\frac{\partial^2 u}{\partial r^2} + \frac{1}{r}\frac{\partial u}{\partial r} - bu = \frac{\partial u}{\partial t} \qquad 0 < r < 1, \qquad t > 0$$

where $b > 0$. Show that the solution is given by

$$u(r, t) = 2e^{-bt} \sum_{n=1}^{\infty} \frac{J_0(k_n r)}{k_n J_1(k_n)} e^{-k_n^2 t}$$

where $J_0(k_n) = 0$ $(n = 1, 2, 3, \ldots)$.

9. Solve the problem described in Sec. 8.4.2 when $f(z) = \sin 3z$.

10. A right circular cylinder is 1 m long and 2 m in diameter. One end and its lateral surface are maintained at a temperature of 0°C, and its other end is at 100°C. Calculate the first three terms in the series solution giving the temperature distribution at an interior point.

11. Show that the solutions of

$$\frac{\partial^2 u}{\partial r^2} + \frac{1}{r}\frac{\partial u}{\partial r} + \frac{\partial^2 u}{\partial z^2} = 0 \qquad 0 < r < 1, \qquad 0 < z < a$$

$$u(r, 0) = 0 \qquad u(r, a) = f(r) \qquad u(1, z) = 0$$

are given by

$$u(r, z) = \sum_{n=1}^{\infty} c_n J_0(k_n r) \frac{\sin k_n z}{\sinh k_n a}$$

where $J_0(k_n) = 0$ $(n = 1, 2, 3, \ldots)$. Find an expression for the constants c_n.

12. Find the solution forms for the boundary-value problem

$$\frac{\partial^2 u}{\partial r^2} + \frac{1}{r}\frac{\partial u}{\partial r} + \frac{\partial^2 u}{\partial z^2} = 0 \qquad 0 < r < 1, \qquad 0 < z < a$$

$$u(r, 0) = f(r) \qquad u(r, a) = 0 \qquad u(1, z) = 0$$

13. Suppose a cylindrical column of unit radius is considered to be of infinite height extending along the z axis. If the lateral surface is maintained at zero temperature and the initial temperature distribution inside the column is prescribed by $f(r, \theta)$, show that the subsequent temperature distribution of the column has the form

$$u(r, \theta, t) = \sum_{m=1}^{\infty} \sum_{n=0}^{\infty} (a_{mn} \cos n\theta + b_{mn} \sin n\theta) J_n(k_{mn} r) e^{-a^2 k_{mn}^2 t}$$

where $J_n(k_{mn}) = 0$ $(m = 1, 2, 3, \ldots, n = 0, 1, 2, \ldots)$.

14. A long hollow cylinder of inner radius α and outer radius β is initially heated to a temperature distribution $u(r, 0) = f(r)$. If both the inner and outer surfaces are maintained at zero temperature, show that the temperature distribution throughout the cylinder has solutions of the form

$$u(r, t) = \sum_{n-1}^{\infty} c_n [Y_0(k_n\alpha)J_0(k_nr) - J_0(k_n\alpha)Y_0(k_nr)]e^{-a^2k_n^2t}$$

where $Y_0(k_n\alpha)J_0(k_n\beta) = J_0(k_n\alpha)Y_0(k_n\beta)$ $(n = 1, 2, 3, \ldots)$.

15. Given the boundary-value problem

$$rR'' + R' + \lambda rR = 0 \qquad R(1) - 0$$

show that $R(r) = 0$ is the only bounded solution when
(a) $\lambda = 0$
(b) $\lambda = -k^2 < 0$

16. Given the boundary-value problem

$$Z'' + \lambda Z = 0 \qquad Z(0) = 0 \qquad Z(\pi) = 0$$

show that $Z(z) = 0$ is the only solution when
(a) $\lambda = 0$
(b) $\lambda = -k^2 < 0$

17. Show that the laplacian $\nabla^2 u$ in spherical coordinates defined by $x = r \cos \theta \sin \phi$, $y = r \sin \theta \sin \phi$, $z = r \cos \phi$ is given by

(a) $\nabla^2 u = \dfrac{\partial^2 u}{\partial r^2} + \dfrac{2}{r}\dfrac{\partial u}{\partial r} + \dfrac{1}{r^2}\dfrac{\partial^2 u}{\partial \phi^2} + \dfrac{\cot \phi}{r^2}\dfrac{\partial u}{\partial \phi} + \dfrac{1}{r^2 \sin^2 \phi}\dfrac{\partial^2 u}{\partial \theta^2}$

(b) $\nabla^2 u = \dfrac{1}{r^2}\left[\dfrac{\partial}{\partial r}\left(r^2 \dfrac{\partial u}{\partial r}\right) + \dfrac{1}{\sin \phi}\dfrac{\partial}{\partial \phi}\left(\sin \phi \dfrac{\partial u}{\partial \phi}\right) + \dfrac{1}{\sin^2 \phi}\dfrac{\partial^2 u}{\partial \theta^2}\right]$

18. Show, by assuming the product form $u(r, \theta, \phi, t) = \psi(r, \theta, \phi)T(t)$, that the heat equation

$$\nabla^2 u = a^{-2}\dfrac{\partial u}{\partial t}$$

reduces to the system of equations
$$\nabla^2 \psi + \lambda \psi = 0$$
$$T' + \lambda a^2 T = 0$$

19. Show, by assuming the product form $u(r, \theta, \phi, t) = \psi(r, \theta, \phi)T(t)$, that the wave equation

$$\nabla^2 u = c^{-2}\dfrac{\partial^2 u}{\partial t^2}$$

reduces to the system of equations

$$\nabla^2\psi + \lambda\psi = 0$$

$$T'' + \lambda c^2 T = 0$$

20. Let $u(r, \phi, t)$ denote the temperature distribution (independent of θ) in a sphere of unit radius, whose surface is maintained at 0°C and whose initial temperature is described by

$$u(r, \phi, 0) = f(r, \phi)$$

   (a) Show that the temperature distribution throughout the sphere at all later times is of the form (see problem 18)

$$u(r, \phi, t) = \sum_{m=1}^{\infty} \sum_{n=0}^{\infty} C_{mn} P_n(\cos \phi) j_n(k_{mn} r) e^{-a^2 k_{mn}^2 t}$$

   where $j_n(k_{mn}) = 0$ $(m = 1, 2, 3, \ldots, n = 0, 1, 2, \ldots)$.
   (b) If the initial temperature distribution is $f(r, \phi) = T_0$, where T_0 is constant, show that the solution in (a) reduces to

$$u(r, t) = \frac{2}{\pi} T_0 \sum_{m=1}^{\infty} (-1)^{m-1}\left(\frac{\sin m\pi r}{mr}\right) e^{-a^2 m^2 \pi^2 t}$$

21. Verify that the separation of variables technique applied to the Helmholtz equation (8.77) by the substitution (8.78) leads to the system of equations (8.79) to (8.81).

22. The velocity potential $u(r, \theta)$ of an acoustic wave scattered by a cylinder of radius b is a solution of

$$\frac{\partial^2 u}{\partial r^2} + \frac{1}{r}\frac{\partial u}{\partial r} + \frac{1}{r^2}\frac{\partial^2 u}{\partial \theta^2} + k^2 u = 0 \qquad b < r < \infty$$

$$u(b, \theta) = -e^{ikb \cos \theta} \qquad \lim_{r\to\infty} \sqrt{r}\left(\frac{\partial u}{\partial r} - iku\right) = 0$$

where the last condition is the *Sommerfeld radiation condition*.
   (a) Show that the separation of variables leads to

$$u(r, \theta) = C_0 H_0^{(1)}(kr) + \sum_{n=1}^{\infty} C_n H_n^{(1)}(kr) \cos n\theta$$

   Hint: $H_n^{(1)}(kr)$ denotes an outgoing wave.
   (b) To satisfy the finite boundary condition, show that

$$C_0 = -\frac{J_0(kb)}{H_0^{(1)}(kb)} \qquad C_n = -\frac{J_n(kb)}{H_n^{(1)}(kb)} \qquad n = 1, 2, 3, \ldots$$

   Hint: Use problem 9a in Exercises 6.2.

8.5   Step-Index Optical Fibers

An optical fiber used in light wave communication systems is a waveguide made of a transparent dielectric whose function is to guide the light over long distances. The fiber consists of an inner cylinder of glass, called the *core*, surrounded by a cylindrical shell of another material of lower refractive index, called the *cladding*. If the core has a uniform refractive index, it is called a **step-index fiber**. For such waveguides, the homogeneous wave equation can be solved in the core and cladding of the guide to obtain expressions for the electric and magnetic field intensities.

If we assume that the electric field has a sinusoidal variation in time at (angular) frequency ω, the wave equation reduces to the *Helmholtz equation*

$$\nabla^2 \mathbf{E} + \kappa^2 \mathbf{E} = 0 \tag{8.88}$$

where $\kappa^2 = \omega^2 \mu\epsilon - \beta^2$ and $\mathbf{E} = (E_x, E_y, E_z)$ is the time-independent *electric field intensity*. Here μ, ϵ, and β are medium constants known, respectively, as the *permeability, permittivity,* and *propagation constant*. The related time-independent *magnetic field intensity* $\mathbf{H} = (H_x, H_y, H_z)$ satisfies the same partial DE. Let us assume that the direction of propagation is the z direction. It can be shown in this case that both E_z and H_z satisfy the two-dimensional form of (8.88) given by

$$\frac{\partial^2 E_z}{\partial x^2} + \frac{\partial^2 E_z}{\partial y^2} + \kappa^2 E_z = 0 \tag{8.89}$$

and

$$\frac{\partial^2 H_z}{\partial x^2} + \frac{\partial^2 H_z}{\partial y^2} + \kappa^2 H_z = 0 \tag{8.90}$$

If the core and cladding of the fiber are circular (see Fig. 8.6), then the proper coordinate system in which to express (8.89) and (8.90) is cylindrical coordinates. By making the change of variables $x = r \cos\theta$, $y = r \sin\theta$, $z = z$, we find that (8.89) becomes (see problem 1 in Exercises 8.4)

$$\frac{\partial^2 E_z}{\partial r^2} + \frac{1}{r}\frac{\partial E_z}{\partial r} + \frac{1}{r^2}\frac{\partial^2 E_z}{\partial \theta^2} + \kappa^2 E_z = 0 \tag{8.91}$$

with an identical equation for H_z.

Let us assume that the core occupies the domain $0 \le r < a$ and that the cladding occupies $a < r < b$. Also we assume that b is large enough to ensure that the cladding field decays exponentially,

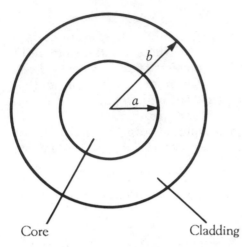

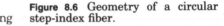

Figure 8.6 Geometry of a circular step-index fiber.

Core   Cladding

approaching zero at the cladding-air interface. The core and cladding are made of different materials, and the behavior of the fields is different in each part of the waveguide. Thus, we need to solve separate problems in each domain. However, the two solutions must agree at the core-cladding interface at $r = a$.

Due to circular symmetry in the fiber, we assume that (8.89) has solutions of the form

$$E_z(r, \theta) = R(r)e^{in\theta} \tag{8.92}$$

where n is a constant restricted to integer values. The direct substitution of (8.92) into (8.89), followed by division by $e^{in\theta}$, leads to the ordinary DE

$$r^2R'' + rR' + (\kappa^2r^2 - n^2)R = 0 \tag{8.93}$$

the general solution of which is

$$R(r) = C_1J_n(\kappa r) + C_2Y_n(\kappa r) \tag{8.94}$$

Energy considerations demand that the field be finite in the core of the fiber. Hence since $Y_n(\kappa r)$ becomes unbounded at $r = 0$, we must select $C_2 = 0$. In the core, therefore, we have the solution

$$E_z = AJ_n(\kappa r)e^{in\theta} \qquad 0 \le r < a \tag{8.95}$$

where A is an arbitrary constant.

The solution (8.94) does not lend itself to an exponential solution form for real κ. However, if we assume that $\kappa = i\gamma$, where $\gamma^2 = \beta^2 - \omega^2\mu\epsilon$, then the general solution of (8.93) becomes

$$R(r) = C_3I_n(\gamma r) + C_4K_n(\gamma r) \tag{8.96}$$

Of the two solutions, only $K_n(\gamma r)$ has an exponential behavior that tends to zero for large arguments. Therefore, we choose $C_0 = 0$, and the solution in the cladding takes the general form

$$E_z = BK_n(\gamma r)e^{in\theta} \qquad a < r < b* \tag{8.97}$$

where B is an arbitrary constant. Similar solutions for the magnetic field component are found to be

$$H_z = CJ_n(\kappa r)e^{in\theta} \qquad 0 \le r < a \tag{8.98}$$

and

$$H_z = DK_n(\gamma r)e^{in\theta} \qquad a < r < b \tag{8.99}$$

where C and D are arbitrary constants.

It can be shown that the transverse components E_r, E_θ, H_r, and H_θ are related to E_z and H_z according to†

$$E_r = -\frac{i}{\kappa^2}\left(\beta\frac{\partial E_z}{\partial r} + \omega\mu\frac{1}{r}\frac{\partial H_z}{\partial\theta}\right) \tag{8.100a}$$

$$E_\theta = -\frac{i}{\kappa^2}\left(\beta\frac{1}{r}\frac{\partial E_z}{\partial\theta} - \omega\mu\frac{\partial H_z}{\partial r}\right) \tag{8.100b}$$

$$H_r = -\frac{i}{\kappa^2}\left(-\beta\frac{\partial H_z}{\partial r} + \omega\epsilon\frac{1}{r}\frac{\partial E_z}{\partial\theta}\right) \tag{8.100c}$$

$$H_\theta = -\frac{i}{\kappa^2}\left(\beta\frac{1}{r}\frac{\partial H_z}{\partial\theta} + \omega\epsilon\frac{\partial E_z}{\partial r}\right) \tag{8.100d}$$

In the core we write $\epsilon = \epsilon_1$, and hence these equations reduce to

$$E_r = -\frac{i}{\kappa^2}\left[A\beta\kappa J_n'(\kappa r) + iC\omega\mu\frac{n}{r}J_n(\kappa r)\right]e^{in\theta} \tag{8.101a}$$

$$E_\theta = -\frac{i}{\kappa^2}\left[iA\beta\frac{n}{r}J_n(\kappa r) - C\kappa\omega\mu J_n'(\kappa r)\right]e^{in\theta} \tag{8.101b}$$

$$H_r = -\frac{i}{\kappa^2}\left[-iA\omega\epsilon_1\frac{n}{r}J_n(\kappa r) + C\kappa\beta J_n'(\kappa r)\right]e^{in\theta} \tag{8.101c}$$

$$H_\theta = -\frac{i}{\kappa^2}\left[A\kappa\omega\epsilon_1 J_n'(\kappa r) + iC\beta\frac{n}{r}J_n(\kappa r)\right]e^{in\theta} \tag{8.101d}$$

---

*An alternate choice to (8.97) is $E_z = BH_n^{(1)}(i\gamma r)e^{in\theta}$, where this is a *Hankel function* of imaginary argument.

† See, for example, A. H. Cherin, *An Introduction to Optical Fibers*, McGraw-Hill, New York, 1983.

The derivation of similar expressions for the cladding where $\epsilon = \epsilon_2$ is left to the exercises (see problem 1 in Exercises 8.5).

At this point constants A, B, C, and D are still unknown. To determine them, it is necessary to apply boundary conditions to the two tangential components of the electric and magnetic fields at the core-cladding interface $r = a$ (see the exercises).

Exercises 8.5

1. Show that the transverse components comparable to (8.101) for the solution in the cladding $(r > a)$ of the waveguide are given by

$$E_r = -\frac{i}{\gamma^2}\left[B\beta\gamma K_n'(\gamma r) + iD\omega\mu\frac{n}{r}K_n(\gamma r)\right]e^{in\theta}$$

$$E_\theta = -\frac{i}{\gamma^2}\left[iB\beta\frac{n}{r}K_n(\gamma r) - D\gamma\omega\mu K_n'(\gamma r)\right]e^{in\theta}$$

$$H_r = -\frac{i}{\gamma^2}\left[-iB\omega\epsilon_2\frac{n}{r}K_n(\gamma r) + D\gamma\beta K_n'(\gamma r)\right]e^{in\theta}$$

$$H_\theta = -\frac{i}{\gamma^2}\left[B\gamma\omega\epsilon_2 K_n'(\gamma r) + iD\beta\frac{n}{r}K_n(\gamma r)\right]e^{in\theta}$$

2. (a) For the special case where $n = 0$ (corresponding to only *meridional rays* propagating in the waveguide), show that Eqs. (8.101b) and (8.101d) for the core reduce to

$$E_\theta = \frac{i}{\kappa}C\omega\mu J_0'(\kappa r)$$

$$H_\theta = -\frac{i}{\kappa}A\omega\epsilon_1 J_0'(\kappa r)$$

   (b) Show that comparable equations in the cladding are

$$E_\theta = \frac{i}{\gamma}D\omega\mu K_0'(\gamma r)$$

$$H_\theta = -\frac{i}{\gamma}B\omega\epsilon_2 K_0'(\gamma r)$$

3. The boundary conditions at the core-cladding interface for the field components E_z and H_z are simply

$$E_z(\text{core}) = E_z(\text{cladding}) \qquad \text{at } r = a$$
$$H_z(\text{core}) = H_z(\text{cladding}) \qquad \text{at } r = a$$

   Show that these conditions lead to, respectively,

$$J_n(\kappa a)A - K_n(\gamma a)C = 0$$
$$J_n(\kappa a)B - K_n(\gamma a)D = 0$$

4. Referring to problem 3, the transverse components E_θ and H_θ must satisfy the additional boundary conditions

$$E_\theta(\text{core}) = E_\theta(\text{cladding}) \qquad \text{at } r = a$$

$$H_\theta(\text{core}) = H_\theta(\text{cladding}) \qquad \text{at } r = a$$

(a)   For the case $n = 0$, use the result of problem 2 to show that the boundary conditions lead to

$$J_0'(\kappa a)B - K_0'(\gamma a)D = 0$$

$$\epsilon_1 \gamma J_0'(\kappa a)A - \epsilon_2 \kappa K_0'(\gamma a)C = 0$$

(b)   Set $n = 0$ in the results of problem 3, and show that the simultaneous solution of the resulting equations with those in (a) leads to the conditions

$$\gamma J_1(\kappa a)K_0(\gamma a) - \kappa J_0(\kappa a)K_1(\gamma a) = 0$$

$$\epsilon_1 \gamma J_1(\kappa a)K_0(\gamma a) - \epsilon_2 \kappa J_0(\kappa a)K_1(\gamma a) = 0$$

[The first of these equations corresponds to the transverse electric (TE) mode for which the longitudinal component $E_z = 0$. The second equation corresponds to the transverse magnetic (TM) mode for which $H_z = 0$.]

The Hypergeometric Function

9.1 Introduction

Because of the many relations connecting the special functions to each other, and to the elementary functions, it is natural to inquire whether more general functions can be developed so that the special functions and elementary functions are merely specializations of these general functions. General functions of this nature have in fact been developed and are collectively referred to as *functions of the hypergeometric type*. There are several varieties of these functions, but the most common are the standard *hypergeometric function* (which we discuss in this chapter) and the *confluent hypergeometric function* (Chap. 10). Still, other generalizations exist, such as *MacRobert's E function* and *Meijer's G function,* for which even *generalized hypergeometric functions* are certain specializations (Chap. 11).

The major development of the theory of the hypergeometric function was carried out by Gauss and published in his famous memoir of 1812, a memoir that is also noted as being the real beginning of rigor in mathematics.* Some important results concerning the hypergeometric function had been developed earlier by Euler and others, but it was Gauss who made the first systematic study of the series that defines this function.

---

* C. F. Gauss, Disquisitiones Generales circa Seriem Infinitam..., *Comment. Soc. Reg. Sci. Gottingensis Recent.,* **2** (1812).

9.2 The Pochhammer Symbol

In dealing with certain product forms, factorials, and gamma functions, it is useful to introduce the abbreviation

$$(a)_0 = 1 \qquad (a)_n = a(a+1)\cdots(a+n-1) \qquad n = 1, 2, 3, \ldots$$

$$(9.1)$$

called the **Pochhammer symbol.** By properties of the gamma function, it follows that this symbol can also be defined by

$$(a)_n = \frac{\Gamma(a+n)}{\Gamma(a)} \qquad n = 0, 1, 2, \ldots \qquad (9.2)$$

Remark: For typographical convenience the symbol $(a)_n$ is sometimes replaced by *Appel's symbol* (a, n).

The Pochhammer symbol $(a)_n$ is important in most of the following material in this text. Because of its close association with the gamma function, it clearly satisfies a large number of identities. Some of the special properties are listed in Theorem 9.1, while other relations are taken up in the exercises (see also the Appendix).

Theorem 9.1. The Pochhammer symbol $(a)_n$ satisfies the identities:

(a) $(1)_n = n!$

(b) $(a+n)(a)_n = a(a+1)_n$

(c) $\begin{pmatrix} -a \\ n \end{pmatrix} = \dfrac{(-1)^n}{n!}(a)_n$

(d) $(a)_{n+k} = (a)_k(a+k)_n = (a)_n(a+n)_k$ (addition formula)

(e) $(a)_{k-n} = (-1)^n(a)_k/(1-a-k)_n$

(f) $(a)_{2n} = 2^{2n}(\tfrac{1}{2}a)_n(\tfrac{1}{2}+\tfrac{1}{2}a)_n$ (duplication formula)

Partial proof: We will prove only parts a, b, and c. The remaining proofs are left to the exercises.
From the definition, it follows that

(a)
$$(1)_n = 1 \times 2 \times \cdots \times n = n!$$

(b)
$$(a+n)(a)_n = a(a+1)\cdots(a+n-1)(a+n)$$
$$= a(a+1)_n$$

$$(c) \qquad \binom{-a}{n} = \frac{-a(-a-1)\cdots(-a-n+1)}{n!}$$

$$= \frac{(-1)^n}{n!} a(a+1)\cdots(a+n-1)$$

$$= \frac{(-1)^n}{n!} (a)_n$$

From the definition, we see that the parameter a can be either positive or negative, but generally we assume $a \neq 0$. An exception to this is the special value $(0)_0 = 1$. If a is a negative integer, we find that (see problem 17 in Exercises 9.2)

$$(-k)_n = \begin{cases} \dfrac{(-1)^n k!}{(k-n)!} & 0 \le n \le k \\ 0 & n > k \end{cases} \qquad (9.3)$$

Theorem 9.1e can be used to give meaning to the Pochhammer symbol for negative index; by setting $k = 0$ we obtain

$$(a)_{-n} = \frac{(-1)^n}{(1-a)_n} \qquad n = 1, 2, 3, \ldots \qquad (9.4)$$

Like the binomial coefficient, the Pochhammer symbol plays a very important role in combinatorial problems, probability theory, and algorithm development. In developing certain relations it is more convenient to use the Pochhammer symbol than the binomial coefficient. The use of this symbol (and the hypergeometric function) in the evaluation of certain series and combinatorial relations is illustrated in Sec. 9.5.

The Pochhammer symbol and binomial coefficient are related directly by the formula given in Theorem 9.1c. A more complex relation between these symbols is developed in the next example.

Example 1: Based on the properties of the Pochhammer symbol listed in Theorem 9.1, show that

$$\binom{a+k-1}{n} = \frac{(-1)^n (1-a)_n (a)_k}{n!\,(a-n)_k} \qquad k = 1, 2, 3, \ldots$$

Solution: From Theorem 9.1c and e, we first obtain

$$\binom{a+k-1}{n} = \frac{(-1)^n}{n!}(1-a-k)_n$$

$$= \frac{(a)_k}{n!\,(a)_{k-n}}$$

Replacing n by $-n$ in Theorem 9.1d, we find

$$(a)_{k-n} = (a)_{-n}(a-n)_k$$
$$= \frac{(-1)^n(a-n)_k}{(1-a)_n}$$

where the last step is a consequence of Eq. (9.4). Combining the above results leads to the desired relation

$$\binom{a+k-1}{n} = \frac{(-1)^n(1-a)_n(a)_k}{n!\,(a-n)_k}$$

Exercises 9.2

In problems 1 to 16, verify the identity.

1. $(-n)_n = (-1)^n n!$

2. $(a-n)_n = (-1)^n(1-a)_n$

3. $(a)_{n+1} = a(a+1)_n$

4. $(a)_{n+k} = (a)_k(a+k)_n$

5. $(a+1)_n - n(a+1)_{n-1} = (a)_n$

6. $(a-1)_n + n(a)_{n-1} = (a)_n$

7. $(n+k)! = n!\,(n+1)$

8. $\Gamma(a+1-n) = \dfrac{(-1)^n\Gamma(a+1)}{(-a)_n}$

9. $(a+n)_{k-n}(a+k)_{n-k} = 1$

10. $(a+k)_{n-k} = (-1)^{n-k}(1-a-n)_{n-k}$

11. $(a)_{k-n} = \dfrac{(-1)^n(a)_k}{(1-a-k)_n}$

12. $(a)_{2n} = 2^{2n}(\tfrac{1}{2}a)_n(\tfrac{1}{2}+\tfrac{1}{2}a)_n$

13. $(2n)! = 2^{2n}(\tfrac{1}{2})_n n!$

14. $(2n+1)! = 2^{2n}(\tfrac{3}{2})_n n!$

15. $\dbinom{2a}{2n} = \dfrac{(-a)_n(\tfrac{1}{2}-a)_n}{(\tfrac{1}{2})_n n!}$

16. $(a)_{3n} = 3^{3n}(\tfrac{1}{3}a)_n(\tfrac{1}{3}+\tfrac{1}{3}a)_n(\tfrac{2}{3}+\tfrac{1}{3}a)_n$

17. Show that $(k = 1, 2, 3, \dots)$

$$(-k)_n = \begin{cases} \dfrac{(-1)^n k!}{(k-n)!} & 0 \le n \le k \\ 0 & n > k \end{cases}$$

18. Show that

$$(1+n)_n = 4^n(\tfrac{1}{2})_n$$

19. Show that

$$\binom{n+a-1}{n} = \frac{(a)_n}{n!}$$

9.3 The Function $F(a, b; c, x)$

The series defined by*

$$1 + \frac{ab}{c}x + \frac{a(a+1)b(b+1)x^2}{c(c+1)\,2!} + \cdots = \sum_{n=0}^{\infty} \frac{(a)_n(b)_n}{(c)_n}\frac{x^n}{n!} \qquad (9.5)$$

is called the **hypergeometric series**. It gets its name from the fact that for $a = 1$ and $c = b$ the series reduces to the elementary *geometric series*

$$1 + x + x^2 + \cdots = \sum_{n=0}^{\infty} x^n \qquad (9.6)$$

Denoting the general term of (9.5) by $u_n(x)$ and applying the ratio test, we see that

$$\lim_{n \to \infty} \left| \frac{u_{n+1}(x)}{u_n(x)} \right| = \lim_{n \to \infty} \left| \frac{(a)_{n+1}(b)_{n+1}x^{n+1}}{(c)_{n+1}(n+1)!} \cdot \frac{(c)_n n!}{(a)_n(b)_n x^n} \right|$$

$$= |x| \lim_{n \to \infty} \left| \frac{(a+n)(b+n)}{(c+n)(n+1)} \right|$$

where we have made use of Theorem 9.1*d*. Completing the limit process reveals that

$$\lim_{n \to \infty} \left| \frac{u_{n+1}(x)}{u_n(x)} \right| = |x| \qquad (9.7)$$

under the assumption that none of a, b, or c is zero or a negative integer. Therefore, we conclude that the series (9.5) converges under these circumstances for all $|x| < 1$ and diverges for all $|x| > 1$. For $|x| = 1$, it can be shown that a sufficient condition for convergence of the series is $c - a - b > 0$.†

The function

$$F(a, b; c; x) = \sum_{n=0}^{\infty} \frac{(a)_n(b)_n}{(c)_n}\frac{x^n}{n!} \qquad |x| < 1 \qquad (9.8)$$

defined by the hypergeometric series is called the **hypergeometric function.** It is also commonly denoted by the symbol

$$_2F_1(a, b; c; x) \equiv F(a, b; c; x) \qquad (9.9)$$

where the 2 and 1 refer to the number of numerator and denominator parameters, respectively, in its series representation. The semicolons

---

* Throughout our discussion the parameters a, b, c are assumed to be real.

† See E. D. Rainville, *Special Functions*, Chelsea, New York, 1971, p. 46.

separate the numerator parameters a and b (which are themselves separated by a comma), the denominator parameter c, and the argument x.

If c is zero or a negative integer, the series (9.8) generally does not exist, and hence the function $F(a, b; c; x)$ is not defined. However, if either a or b (or both) is zero or a negative integer, the series is finite and thus converges for *all* x. That is, if $a = -m$ ($m = 0, 1, 2, \dots$), then $(-m)_n = 0$ when $n > m$, and in this case (9.8) reduces to the hypergeometric polynomial defined by

$$F(-m, b; c; x) = \sum_{n=0}^{m} \frac{(-m)_n (b)_n}{(c)_n} \frac{x^n}{n!} \qquad -\infty < x < \infty \qquad (9.10)$$

9.3.1 Elementary properties

There are several properties of the hypergeometric function that are immediate consequences of its definition (9.8). First, we note the *symmetry property* of the parameters a and b, that is,

$$F(a, b; c; x) = F(b, a; c; x) \qquad (9.11)$$

Second, by differentiating the series (9.8) termwise, we find that

$$\frac{d}{dx} F(a, b; c; x) = \underbrace{\sum_{n=1}^{\infty} \frac{(a)_n (b)_n}{(c)_n} \frac{x^{n-1}}{(n-1)!}}_{n \to n+1}$$

$$= \sum_{n=0}^{\infty} \frac{(a)_{n+1} (b)_{n+1}}{(c)_{n+1}} \frac{x^n}{n!}$$

$$= \frac{ab}{c} \sum_{n=0}^{\infty} \frac{(a+1)_n (b+1)_n}{(c+1)_n} \frac{x^n}{n!}$$

hence

$$\frac{d}{dx} F(a, b; c; x) = \frac{ab}{c} F(a+1, b+1; c+1; x) \qquad (9.12)$$

Repeated application of (9.12) leads to the general formula (see problem 1 in Exercises 9.3)

$$\frac{d^k}{dx^k} F(a, b; c; x) = \frac{(a)_k (b)_k}{(c)_k} F(a+k, b+k; c+k; x)$$

$$k = 1, 2, 3, \dots \qquad (9.13)$$

The parameters a, b, and c in the definition of the hypergeometric function play much the same role in the relationships of this function as the parameter n or p did for the Legendre polynomials and Bessel functions. The usual nomenclature for the hypergeometric functions in which one parameter changes by $+1$ or -1 is *contiguous functions*. There are six contiguous functions, defined by $F(a \pm 1, b; c; x)$, $F(a, b \pm 1; c; x)$, and $F(a, b; c \pm 1; x)$. Gauss was the first to show that between $F(a, b; c; x)$ and any two contiguous functions there exists a linear relation with coefficients at most linear in x. The 6 contiguous functions, taken 2 at a time, lead to a total of 15 recurrence relations of this kind, that is, $\binom{6}{2} = 15.$*

To derive one of the 15 recurrence relations, we first observe that

$$x \frac{d}{dx} F(a, b; c; x) + aF(a, b; c; x)$$

$$= \sum_{n=0}^{\infty} \frac{(a)_n (b)_n}{(c)_n} \frac{nx^n}{n!} + \sum_{n=0}^{\infty} \frac{a(a)_n (b)_n}{(c)_n} \frac{x^n}{n!}$$

$$= \sum_{n=0}^{\infty} \frac{(a+n)(a)_n (b)_n}{(c)_n} \frac{x^n}{n!}$$

$$= a \sum_{n=0}^{\infty} \frac{(a+1)_n (b)_n}{(c)_n} \frac{x^n}{n!}$$

from which we deduce

$$x \frac{d}{dx} F(a, b; c; x) + aF(a, b; c; x) = aF(a+1, b; c; x) \qquad (9.14)$$

Similarly, from the symmetry property (9.11),

$$x \frac{d}{dx} F(a, b; c; x) + bF(a, b; c; x) = bF(a, b+1; c; x) \qquad (9.15)$$

and by subtracting (9.15) from (9.14), it follows at once that

$$(a-b)F(a, b; c; x) = aF(a+1, b; c; x) - bF(a, b+1; c; x)$$

$$(9.16)$$

---

*For a listing of all 15 relations, see A. Erdelyi et al., *Higher Transcendental Functions*, vol. 1, McGraw-Hill, New York, 1953, pp. 103–104.

which is one of the simplest recurrence relations involving the contiguous functions. Some of the other recurrence relations are taken up in the exercises.

9.3.2   Integral representation

To derive an integral representation for the hypergeometric function, we start with the beta-function relation (see Sec. 2.4)

$$B(n+b, c-b) = \int_0^1 t^{n+b-1}(1-t)^{c-b-1}\,dt \qquad c>b>0 \quad (9.17)$$

from which we deduce (for $n = 0, 1, 2, \dots$)

$$\frac{(b)_n}{(c)_n} = \frac{\Gamma(c)}{\Gamma(b)\Gamma(c-b)} \int_0^1 t^{n+b-1}(1-t)^{c-b-1}\,dt \qquad (9.18)$$

The substitution of (9.18) into (9.8) yields

$$F(a, b; c; x) = \frac{\Gamma(c)}{\Gamma(b)\Gamma(c-b)} \sum_{n=0}^{\infty} \frac{(a)_n}{n!} x^n \int_0^1 t^{n+b-1}(1-t)^{c-b-1}\,dt$$

$$= \frac{\Gamma(c)}{\Gamma(b)\Gamma(c-b)} \int_0^1 t^{b-1}(1-t)^{c-b-1}\left[\sum_{n=0}^{\infty} \frac{(a)_n}{n!}(xt)^n\right] dt \quad (9.19)$$

where we have reversed the order of integration and summation. Now, using the relation (from Theorem 9.1)

$$\frac{(a)_n}{n!} = \binom{-a}{n}(-1)^n \qquad (9.20)$$

we recognize the series in (9.19) as a binomial series which has the sum

$$\sum_{n=0}^{\infty} \frac{(a)_n}{n!}(xt)^n = \sum_{n=0}^{\infty} \binom{-a}{n}(-xt)^n = (1-xt)^{-a} \qquad (9.21)$$

provided $|xt| < 1$. Hence, (9.19) gives us the **integral representation**

$$F(a, b; c; x) = \frac{\Gamma(c)}{\Gamma(b)\Gamma(c-b)} \int_0^1 t^{b-1}(1-t)^{c-b-1}(1-xt)^{-a}\,dt$$

$$c>b>0 \quad (9.22)$$

Although (9.22) was derived under the assumption that $|xt| < 1$, it can be shown that the integral converges also for $x = 1$.* The convergence of (9.22) for $x = 1$ is important in our proof of the following useful theorem.

Theorem 9.2.   For $c \neq 0, -1, -2, \ldots$ and $c - a - b > 0$,

$$F(a, b; c; 1) = \frac{\Gamma(c)\Gamma(c - a - b)}{\Gamma(c - a)\Gamma(c - b)}$$

Proof: We prove the theorem only with the added restriction $c > b > 0$, although it is valid without this restriction. We simply set $x = 1$ in (9.22) to get

$$F(a, b; c; 1) = \frac{\Gamma(c)}{\Gamma(b)\Gamma(c - b)} \int_0^1 t^{b-1}(1 - t)^{c-b-1}(1 - t)^{-a}\, dt$$

$$= \frac{\Gamma(c)}{\Gamma(b)\Gamma(c - b)} \int_0^1 t^{b-1}(1 - t)^{c-a-b-1}\, dt$$

which, evaluated as a beta integral, yields our result

$$F(a, b; c; 1) = \frac{\Gamma(c)\Gamma(b)\Gamma(c - a - b)}{\Gamma(b)\Gamma(c - b)\Gamma(c - a)}$$

$$= \frac{\Gamma(c)\Gamma(c - a - b)}{\Gamma(c - a)\Gamma(c - b)}$$

9.3.3   The hypergeometric equation

The linear second-order DE

$$x(1 - x)y'' + [c - (a + b + 1)x]y' - aby = 0 \qquad (9.23)$$

is called the **hypergeometric equation** of Gauss. It is so named because the function

$$y_1 = F(a, b; c; x) \qquad c \neq 0, -1, -2, \ldots \qquad (9.24)$$

is a solution. To verify that (9.24) is indeed a solution, we can substitute the series for $F(a, b; c; x)$ directly into (9.23).

Examination of the coefficient of y'' reveals that both $x = 0$ and $x = 1$ are (finite) singular points of the equation. Therefore, to find a

---

* See E. D. Rainville, *Special Functions*, Chelsea, New York, 1971, pp. 48–49.

second series solution about $x = 0$ would normally require use of the Frobenius method.* Under special restrictions on the parameter c, however, we can produce a second (linearly independent) solution of (9.23) without resorting to this more general method. We simply make the change of dependent variable

$$y = x^{1-c}z \tag{9.25}$$

from which we calculate

$$y' = x^{1-c}z' + (1-c)x^{-c}z \tag{9.26a}$$

$$y'' = x^{1-c}z'' + 2(1-c)x^{-c}z' - c(1-c)x^{-c-1}z \tag{9.26b}$$

The substitution of (9.25), (9.26a), and (9.26b) into (9.23) leads to (upon algebraic simplification).

$$x(1-x)z'' + [2 - c - (a+b-2c+3)x]z'$$
$$- (1+a-c)(1+b-c)z = 0 \tag{9.27}$$

which we recognize as another form of (9.23). Hence Eq. (9.27) has the solution

$$z = F(1+a-c, 1+b-c; 2-c; x) \qquad c \neq 2, 3, 4, \ldots \tag{9.28}$$

and so we deduce that

$$y_2 = x^{1-c}F(1+a-c, 1+b-c; 2-c; x) \qquad c \neq 2, 3, 4, \ldots \tag{9.29}$$

is a second solution of (9.23). For $c = 2, 3, 4, ..$, the hypergeometric function in (9.29) does not usually exist, while for $c = 1$ the solutions (9.29) and (9.24) are identical. However, if we restrict c to $c \neq 0, \pm 1, \pm 2, \ldots$, then (9.29) is linearly independent of (9.24) and

$$y = C_1 F(a, b; c; x) + C_2 x^{1-c}F(1+a-c, 1+b-c; 2-c; x)$$
$$\tag{9.30}$$

is a general solution of Eq. (9.23).

To cover the cases when $c = 2, 3, 4, \ldots$, a *hypergeometric function of the second kind* can be introduced (see problem 28 in Exercises 9.3). However, beyond its connection as a solution to the hypergeometric equation of Gauss, the hypergeometric function of the second kind has limited usefulness in applications.

---

*For an introductory discussion of the Frobenius method, see L. C. Andrews, *Introduction to Differential Equations with Boundary Value Problems*, Harper-Collins, New York, 1991, chap. 7.

Remark: Actually $y_1 = F(a, b; c; x)$ and $y_2 = x^{1-c}F(1 + a - c, 1 + b - c; 2 - c; x)$ are only two of a total of 24 solutions of Eq. (9.23) that can be expressed in terms of the hypergeometric function. For a listing of all 24 solutions, see W. W. Bell, *Special Functions for Scientists and Engineers,* Van Nostrand, London, 1968, pp. 208–209.

Exercises 9.3

1. Show that (for $k = 1, 2, 3, \ldots$)

$$\frac{d^k}{dx^k} F(a, b; c; x) = \frac{(a)_k(b)_k}{(c)_k} F(a + k, b + k; c + k; x)$$

2. Show that (for $k = 1, 2, 3, \ldots$)

   (a) $\dfrac{d}{dx} [x^a F(a, b; c; x)] = ax^{a-1}F(a + 1, b; c; x)$

   (b) $\dfrac{d^k}{dx^k} [x^{a-1+k}F(a, b; c; x)] = (a)_k x^{a-1}F(a + k, b; c; x)$

   (c) $\dfrac{d^x}{dx^k} [x^{c-1}F(a, b; c; x)] = (c - k)_k x^{c-1-k}F(a, b; c - k; x)$

In problems 3 to 6, verify the differentiation formula.

3. $x \dfrac{d}{dx} F(a, b; c; x) + (c - 1)F(a, b; c; x) = (c - 1)F(a, b; c - 1; x)$

4. $x \dfrac{d}{dx} F(a - 1, b; c; x) = (a - 1)F(a, b; c; x)$

$$- (a - 1)F(a - 1, b; c; x)$$

5. $(1 - x)x \dfrac{d}{dx} F(a, b; c; x) = (a + b - c)xF(a, b; c; x)$

$$+ c^{-1}(c - a)(c - b)xF(a, b; c + 1; x)$$

6. $x \dfrac{d}{dx} F(a - 1, b; c; x) = (a - 1)xF(a, b; c; x)$

$$- c^{-1}(a - 1)(c - b)xF(a, b; c + 1; x)$$

In problems 7 to 13, verify the given contiguous relation by using the results of problems 3 to 6 or by series representations.

7. $(b - a)(1 - x)F(a, b; c; x) = (c - a)F(a - 1, b; c; x)$

$$-(c - b)F(a, b - 1; c; x)$$

8. $(1 - x)F(a, b; c; x) = F(a - 1, b; c; x)$

$$- c^{-1}(c - b)xF(a, b; c + 1; x)$$

9. $(1-x)F(a, b; c; x) = F(a, b-1; c; x)$
$$-c^{-1}(c-a)xF(a, b; c+1; x)$$

10. $(c-a-b)F(a, b; c; x) + a(1-x)F(a+1, b; c; x)$
$$= (c-b)F(a, b-1; c; x)$$

11. $(c-a-b)F(a, b; c; x) + b(1-x)F(a, b+1; c; x)$
$$= (c-a)F(a-1, b; c; x)$$

12. $(c-b-1)F(a, b; c; x) + bF(a, b+1; c; x)$
$$= (c-1)F(a, b; c-1; x)$$

13. $[2b-c+(a-b)x]F(a, b; c; x) = b(1-x)F(a, b+1; c; x)$
$$-(c-b)F(a, b-1; c; x)$$

In problems 14 and 15, verify the formula by direct substitution of the series representations.

14. $F(a, b+1; c; x) - F(a, b; c; x) = \dfrac{ax}{c}F(a+1, b+1; c+1; x)$

15. $F(a, b; c; x) - F(a, b; c-1; x)$
$$= -\dfrac{abx}{c(c-1)}F(a+1, b+1; c+1; x)$$

In problems 16 and 17, use termwise integration to derive the given integral representation.

16. $F(a, b; c; x) = \dfrac{\Gamma(c)}{\Gamma(d)\Gamma(c-d)} \displaystyle\int_0^1 t^{d-1}(1-t)^{c-d-1}F(a, b; d; xt)\, dt,$
$$c > d > 0$$

17. $F(a, b; c+1; x) = c \displaystyle\int_0^1 F(a, b; c; xt)t^{c-1}\, dt, \ c > 0$

18. Show that $(s > 0)$

(a) $\displaystyle\int_0^\infty e^{-st}F[a, b; 1; x(1-e^{-t})]\, dt = \dfrac{1}{s}F(a, b; s+1; x)$

(b) $\displaystyle\int_0^\infty e^{-st}F(a, b; 1; 1-e^{-t})\, dt = \dfrac{\Gamma(s)\Gamma(s+1-a-b)}{\Gamma(s+1-a)\Gamma(s+1-b)}$

   Hint: Set $x = 1$ in (a).

19. Show that (for $n = 0, 1, 2, ..$)

(a) $F(-n, b; c; 1) = \dfrac{(c-b)_n}{(c)_n}$

(b) $F(-n, a+n; c; 1) = (-1)^n \dfrac{(1+a-c)_n}{(c)_n}$

(c) $F(-n, 1-b-n; c; 1) = \dfrac{(c+b-1)_{2n}}{(c)_n(c+b-1)_n}$

20. Show that

$$F(-\tfrac{1}{2}n, \tfrac{1}{2} - \tfrac{1}{2}n; b + \tfrac{1}{2}; 1) = \frac{2^n (b)_n}{(2b)_n} \qquad n = 0, 1, 2, \ldots$$

21. Using the result of problem 19a, show that (for $p = 0, 1, 2, \ldots$)

(a) $F(-p, a + n + 1; a + 1; 1) = \begin{cases} 0 & 0 \leq n \leq p - 1 \\ \dfrac{(-1)^p p!}{(a+1)_p} & n = p \end{cases}$

(b) $F(-p, a + n + 2; a + 1; 1)$

$$= \begin{cases} 0 & 0 \leq n \leq p - 2 \\ \dfrac{(-1)^p (n+1)!}{(a+1)_p (n+1-p)!} & n = p - 1, p \end{cases}$$

22. Given the generating function

$$w(x, t) = (1 - t)^{b-c}(1 - t + xt)^{-b} \qquad c \neq 0, -1, -2, \ldots$$

show that

$$w(x, t) = \sum_{n=0}^{\infty} \frac{(c)_n}{n!} F(-n, b; c; x) t^n$$

where $F(-n, b; c; x)$ denotes the *hypergeometric polynomials* defined by Eq. (9.10).

23. Show that for $|x| < 1$ and $|x/(1-x)| < 1$,

$$(1 - x)^{-a} F\left(a, c - b; c; \frac{-x}{1-x}\right) = F(a, b; c; x)$$

24. By substituting $y = x/(x - 1)$ in problem 23, deduce that

(a) $F(a, c - b; c; x) = (1 - x)^{b-c} F\left(c - a, c - b; c; \dfrac{-x}{1-x}\right)$

(b) $F\left(a, c - b; c; \dfrac{-x}{1-x}\right) = (1 - x)^{c-b} F(c - a, c - b; c; x)$

(c) $F(a, b; c; x) = (1 - x)^{c-a-b} F(c - a, c - b; c; x)$

25. Show that

$$(1 - x)^{-a} F\left[\frac{1}{2}a, \frac{1}{2} + \frac{1}{2}a - b; 1 + a - b; \frac{-4x}{(1-x)^2}\right]$$

$$= F(a, b; 1 + a - b; x)$$

26. Use problems 23 to 25 to deduce that

(a) $\quad F(a, b; 1 + a - b; -1) = \dfrac{\Gamma(1 + a - b)\Gamma(1 + \frac{1}{2}a)}{\Gamma(1 + a)\Gamma(1 + \frac{1}{2}a - b)}$

(b) $\quad F\left(a, 1 - a; c; \dfrac{1}{2}\right) = \dfrac{\Gamma(\frac{1}{2}c)\Gamma(\frac{1}{2}c + \frac{1}{2})}{\Gamma(\frac{1}{2}a + \frac{1}{2}c)\Gamma(\frac{1}{2} - \frac{1}{2}a + \frac{1}{2}c)}$

(c) $\quad F\left(2a, 2b; a + b + \dfrac{1}{2}; \dfrac{1}{2}\right) = \dfrac{\Gamma(a + b + \frac{1}{2})\sqrt{\pi}}{\Gamma(a + \frac{1}{2})\Gamma(b + \frac{1}{2})}$

27. By assuming a power series solution of the form

$$y = \sum_{n=0}^{\infty} A_n x^n$$

show that $y = F(a, b; c; x)$ is a solution of the hypergeometric equation

$$x(1 - x)y'' + [c - (a + b + 1)x]y' - aby = 0$$

28. The *hypergeometric function of the second kind* is defined by

$$G(a, b; c; x) = \dfrac{\Gamma(1 - c)}{\Gamma(a - c + 1)\Gamma(b - c + 1)} F(a, b; c; x)$$

$$+ \dfrac{\Gamma(c - 1)}{\Gamma(a)\Gamma(b)} x^{1-c} F(1 + a - c, 1 + b - c; 2 - c; x)$$

(a) Show that $G(a, b; c; x)$ is a solution of the hypergeometric equation in problem 27, $c \neq 0, \pm 1, \pm 2, \ldots$.

(b) Show that $G(a, b; c; x) = x^{1-c} G(1 + a - c, 1 + b - c; 2 - c; x)$.

29. Show that the wronksian of $F(a, b; c; x)$ and $G(a, b; c; x)$ is given by (see problem 28)

$$W(F, G)(x) = -\dfrac{\Gamma(c)}{\Gamma(a)\Gamma(b)} x^{-c}(1 - x)^{c - a - b - 1}$$

30. Derive the generating-function relation

$$(1 - xt)^{-a} F\left[\dfrac{1}{2}a, \dfrac{1}{2}a + \dfrac{1}{2}; 1, \dfrac{t^2(x^2 - 1)}{(1 - xt)^2}\right] = \sum_{n=0}^{\infty} \dfrac{(a)_n P_n(x)}{n!} t^n$$

where $P_n(x)$ is the nth Legendre polynomial.

9.4   Relation to Other Functions

The hypergeometric function is important in many areas of mathematical analysis and its applications. Partly this is a consequence of the fact that so many elementary and special functions are simply special cases of the hypergeometric function. For example, the specialization

$$F(1, b; b; x) = \sum_{n=0}^{\infty} \frac{(1)_n}{n!} x^n = \sum_{n=0}^{\infty} x^n$$

reveals that

$$F(1, b; b; x) = (1 - x)^{-1} \qquad (9.31)$$

Similarly, it can be established that

$$\arcsin x = xF({}^{1}\!/_{2}, {}^{1}\!/_{2}; {}^{3}\!/_{2}; x^2) \qquad (9.32)$$

and

$$\ln(1 - x) = -xF(1, 1; 2; x) \qquad (9.33)$$

Example 2: Show that $\arcsin x = xF({}^{1}\!/_{2}, {}^{1}\!/_{2}; {}^{3}\!/_{2}; x^2)$.

Solution: From the calculus, we recall that

$$\arcsin x = \sum_{n=0}^{\infty} \frac{(2n)! \, x^{2n+1}}{2^{2n}(n!)^2(2n+1)}$$

To recognize this series as a hypergeometric series, we need to express the coefficient of $x^{2n+1}/n!$ in terms of Pochhammer symbols. Thus, using the results of problems 13 and 14 in Exercises 9.2, we have

$$(2n)! = 2^{2n}({}^{1}\!/_{2})_n n!$$

$$(2n + 1) = \frac{(2n + 1)!}{(2n)!} = \frac{({}^{3}\!/_{2})_n}{({}^{1}\!/_{2})_n}$$

and making these substitutions leads to

$$\arcsin x = x \sum_{n=0}^{\infty} \frac{({}^{1}\!/_{2})_n({}^{1}\!/_{2})_n}{({}^{3}\!/_{2})_n} \frac{x^{2n}}{n!}$$

from which we deduce

$$\arcsin x = xF({}^{1}\!/_{2}, {}^{1}\!/_{2}; {}^{3}\!/_{2}; x^2)$$

The verification of (9.33) along with several other such relations involving elementary functions is left to the exercises.

A more involved relationship to establish is given by

$$P_n(x) = F\left(-n, n + 1; 1; \frac{1 - x}{2}\right) \qquad (9.34)$$

where $P_n(x)$ is the nth Legendre polynomial. To prove (9.34), we first observe that

$$(1 - 2xt + t^2)^{-1/2} = [(1 - t)^2 - 2t(x - 1)]^{-1/2}$$

$$= (1 - t)^{-1}\left[1 - \frac{2t(x - 1)}{(1 - t)^2}\right]^{-1/2} \tag{9.35}$$

and thus deduce the relation

$$\sum_{n=0}^{\infty} P_n(x)t^n = (1 - 2xt + t^2)^{-1/2}$$

$$= (1 - t)^{-1}\left[1 - \frac{2t(x - 1)}{(1 - t)^2}\right]^{-1/2}$$

$$= \sum_{k=0}^{\infty} \binom{-\frac{1}{2}}{k} \frac{(-1)^k (2t)^k (x - 1)^k}{(1 - t)^{2k+1}} \tag{9.36}$$

Our objective now is to recognize the right-hand side of (9.36) as a power series in t which has the coefficient $F[-n, n + 1; 1; (1 - x)/2]$. To obtain powers of t, we further expand $(1 - t)^{-2k-1}$ in a binomial series and interchange the order of summation. Hence,

$$\sum_{n=0}^{\infty} P_n(x)t^n = \sum_{m=0}^{\infty} \sum_{k=0}^{\infty} \binom{-\frac{1}{2}}{k}\binom{-2k - 1}{m}(-1)^{m+k} 2^k (x - 1)^k t^{k+m}$$

$$= \sum_{n=0}^{\infty} \sum_{k=0}^{n} \binom{-\frac{1}{2}}{k}\binom{-2k - 1}{n - k}(-1)^n 2^k (x - 1)^k t^n \tag{9.37}$$

where the last step is a result of the index change $m = n - k$. Next, from Theorem 9.1c we can write

$$\binom{-\frac{1}{2}}{k}\binom{-2k - 1}{n - k} = \frac{(-1)^n (\frac{1}{2})_k (2k + 1)_{n-k}}{k!(n - k)!} \tag{9.38}$$

but from problems 7 and 13 in Exercises 9.2, we further have

$$(2k + 1)_{n-k} = \frac{(n + k)!}{(2k)!} = \frac{n!(n + 1)_k}{2^{2k}(\frac{1}{2})_k k!} \tag{9.39}$$

Finally, setting $a = 1$ in problem 11 in Exercises 9.2 leads to

$$(n - k)! = \frac{(-1)^k n!}{(-n)_k} \tag{9.40}$$

so by combining the results of (9.38), (9.39), and (9.40), we find that (9.37) becomes

$$\sum_{n=0}^{\infty} P_n(x)t^n = \sum_{n=0}^{\infty} \left[\sum_{k=0}^{n} \frac{(-n)_k(n+1)_k}{(1)_k k!} \left(\frac{1-x}{2}\right)^k \right] t^n$$

$$= \sum_{n=0}^{\infty} F\left(-n, n+1; 1; \frac{1-x}{2}\right) t^n \qquad (9.41)$$

from which (9.34) follows by comparison of like terms.

9.4.1 Legendre functions

The relation (9.34) between the nth Legendre polynomial and the hypergeometric function provides us with a natural way of introducing the more general function

$$P_v(x) = F\left(-v, v+1; 1; \frac{1-x}{2}\right) \qquad (9.42)$$

where v is not restricted to integer values. We call $P_v(x)$ a **Legendre function of the first kind** of degree v; it is not a polynomial except in the special case when $v = n$ ($n = 0, 1, 2, \ldots$). A *Legendre function of the second kind*, denoted $Q_v(x)$, also can be defined in terms of the hypergeometric function, although we do not discuss it.*

The function $P_v(x)$ has many properties in common with the Legendre polynomial $P_n(x)$. For example, by setting $x = 1$ in (9.42), we obtain

$$P_v(1) = F(-v, v+1; 1; 0) = 1 \qquad (9.43)$$

The substitution of $x = 0$ in (9.42) leads to

$$P_v(0) = F(-v, v+1; 1; \tfrac{1}{2}) \qquad (9.44)$$

and by using the relation (see problem 26b in Exercises 9.3)

$$F\left(a, 1-a; c; \frac{1}{2}\right) = \frac{\Gamma(\tfrac{1}{2}c)\Gamma(\tfrac{1}{2}c + \tfrac{1}{2})}{\Gamma(\tfrac{1}{2}a + \tfrac{1}{2}c)\Gamma(\tfrac{1}{2} - \tfrac{1}{2}a + \tfrac{1}{2}c)} \qquad (9.45)$$

we deduce that

$$P_v(0) = \frac{\sqrt{\pi}}{\Gamma(\tfrac{1}{2} - \tfrac{1}{2}v)(\Gamma(\tfrac{1}{2}v + 1)} \qquad (9.46)$$

---

* See T. M. MacRobert, *Spherical Harmonics*, Pergamon, Oxford, 1967, chap. 6.

Recalling the identity

$$\Gamma(x)\Gamma(1-x) = \frac{\pi}{\sin \pi x} \tag{9.47}$$

we can express (9.46) in the alternative form

$$P_\nu(0) = \frac{\Gamma(\tfrac{1}{2}\nu + \tfrac{1}{2})}{\sqrt{\pi}\,\Gamma(\tfrac{1}{2}\nu + 1)} \cos \frac{\nu\pi}{2} \tag{9.48}$$

When ν is a nonnegative integer, we find that (9.48) reduces to the results that we previously derived for the Legendre polynomials (see problem 22 in Exercises 9.4).

Various recurrence formulas for $P_\nu(x)$ can be derived by expressing this function in its series representation or by using properties of the hypergeometric function. For example, it can be verified that

$$(\nu + 1)P_{\nu+1}(x) - (2\nu + 1)xP_\nu(x) + \nu P_{\nu-1}(x) = 0 \tag{9.49}$$

$$P'_{\nu+1}(x) - xP'_\nu(x) = (\nu + 1)P_\nu(x) \tag{9.50}$$

$$xP'_\nu(x) - P'_{\nu-1}(x) = \nu P_\nu(x) \tag{9.51}$$

and so forth.

The Legendre functions $P_\nu(x)$ are important for theoretical purposes in the general study of spherical harmonics. Their properties are important also from a more practical point of view, since these functions are prominent in solving Laplace's equation in various coordinate systems, such as toroidal coordinates.

Exercises 9.4

In problems 1 to 8, compare series to deduce the result.

1. $1 = F(0, b; c; x)$
2. $(1 - x)^{-a} = F(a, b; b; x)$
3. $\ln(1 - x) = -xF(1, 1; 2; x)$
4. $\ln\dfrac{1+x}{1-x} = 2xF\left(\dfrac{1}{2}, 1; \dfrac{3}{2}; x^2\right)$
5. $\arctan x = xF(\tfrac{1}{2}, 1; \tfrac{3}{2}; -x^2)$
6. $(1 + x)(1 - x)^{-2a-1} = F(2a, a + 1; a; x)$
7. $\tfrac{1}{2}(1 + \sqrt{x})^{-2a} + \tfrac{1}{2}(1 - \sqrt{x})^{-2a} = F(a, a + \tfrac{1}{2}; \tfrac{1}{2}; x)$
8. $\left(\dfrac{1 + \sqrt{1-x}}{2}\right)^{1-2a} = F\left(a - \dfrac{1}{2}, a; 2a; x\right)$

9. $\dfrac{1}{\sqrt{1-x}}\left(\dfrac{1+\sqrt{1-x}}{2}\right)^{1-2a} = F\left(a, a + \dfrac{1}{2}; 2a; x\right)$

10. You are given

$$K(x) = \int_0^{\pi/2} (1 - x^2 \sin^2 \phi)^{-1/2}\, d\phi$$

$$E(x) = \int_0^{\pi/2} (1 - x^2 \sin^2 \psi)^{1/2}\, d\psi$$

Show that

(a) $K(x) = \frac{1}{2}\pi F(\frac{1}{2}, \frac{1}{2}; 1; x^2)$
(b) $E(x) = \frac{1}{2}\pi F(-\frac{1}{2}, \frac{1}{2}; 1; x^2)$

11. Show that the *associated Legendre functions*

$$P_n^m(x) = (1 - x^2)^{m/2}\, \dfrac{d^m}{dx^m} P_n(x)$$

satisfy the relation

$$P_n^m(x) = \dfrac{(n + m)!}{2^m (n - m)!\, m!}\, (1 - x^2)^{m/2}$$

$$\times F\left(m - n, m + n + 1; m + 1; \dfrac{1-x}{2}\right)$$

12. Show that the *Chebyshev polynomials of the first kind*

$$T_n(x) = \sum_{k=0}^{[n/2]} \dfrac{(-1)^k (n - k - 1)!}{k!\,(n - 2k)!}\, (2x)^{n-2k} \qquad n \ge 1$$

satisfy the relation

$$T_n(x) = F\left(-n, n; \dfrac{1}{2}; \dfrac{1-x}{2}\right)$$

13. Show that the *Chebyshev polynomials of the second kind*

$$U_n(x) = \sum_{k=0}^{[n/2]} \binom{n - k}{k}(-1)^k (2x)^{n-2k}$$

satisfy the relation

$$U_n(x) = (n + 1)F\left(-n, n + 2; \dfrac{3}{2}; \dfrac{1-x}{2}\right)$$

14. Show that the *Gegenbauer polynomials*

$$C_n^\lambda(x) = \sum_{k=0}^{[n/2]} \binom{-\lambda}{n - k}\binom{n - k}{k}(-1)^n (2x)^{n-2k}$$

satisfy the relations

(a) $C_{2n}^{\lambda}(x) = (-1)^n \dfrac{(\lambda)_n}{n!} F\left(-n, \lambda+n; \dfrac{1}{2}; x^2\right)$

(b) $C_{2n+1}^{\lambda}(x) = (-1)^n \dfrac{(\lambda)_{n+1}}{n!} F\left(-n, \lambda+n; \dfrac{3}{2}; x^2\right)$

(c) $C_n^{\lambda}(x) = \dbinom{n+2\lambda-1}{n} F\left(-n, 2\lambda+n; \lambda+\dfrac{1}{2}; \dfrac{1-x}{2}\right)$

15. Show that the *Jacobi polynomials*

$$P_n^{(a,\,b)}(x) = \frac{1}{2^n} \sum_{k=0}^{n} \binom{n+a}{n-k}\binom{n+b}{n-k}(x+1)^{n-k}(x-1)^k$$

satisfy the relations

(a) $P_n^{(a,\,b)}(x) = (-1)^n \dbinom{n+b}{b} F\left(-n, n+a+b+1; 1+b; \dfrac{1+x}{2}\right)$

(b) $P_n^{(a,\,b)}(x) = \dbinom{n+a}{a} F\left(-n, n+a+b+1; 1+a; \dfrac{1-x}{2}\right)$

16. Given the *incomplete beta function*

$$B_x(p,q) = \int_0^x t^{p-1}(1-t)^{q-1}\,dt \qquad p, q > 0$$

show that

(a) $B_x(p,q) = \dfrac{x^p}{p} F(p, 1-q; 1+p; x)$

(b) $B_1(p,q) = \dfrac{\Gamma(p)\Gamma(q)}{\Gamma(p+q)}$

17. Show that

$$\sum_{k=0}^{n-1} \binom{a}{k} x^k = (1+x)^a - \binom{a}{n} x^n F(n-a, 1; n+1; -x)$$

18. Verify that

$$P_\nu(x) = \sum_{k=0}^{\infty} \frac{(-\nu)_k (\nu+1)_k}{(k!)^2}\left(\frac{1-x}{2}\right)^k$$

In problems 19 to 21, use the series representation in problem 18 to deduce the given recurrence formula.

19. $(\nu+1)P_{\nu+1}(x) - (2\nu+1)xP_\nu(x) + \nu P_{\nu-1}(x) = 0$

20. $P'_{v+1}(x) - xP'_v(x) = (v+1)P_v(x)$

21. $xP'_v(x) - P'_{v-1}(x) = vP_v(x)$

22. Using the relation (9.48), show that (for $k = 0, 1, 2, \ldots$)
    (a) $P_{2k+1}(0) = 0$

    (b) $P_{2h}(0) = \dfrac{(-1)^k}{k!} \left(\dfrac{1}{2}\right)_k$

23. Show that

$$P'_v(0) = \frac{2\Gamma(\tfrac{1}{2}v + 1)}{\sqrt{\pi}\,\Gamma(\tfrac{1}{2}v + \tfrac{1}{2})} \sin \frac{v\pi}{2}$$

24. Show that $P_v(x) = P_{-v-1}(x)$.
    Hint: Recall that $F(a, b; c; x) = F(b, a; c; x)$.

25. By making the substitution $x = 1 - 2z$ in the generalized form of Legendre's equation

$$(1 - x^2)y'' - 2xy' + v(v+1)y = 0$$

show that it transforms to Gauss' hypergeometric equation and thus deduce that $y = F[-v, v+1; 1; (1-x)/2)]$ is one solution of the generalized Legendre equation.

26. Show that

$$\frac{1}{k!} = \frac{2}{\pi(\tfrac{1}{2})_k} \int_0^{\pi/2} \sin^{2k} \phi\, d\phi \qquad k = 0, 1, 2, \ldots$$

and then, by expressing $P_v(x)$ in its series representation (problem 18), deduce that

$$P_v(x) = \frac{2}{\pi} \int_0^{\pi/2} F\left(-v, v+1; \frac{1}{2}; \frac{1-x}{2} \sin^2 \phi\right) d\phi$$

9.5 Summing Series and Evaluating Integrals

The hypergeometric function obviously has many areas of application because of its connection with other functions such as the inverse trigonometric functions, logarithmic functions, and the Legendre polynomials. However, it is also a useful tool in the evaluation or recognition of various series, both finite and infinite.

Example 3: Prove the combinatorial formula

$$\sum_{k=0}^{n} (-1)^k \binom{n}{k}\binom{a+k}{m} = (-1)^n \binom{a}{m-n} \qquad m = 0, 1, 2, \ldots$$

Solution: From Theorem 9.1c and Example 1,

$$\binom{n}{k} = \frac{(-1)^k}{k!}(-n)_k$$

$$\binom{a+k}{m} = \frac{(-1)^m(-a)_m(a+1)_k}{m!\,(a+1-m)_k}$$

and therefore

$$\sum_{k=0}^{n}(-1)^k\binom{n}{k}\binom{a+k}{m} = \frac{(-1)^m(-a)_m}{m!}\sum_{k=0}^{n}\frac{(-n)_k(a+1)_k}{(a+1-m)_k k!}$$

$$= \frac{(-1)^m(-a)_m}{m!}F(-n,a+1;a+1-m;1)$$

Recalling Theorem 9.2,

$$F(-n,a+1;a+1-m;1) = \frac{\Gamma(a+1-m)\Gamma(n-m)}{\Gamma(a+1+n-m)\Gamma(-m)}$$

$$= \frac{(-1)^n m!}{(m-n)!\,(a+1-m)_n}$$

where the last step follows from Eq. (9.1) and problem 11 in Exercises 2.2. Theorem 9.1e leads to

$$(-a)_m = (-1)^n(-a)_{m-n}(a+1-m)_n$$

and thus by combining results, we obtain

$$\sum_{k=0}^{n}(-1)^k\binom{n}{k}\binom{a+k}{m} = \frac{(-1)^m}{(m-n)!}(-a)_{m-n}$$

$$= (-1)^n\binom{a}{m-n}$$

following another application of Theorem 9.1c.

The hypergeometric function is useful also in the evaluation of certain integrals, as illustrated in the next example.

Example 4: Show that for $a > -1$

$$\int_0^\infty x^a [L_p^{(a)}(x)]^2 e^{-x}\,dx = \frac{\Gamma(p+a+1)}{p!} \qquad p = 0, 1, 2, \ldots$$

where $L_p^{(a)}(x)$ is the associated Laguerre polynomial.

Solution: By writing

$$L_p^{(a)}(x) = \sum_{n=0}^{P} (-1)^n \binom{p+a}{p-n} \frac{x^n}{n!}$$

$$= \binom{p+a}{p} \sum_{n=0}^{P} \frac{(-p)_n}{(a+1)_p} \frac{x^n}{n!}$$

we have

$$\int_0^\infty x^a [L_p^{(a)}(x)]^2 e^{-x}\, dx$$

$$= \binom{p+a}{p}^2 \sum_{n=0}^{P} \frac{(-p)_n}{(a+1)_n n!} \sum_{k=0}^{P} \frac{(-p)_k}{(a+1)_k k!} \int_0^\infty x^{a+n+k} e^{-x}\, dx$$

This last integral can be evaluated by using properties of the gamma function and two applications of problem 7 in Exercises 9.2, to get

$$\int_0^\infty x^{a+n+k} e^{-x}\, dx = \Gamma(a+n+k+1)$$

$$= \Gamma(a+n+1)(a+n+1)_k$$

$$= \Gamma(a+1)(a+1)_n (a+n+1)_k$$

Hence

$$\int_0^\infty x^a [L_p^{(a)}(x)]^2 e^{-x}\, dx$$

$$= \binom{p+a}{p}^2 \Gamma(a+1) \sum_{n=0}^{P} \frac{(-p)_n}{n!} \sum_{k=0}^{P} \frac{(-p)_k (a+n+1)_k}{(a+1)_k k!}$$

$$= \binom{p+a}{p}^2 \Gamma(a+1) \sum_{n=0}^{P} \frac{(-p)_n}{n!} F(-p, a+n+1; a+1; 1)$$

and by using the result of problem 21a in Exercises 9.3, see that

$$F(-p, a+n+1; a+1; 1) = \begin{cases} 0 & 0 \le n \le p-1 \\ \dfrac{(-1)^p p!}{(a+1)_p} & n = p \end{cases}$$

Finally, substitution of this last expression for the hypergeometric function leads us to our intended result,

$$\int_0^\infty x^a [L_p^{(a)}(x)]^2 e^{-x}\, dx = \left(\frac{p+a}{p}\right)^2 \Gamma(a+1)\frac{(-p)_p(-1)^p p!}{(a+1)_p p!}$$

$$= \left(\frac{p+a}{p}\right)^2 \frac{\Gamma(a+1)p!}{(a+1)_p}$$

$$= \frac{\Gamma(p+a+1)}{p!}$$

9.5.1 Action-angle variables

Action-angle variables are introduced in periodic motions of conservative hamiltonian systems in order to simplify the solution. Consider, for example, the specific *hamiltonian*

$$H(p,q) = \frac{p^2}{2m} + V(q) \tag{9.52}$$

in which q and p denote the position and momentum, respectively, and $V(q)$ is the potential energy. If the total energy of this system is equal to E, the momentum is then given by

$$p(q,E) = \pm \sqrt{2m[E - V(q)]} \tag{9.53}$$

and this in turn leads to the *action integral* defined by $[V(q_i) = E, i = 1, 2]$

$$I(E) = \frac{1}{\pi}\int_{q_1}^{q_2} \sqrt{2m[E - V(q)]}\, dq \tag{9.54}$$

Example 5: Evaluate the action integral (9.54) for the special case in which $V(q) = U\tan^2 q$ $(U > 0)$.

Solution: Here we find that $\tan^2 q_2 = E/U$, $q_1 = -q_2$, so

$$I = \frac{1}{\pi}\int_{q_1}^{q_2} \sqrt{2m(E - U\tan^2 q)}\, dq$$

$$= \frac{2}{\pi}\sqrt{2mU}\int_0^{\tan^{-1} a} \sqrt{a^2 - \tan^2 q}\, dq$$

where $a = E/U$. To evaluate the above integral, first we assume that

$a < 1$ and make the change of variable $ax = \tan q$, which leads to

$$\int_0^{\tan^{-1} a} \sqrt{a^2 - \tan^2 q}\, dq = a^2 \int_0^1 \frac{\sqrt{1 - x^0}}{1 + a^2 x^2}\, dx$$

$$= a^2 \sum_{n=0}^{\infty} (-1)^n a^{2n} \int_0^1 x^{2n} \sqrt{1 - x^2}\, dx$$

where we have expanded $(1 + a^2 x^2)^{-1}$ in a binomial series and interchanged the order of integration and summation. By substituting $t = x^2$ in this last integral, we can use properties of the beta function (Sec. 2.4) to deduce that

$$\int_0^1 x^{2n} \sqrt{1 - x^2}\, dx = \tfrac{1}{2} \int_0^1 t^{n - 1/2} (1 - t)^{1/2}\, dt$$

$$= \frac{\Gamma(n + \tfrac{1}{2}) \Gamma(\tfrac{3}{2})}{2 \Gamma(n + 2)}$$

Expressing all terms of the above series in terms of Pochhammer symbols, we obtain

$$\int_0^{\tan^{-1} a} \sqrt{a^2 - \tan^2 q}\, dq = \frac{a^2}{\pi} \sum_{n=0}^{\infty} \frac{(\tfrac{1}{2})_n (1)_n}{(2)_n} \frac{(-a^2)^n}{n!}$$

$$= \frac{a^2}{\pi} F\left(\frac{1}{2}, 1; 2; -a^2\right)$$

$$= \frac{a^2}{\pi} \left(\frac{2}{1 + \sqrt{1 + a^2}}\right)$$

where the last step follows from problem 8 in Exercises 9.4. By simplification of this last expression, we get

$$\int_0^{\tan^{-1} a} \sqrt{a^2 - \tan^2 \theta}\, d\theta = \frac{\pi}{2} (\sqrt{1 + a^2} - 1)$$

and finally returning to the original expression, we deduce that

$$I = \sqrt{2m} \, (\sqrt{E + U} - \sqrt{U})$$

Although we made the assumption $E < U$, the final result is valid also for $E \geq U$.

Exercises 9.5

1. Show that

$$\sum_{k=0}^{n} \frac{(-n)_k (b)_k}{(c)_k k!} = \frac{(c-b)_n}{(c)_n}$$

2. Show that

$$\sum_{k=0}^{n} \frac{(a)_k (b)_{n-k}}{(n-k)! k!} = \frac{(a+b)_n}{n!}$$

3. Show that

$$\sum_{k=1}^{n} \frac{(a)_k (b)_{n-k}}{(n-k)! (k-1)!} = \frac{a(a+b+1)_{n-1}}{(n-1)!} \qquad n \geq 1$$

In problems 4 to 21, verify the given identity.

4. $\displaystyle\sum_{k=0}^{n} \binom{n}{k}^2 = \binom{2n}{n}$

5. $\displaystyle\sum_{k=0}^{n} \binom{2n+1}{k} = 2^{2n}$

6. $\displaystyle\sum_{k=0}^{n} \binom{k}{m} = \binom{n+1}{m+1}$

7. $\displaystyle\sum_{k=0}^{n} (-1)^k \binom{m}{k} = (-1)^n \binom{m-1}{n}, \qquad m > n$

8. $\displaystyle\sum_{k=1}^{n} k \binom{n}{k}^2 = \frac{(2n-1)!}{[(n-1)!]^2}$

9. $\displaystyle\sum_{k=0}^{n} \binom{r}{k}\binom{m}{n-k} = \binom{r+m}{n}$

10. $\displaystyle\sum_{k=0}^{n} \binom{r}{k}\binom{m}{n+k} = \binom{r+m}{r+n}, \qquad r \geq 0$

11. $\displaystyle\sum_{k=0}^{n} \binom{2n}{2k}\binom{2k}{k}\binom{2n-2k}{n-k} = \binom{2n}{n}^2$

12. $\displaystyle\sum_{k=0}^{n} (-1)^k \binom{a+k-1}{k}\binom{a}{n-k} = 0, \qquad n \geq 1$

13. $\displaystyle\sum_{k=0}^{n} \binom{n-k}{m}\binom{p+k}{q} = \binom{n+p+1}{m+q+1}, \qquad p, q = 0, 1, 2, \ldots$

14. $\displaystyle\sum_{k=0}^{n} \binom{m}{k}\binom{r+k}{n}(-1)^k = (-1)^m \binom{r}{n-m}, \qquad 0 \leq m \leq n$

15. $\displaystyle\sum_{k=0}^{n}\binom{2n}{2k}=2^{2n-1}$, $\quad n>0$

16. $\displaystyle\sum_{k=0}^{n}(-1)^k\binom{2n}{k}^2=(-1)^n\binom{2n}{n}$

Hint: Use problem 26 in Exercises 9.3.

17. $\displaystyle\sum_{k=0}^{n-p}\binom{n}{k}\binom{n}{p+h}=\frac{(2n)!}{(n\ p)!\,(n+p)!}$, $\quad 0\le p\le n$

18. $\displaystyle\sum_{k=0}^{n}(-1)^k\binom{2n}{2k}\binom{-\frac{1}{2}}{k}=\frac{2^{2n}\Gamma(2n+\frac{1}{2})}{\sqrt{\pi}\,(2n)!}$ $\quad n=0,1,2,\dots$

19. $\displaystyle\sum_{n=0}^{\infty}\frac{1\times3\times5\times\cdots\times(2n-1)}{2\times4\times6\times\cdots\times(2n)}\left(\frac{1}{2}\right)^n=\sqrt{\frac{2}{3}}$

20. $\displaystyle\sum_{n=0}^{\infty}\frac{(n!)^2 2^n}{(2n+1)!}=\frac{\pi}{2}$

Hint: Use problem 26 in Exercises 9.3.

21. $\displaystyle\sum_{n=0}^{\infty}\frac{[(2n)!]^2}{(2n+1)!\,(n!)^2}\left(\frac{1}{8}\right)^n=\frac{\pi\sqrt{2}}{4}$

Hint: Use problem 26 in Exercises 9.3.

22. Show that

(a) $\displaystyle J_0(x)\cos x=\frac{1}{\sqrt{\pi}}\sum_{n=0}^{\infty}\frac{(-1)^n\Gamma(2n+\frac{1}{2})}{[(2n)!]^2}(2x)^{2n}$

(b) $\displaystyle J_0(x)\sin x=\frac{1}{\sqrt{\pi}}\sum_{n=0}^{\infty}\frac{(-1)^n\Gamma(2n+\frac{3}{2})}{[(2n+1)!]^2}(2x)^{2n+1}$

Hint: See problem 18.

23. Show that $(a>-1)$

$$\int_0^\infty x^{a+1}[L_p^{(a)}(x)]^2 e^{-x}\,dx=\frac{\Gamma(a+p+1)}{p!}(2p+a+1)$$

24. Show that* $(m=0,1,2,\dots)$

$$\int_0^a x^m [L_p^{(m)}(x)]^2 e^{-x}\,dx=1-\binom{p+m}{p}e^{-a}\sum_{n=0}^{p}\frac{(-p)_n}{n!}$$

$$\times\sum_{k=0}^{p}\frac{(-p)_k(m+n+1)_k}{(m+1)_k k!}e_{m+n+k}^a$$

where $e_n^x=\sum_{k=0}^{n}(x^k/k!)$.

---

* The evaluation of this integral is important in determining the total energy contained within the spot size of a Laguerre gaussian beam. See R. L. Phillips and L. C. Andrews, "Spot Size and Divergence for Laguerre Gaussian Beams of Any Order," *Applied Optics*, **22**(5): 643–644, March 1983.

In problems 25 to 27, verify the given identity.

25. $J_0(ax)J_0(bx) = \sum_{n=0}^{\infty} \dfrac{(-1)^n(ax/2)^{2n}}{(n!)^2} F\left(-n, -n; 1; \dfrac{b^2}{a^2}\right)$

26. $J_p(ax)J_\mu(bx) = \dfrac{(ax/2)^p(bx/2)^\mu}{\Gamma(\mu+1)}$

$\times \sum_{n=0}^{\infty} \dfrac{(-1)^n(ax/2)^{2n}}{n!\,\Gamma(\mu+n+1)} F\left(-n, -p-\mu-n; \mu+1; \dfrac{b^2}{a^2}\right)$

27. $[J_p(x)]^2 = \sum_{n=0}^{\infty} \dfrac{(-1)^n\Gamma(2p+2n+1)(x/2)^{2n+2p}}{n!\,\Gamma(2p+n+1)[\Gamma(p+n+1)]^2}$

28. Show that

$$[F(a,b;c;x)]^2 = \sum_{n=0}^{\infty} \dfrac{(1/2)_n(a)_n(b)_n(c-a)_n(c-b)_n}{(c)_n(c)_{2n}(c+n-1/2)_n} \cdot \dfrac{x^{2n}}{n!}$$

$$\times F(2a+2n, 2b+2n; 2c+4n; x)$$

29. Show that

$$\int_0^{\pi/2} [J_1(x \sin\theta)]^2 \csc\theta\, d\theta = \dfrac{1}{2} - \dfrac{J_1(2x)}{2x}$$

Hint: Use problem 27.

30. Show that $(n = 0, 1, 2, \dots)$

$$\dfrac{1}{\pi}\int_0^{\pi} J_0(2x\sin\phi)\cos 2n\phi\, d\phi = [J_n(x)]^2$$

Hint: Use problem 27.

31. Show that for $a > b > 0$, $p - \mu + 1 > 0$.

$$\int_0^{\infty} e^{-ax}x^{-\mu}J_p(bx)\, dx = \left(\dfrac{b}{2a}\right)^p \dfrac{a^{\mu-1}\Gamma(p-\mu+1)}{\Gamma(p+1)}$$

$$\times F\left(\dfrac{p-\mu+1}{2}, \dfrac{p-\mu+2}{2}; p+1; -\dfrac{b^2}{a^2}\right)$$

32. For $\mu = -p$, show that the result of problem 31 reduces to

$$\int_0^{\infty} e^{-ax}x^pJ_p(bx)\, dx = \left(\dfrac{b}{2}\right)^p \dfrac{\Gamma(2p+1)}{\Gamma(p+1)(a^2+b^2)^{p+1/2}}$$

33. For $\mu = 1$, show that the result of problem 31 reduces to

$$\int_0^{\infty} e^{-ax}x^{-1}J_p(bx)\, dx = \dfrac{1}{p}\left(\dfrac{\sqrt{a^2+b^2}-a}{b}\right)^p$$

Hint: Use problem 8 in Exercises 9.4.

34. For $\mu = 0$, show that the result of problem 31 reduces to

$$\int_0^{\infty} e^{-ax}J_p(bx)\, dx = \dfrac{1}{p}\dfrac{(\sqrt{a^2+b^2}-a)^p}{b^p\sqrt{a^2+b^2}}$$

Hint: Use problem 9 in Exercises 9.4.

10

The Confluent
Hypergeometric Functions

10.1 Introduction

Whereas Gauss was largely responsible for the systematic study of the hypergeometric function, E. E. Kummer (1810–1893) is the person most associated with developing properties of the related *confluent hypergeometric function*. Kummer published his work on this function in 1836,* and since that time it has been commonly referred to as *Kummer's function*. Like the hypergeometric function, the confluent hypergeometric function is related to a large number of other functions.

Kummer's function satisfies a second-order linear differential equation called the *confluent hypergeometric equation*. A second solution of this DE leads to the definition of the *confluent hypergeometric function of the second kind*, which is also related to many other functions. At the beginning of the twentieth century (1904), Whittaker introduced another pair of confluent hypergeometric functions that now bears his name.† The *Whittaker functions* arise as solutions of the confluent hypergeometric equation after a transformation to *Liouville's standard form of the DE*.

---

* E. E. Kummer, "Über die Hypergeometrische Reihe $F(a;b;x)$," *J. Reine Angew. Math.*, **15**: 39–83, 127–172, 1836.
† E. T. Whittaker, "An Expression of Certain Known Functions as Generalized Hypergeometric Series," *Bull. Amer. Math. Soc.*, **10**: 125–134, 1904.

10.2 The Functions $M(a; c; x)$ and $U(a; c; x)$

Perhaps even more important in applications than the hypergeometric function is the related function

$$M(a; c; x) = \sum_{n=0}^{\infty} \frac{(a)_n x^n}{(c)_n n!} \qquad -\infty < x < \infty \qquad (10.1)$$

called the **confluent hypergeometric function**.* It is related to the hypergeometric function according to

$$M(a; c; x) = \lim_{b \to \infty} F\left(a, b; c; \frac{x}{b}\right) \qquad (10.2)$$

To see this, we note that

$$\lim_{b \to \infty} F\left(a, b; c; \frac{x}{b}\right) = \lim_{b \to \infty} \sum_{n=0}^{\infty} \frac{(a)_n (b)_n}{(c)_n} \frac{(x/b)^n}{n!}$$

$$= \sum_{n=0}^{\infty} \frac{(a)_n x^n}{(c)_n n!} \lim_{b \to \infty} \frac{(b)_n}{b^n}$$

where clearly $(b)_n/b^n = b(b+1) \cdots (b+n-1)/b^n \to 1$ as $b \to \infty$.

Remark: The function $M(a; c; x)$ is also designated by $\Phi(a; c; x)$ or $_1F_1(a; c; x)$, and commas are sometimes used in place of semicolons.

As was the case for the hypergeometric function, the series (10.1) is normally not defined for $c = 0, -1, -2, \ldots$, and if a is a negative integer, the series truncates. By application of the ratio test, it can be shown that the confluent hypergeometric series (10.1) converges for all (finite) x (see problem 1 in Exercises 10.2).

10.2.1 Elementary properties of $M(a; c; x)$

Because of the similarity of definition to that of $F(a, b; c; x)$, the function $M(a; c; x)$ obviously has many properties analogous to those of the hypergeometric function. For example, it is easy to show that

$$\frac{d}{dx} M(a; c; x) = \frac{a}{c} M(a+1; c+1; x) \qquad (10.3)$$

---

*The term *confluent* refers to the fact that, due to the transformation (10.2), two singularities in the hypergeometric differential equation (at 1 and infinity) are merged into one singularity (at infinity) in the confluent hypergeometric differential equation. See Sec. 10.2.2.

whereas in general

$$\frac{d^k}{dx^k} M(a;c;x) = \frac{(a)_k}{(c)_k} M(a+k;c+k;x) \qquad k = 1, 2, 3, \ldots \quad (10.4)$$

The function $M(a;c;x)$ also satisfies recurrence relations involving the contiguous functions $M(a \pm 1; c; x)$ and $M(a; c \pm 1; x)$. From these four contiguous functions, taken two at a time, we find six recurrence relations with coefficients at most linear in x:

$$(c - a - 1)M(a;c;x) + aM(a+1;c;x)$$
$$- (c - 1)M(a;c - 1;x) \quad (10.5)$$

$$cM(a;c;x) - cM(a-1;c;x) = xM(a;c+1;x) \quad (10.6)$$

$$(a - 1 + x)M(a;c;x) + (c - a)M(a-1;c;x)$$
$$= (c - 1)M(a;c - 1;x) \quad (10.7)$$

$$c(a + x)M(a;c;x) - acM(a+1;c;x)$$
$$= (c - a)xM(a;c + 1;x) \quad (10.8)$$

$$(c - a)M(a - 1;c;x) + (2a - c + x)M(a;c;x)$$
$$= aM(a + 1;c;x) \quad (10.9)$$

$$c(c - 1)M(a;c - 1;x) - c(c - 1 + x)M(a;c;x)$$
$$= (a - c)xM(a;c + 1;x) \quad (10.10)$$

The verification of these relations is left to the exercises.

To obtain an integral representation of $M(a;c;x)$, we first recall the identity [Eq. (9.18) in Sec. 9.3.2]

$$\frac{(a)_n}{(c)_n} = \frac{\Gamma(c)}{\Gamma(a)\Gamma(c - a)} \int_0^1 t^{a+n-1}(1 - t)^{c-a-1}\, dt \qquad c > a > 0 \quad (10.11)$$

for $n = 0, 1, 2, \ldots$. Thus it follows from (10.1) that

$$M(a;c;x) = \frac{\Gamma(c)}{\Gamma(a)\Gamma(c - a)} \sum_{n=0}^{\infty} \frac{x^n}{n!} \int_0^1 t^{a+n-1}(1 - t)^{c-a-1}\, dt$$

$$= \frac{\Gamma(c)}{\Gamma(a)\Gamma(c - a)} \int_0^1 t^{a-1}(1 - t)^{c-a-1}\left[\sum_{n=0}^{\infty} \frac{(xt)^n}{n!}\right] dt \quad (10.12)$$

where we have interchanged the order of integration and summation.

Recognizing the infinite sum in (10.12) as that of an exponential, we deduce the **integral representation**

$$M(a;c;x) = \frac{\Gamma(c)}{\Gamma(a)\Gamma(c-a)} \int_0^1 e^{xt} t^{a-1}(1-t)^{c-a-1} dt \qquad c > a > 0$$

(10.13)

The integral formula (10.13) can now be used to derive a very important result concerning confluent hypergeometric functions. We simply make the change of variable $t = 1 - u$ to get

$$M(a;c;x) = \frac{\Gamma(c)}{\Gamma(a)\Gamma(c-a)} e^x \int_0^1 e^{-xu} u^{c-a-1}(1-u)^{a-1} du \quad (10.14)$$

which implies

$$M(a;c;x) = e^x M(c-a;c;-x) \tag{10.15}$$

known as **Kummer's transformation.** Even though (10.13) requires that $c > a > 0$, the result (10.15) is valid for all values of the parameters for which the confluent hypergeometric function is defined.

10.2.2 Confluent hypergeometric equation and $U(a; c; x)$

The hypergeometric function $y = F(a, b; c; t)$ is a solution of Gauss' equation

$$t(1-t)\frac{d^2y}{dt^2} + [c - (a+b+1)t]\frac{dy}{dt} - aby = 0 \tag{10.16}$$

By making the change of variable $t = x/b$, (10.16) becomes

$$x\left(1 - \frac{x}{b}\right)y'' + \left(c - x - \frac{a+1}{b}x\right)y' - ay = 0$$

and then allowing $b \to \infty$, we find

$$xy'' + (c - x)y' - ay = 0 \tag{10.17}$$

Now since

$$M(a;c;x) = \lim_{b\to\infty} F\left(a, b; c; \frac{x}{b}\right)$$

it follows that $y_1 = M(a;c;x)$ is a solution of Eq. (10.17), which is called the **confluent hypergeometric equation.**

By making the change of variable $y = x^{1-c}z$, we find that (10.17) becomes (after simplification)

$$xz'' + (2 - c - x)z' - (1 + a - c)z = 0 \qquad (10.18)$$

Thus, by comparing (10.18) with (10.17), it is clear that

$$z = M(1 + a - c; 2 - c; x) \qquad c \neq 2, 3, 4, \ldots$$

is a solution of (10.18), and hence

$$y_2 = x^{1-c}M(1 + a - c; 2 - c; x) \qquad c \neq 2, 3, 4, \ldots \qquad (10.19)$$

is a second solution of Eq. (10.17). Furthermore, if c is not an integer (positive, zero, or negative), then y_2 is linearly independent of $y_1 = M(a; c; x)$, and in this case the general solution of (10.17) is

$$y = C_1 M(a; c; x) + C_2 x^{1-c}M(1 + a - c; 2 - c; x)$$
$$c \neq 0, \pm 1, \pm 2, \ldots \qquad (10.20)$$

where C_1 and C_2 are any constants.

To remove the restriction $c \neq 1, 2, 3, \ldots$ in the general solution (10.20), we introduce the function $(c \neq 0, -1, -2, \ldots)$*

$$U(a; c; x) = \frac{\pi}{\sin c\pi} \left[\frac{M(a; c; x)}{\Gamma(1 + a - c)\Gamma(c)} - \frac{x^{1-c}M(1 + a - c; 2 - c; x)}{\Gamma(a)\Gamma(2 - c)} \right]$$
$$(10.21)$$

called the **confluent hypergeometric function of the second kind.** For nonintegral values of c, $U(a; c; x)$ is surely a solution of (10.17), since it is simply a linear combination of two solutions. For $c = 1, 2, 3, \ldots$, we find that (10.21) assumes the indeterminate form $0/0$, and in this case we define (analogous to Bessel functions of the second kind)

$$U(a; n + 1; x) = \lim_{c \to n+1} U(a; c; x) \qquad n = 0, 1, 2, \ldots \qquad (10.22)$$

which can also be shown to be a solution of (10.17).

To investigate the behavior of $U(a; c; x)$ when a is a nonpositive integer, we set $a = -n$ $(n = 0, 1, 2, \ldots)$ in (10.21) to find

$$U(-n; c; x) = \frac{\pi}{\sin \pi c} \frac{M(-n; c; x)}{\Gamma(1 - n - c)\Gamma(c)}$$

---

* See also problem 20 in Exercises 10.2.

which, by use of the identity $\Gamma(x)\Gamma(1-x) = \pi/\sin \pi x$, becomes

$$U(-n;c;x) = (-1)^n(c)_n M(-n;c;x) \qquad n = 0, 1, 2, \ldots \quad (10.23)$$

Hence, the functions $U(a;c;x)$ and $M(a;c;x)$ are clearly linearly dependent for $a = 0, -1, -2, \ldots$ and therefore do not constitute a fundamental set of solutions of (10.17) in this case. Nonetheless, for both $a, c \neq 0, -1, -2, \ldots$, it can be shown that $U(a;c;x)$ and $M(a;c;x)$ are linearly independent functions, and in this case the general solution of (10.17) is (see problem 22, in Exercises 10.2)

$$y = C_1 M(a;c;x) + C_2 U(a;c;x) \qquad a, c \neq 0, -1, -2, \ldots \quad (10.24)$$

The function $U(a;c;x)$ has many properties like $M(a;c;x)$. Directly from its definition (10.21), we first note that (see problem 21 in Exercises 10.2)*

$$U(a;c;x) = x^{1-c} U(1 + a - c; 2 - c; x) \qquad (10.25)$$

while the derivative relations are readily found to be (see problem 23 in Exercises 10.2)

$$\frac{d}{dx} U(a;c;x) = -aU(a+1;c+1;x) \qquad (10.26)$$

and

$$\frac{d^k}{dx^k} U(a;c;x) = (-1)^k (a)_k U(a+k;c+k;x) \qquad k = 1, 2, 3, \ldots \quad (10.27)$$

Although more difficult to show, it has the **integral representation**

$$U(a;c;x) = \frac{1}{\Gamma(a)} \int_0^\infty e^{-xt} t^{a-1} (1+t)^{c-a-1} \, dt \qquad a > 0, x > 0 \quad (10.28)$$

Some additional properties are taken up in the exercises.

10.2.3 Asymptotic formulas

From the series representation (10.1) of $M(a;c;x)$, it follows directly that for small values of x

$$M(a;c;x) \sim 1 \qquad x \to 0 \qquad (10.29)$$

---

* From (10.25), it follows that $U(a;c;x)$ is defined also for $c = 0, -1, -2, \ldots$.

For $U(a;c;x)$ it can be shown that

$$U(a;c;x) \sim \frac{\Gamma(c-1)}{\Gamma(a)} x^{1-c} \qquad c>1, \qquad x \to 0^+ \qquad (10.30a)$$

$$U(a;1;x) \sim -\frac{1}{\Gamma(a)} [\ln x + \psi(a)] \qquad x \to 0^+ \qquad (10.30b)$$

the details of which we leave to the exercises (see problem 27 in Exercises 10.2).

To derive an asymptotic formula for $M(a;c;x)$ that is valid for large x, we begin with the integral representation [see (10.14)]

$$M(a;c;x) = \frac{\Gamma(c)}{\Gamma(a)\Gamma(c-a)} e^x \int_0^1 e^{-xu} u^{c-a-1} (1-u)^{a-1} \, du$$

This integral can further be expressed as the difference of two integrals by writing

$$\int_0^1 e^{-xu} u^{c-a-1} (1-u)^{a-1} \, du = \int_0^\infty e^{-xu} u^{c-a-1} (1-u)^{a-1} \, du$$

$$- \int_1^\infty e^{-xu} u^{c-a-1} (1-u)^{a-1} \, du$$

Next we make the substitution $s = xu$ in the first integral on the right and the substitution $t = x(u-1)$ in the second integral on the right. This action yields

$$\int_0^1 e^{-xu} u^{c-a-1} (1-u)^{a-1} \, du = x^{a-c} \int_0^\infty e^{-s} s^{c-a-1} \left(1 - \frac{s}{x}\right)^{a-1} ds$$

$$- \frac{e^{-x}}{x} \int_0^\infty e^{-t} \left(1 + \frac{t}{x}\right)^{c-a-1} \left(-\frac{t}{x}\right)^{a-1} dt \quad (10.31)$$

Hence, for $x \gg s$ and $x \to \infty$, we make the approximations

$$\left(1 - \frac{s}{x}\right)^{a-1} \simeq 1 \qquad \frac{e^{-x}}{x} \simeq 0$$

and find that (10.31) leads to

$$\int_0^1 e^{-xu} u^{c-a-1} (1-u)^{a-1} \, du \sim x^{a-c} \int_0^\infty e^{-s} s^{c-a-1} \, ds$$

$$\sim x^{a-c} \Gamma(c-a)$$

Substituting this result into (10.14), we deduce the **asymptotic formula**

$$M(a;c;x) \sim \frac{\Gamma(c)}{\Gamma(a)} x^{a-c} e^{x} \qquad x \to \infty \tag{10.32}$$

for a, $c \neq 0$, $-1, -2, \ldots$. If, instead of approximating the term $(1 - s/x)^{a-1}$ by unity, we choose to expand it in the binomial series

$$\left(1 - \frac{s}{x}\right)^{a-1} = \sum_{n=0}^{\infty} \binom{a-1}{n} (-1)^{n} \left(\frac{s}{x}\right)^{n} \qquad s < x$$

then we obtain the full asymptotic series given in problem 28 in Exercises 10.2. Finally, by using Kummer's transformation (10.15), it readily follows from (10.32) that

$$M(a;c;-x) \sim \frac{\Gamma(c)}{\Gamma(c-a)} x^{-a} \qquad x \to \infty \tag{10.33}$$

Last, if we utilize the integral representation (10.28) for $U(a;c;x)$, it follows in a like manner that (see problems 29 and 30 in Exercises 10.2)

$$U(a;c;x) \sim x^{-a} \qquad x \to \infty \tag{10.34}$$

Exercises 10.2

1. By applying the ratio test to Eq. (10.1), show that the confluent hypergeometric series converges for all x.

2. Show that

   (a) $\dfrac{d}{dx} M(a;c;x) = \dfrac{a}{c} M(a+1;c+1;x)$

   (b) $\dfrac{d^{k}}{dx^{k}} M(a;c;x) = \dfrac{(a)_{k}}{(c)_{k}} M(a+k;c+k;x), k = 1, 2, 3, \ldots$

In problems 3 to 7, verify the differentiation formula.

3. $x \dfrac{d}{dx} M(a;c;x) + aM(a;c;x) = aM(a+1;c;x)$

4. $x \dfrac{d}{dx} M(a;c;x) + (c-a-x)M(a;c;x) = (c-a)M(a-1;c;x)$

5. $c \dfrac{d}{dx} M(a;c;x) - cM(a;c;x) = (a-c)M(a;c+1;x)$

6. $x\dfrac{d}{dx}M(a;c;x)+(c-1)M(a;c;x)=(c-1)M(a;c-1;x)$

7. $x\dfrac{d}{dx}M(a;c;x)+(c-1-x)M(a;c;x)$

$\quad=(c-1)M(a-1;c-1;x)$

In problems 8 to 13, verify the contiguous relation by using the results of problems 3 to 7 or by using series representations.

8. $(c-a-1)M(a;c;x)+aM(a+1;c;x)=(c-1)M(a;c-1;x)$

9. $cM(a;c;x)-cM(a-1;c;x)=xM(a;c+1;x)$

10. $(a\quad 1\mid x)M(a;c;x)+(c-a)M(a-1;c;x)$

$\quad=(c-1)M(a;c-1;x)$

11. $c(a+x)M(a;c;x)-acM(a+1;c;x)=(c-a)xM(a;c+1;x)$

12. $(c-a)M(a-1;c;x)+(2a-c+x)M(a;c;x)=aM(a+1;c;x)$

13. $c(c-1)M(a;c-1;x)+c(c-1+x)M(a;c;x)$

$\quad=(a-c)xM(a;c+1;x)$

14. Show that

$$M(a+1;c;x)-M(a;c;x)=\frac{x}{c}M(a+1;c+1;x)$$

15. Show that

$$M(a;c;x)=\frac{c-a}{c}M(a;c+1;x)+\frac{a}{c}M(a+1;c+1;x)$$

In problems 16 to 18, derive the integral relation.

16. $M(a;c;x)=\dfrac{\Gamma(c)2^{1-c}}{\Gamma(a)\Gamma(c-a)}e^{x/2}\displaystyle\int_{-1}^{1}e^{-xt/2}(1+t)^{c-a-1}(1-t)^{a-1}\,dt,$

$c>a>0$

17. $M(a;c;x)=\dfrac{\Gamma(c)}{\Gamma(c-a)}e^{x}x^{(1-c)/2}\displaystyle\int_{0}^{\infty}e^{-t}t^{(c-1)/2-a}J_{c-1}(2\sqrt{xt})\,dt,$

$c>a>0,\ x>0$

18. $\displaystyle\int_{0}^{\infty}e^{-st}M(a;c;t)\,dt=\dfrac{1}{s}F\left(a,1;c;\dfrac{1}{s}\right),\ s>1$

19. By substituting its series representation, show directly that $y=M(a;c;x)$ is a solution of

$$xy''+(c-x)y'-ay=0$$

20. Show that for $c \neq 1, 2, 3, \ldots$ the confluent hypergeometric function of the second kind (10.21) can be written as

$$U(a;c;x) = \frac{\Gamma(1-c)}{\Gamma(1+a-c)} M(a;c;x)$$

$$+ \frac{\Gamma(c-1)}{\Gamma(a)} x^{1-c} M(1+a-c; 2-c; x)$$

21. Verify the Kummer relation

$$U(a;c;x) = x^{1-c} U(1+a-c; 2-c; x)$$

22. Show that the wronskian of the confluent hypergeometric functions is given by

$$W(M, U)(x) = -\frac{\Gamma(c)}{\Gamma(a)} x^{-c} e^{x}$$

    Hint: See problem 7 in Exercises 6.6.

23. Show that

    (a) $\dfrac{d}{dx} U(a;c;x) = -aU(a+1;c+1;x)$

    (b) $\dfrac{d^{k}}{dx^{k}} U(a;c;x) = (-1)^{k} (a)_{k} U(a+k; c+k; x), \; k = 1, 2, 3, \ldots$

24. Show that $U(a;c;x)$ has (among others) the contiguous relations
    (a) $U(a;c;x) - aU(a+1;c;x) = U(a;c-1;x)$
    (b) $(c-a)U(a;c;x) + U(a-1;c;x) = xU(a;c+1;x)$

25. From the well-known result of calculus

$$f(x+y) = \sum_{n=0}^{\infty} f^{(n)}(x) \frac{y^{n}}{n!} \qquad |y| < \rho$$

    derive the addition formulas

    (a) $M(a;c;x+y) = \displaystyle\sum_{n=0}^{\infty} \frac{(a)_{n} y^{n}}{(c)_{n} n!} M(a+n; c+n; x)$

    (b) $U(a;c;x+y) = \displaystyle\sum_{n=0}^{\infty} \frac{(a)_{n}}{n!} (-1)^{n} y^{n} U(a+n; c+n; x)$

26. From the result of problem 25, deduce the multiplication formulas

    (a) $M(a;c;xy) = \displaystyle\sum_{n=0}^{\infty} \frac{(a)_{n} x^{n} (y-1)^{n}}{(c)_{n} n!} M(a+n; c+n; x)$

    (b) $U(a;c;xy) = \displaystyle\sum_{n=0}^{\infty} \frac{(a)_{n} x^{n} (1-y)^{n}}{n!} U(a+n; c+n; x)$

27. For small arguments, show that

(a) $\quad U(a;c;x) \sim \dfrac{\Gamma(c-1)}{\Gamma(a)} x^{1-c}, c>1, x\to0^+$

(b) $\quad U(a;1;x) \sim -\dfrac{1}{\Gamma(a)}[\ln x + \psi(a)], x\to0^+$

28. Starting with Eq. (10.31), show by expanding $(1-s/x)^{a-1}$ in a binomial series that

$$M(a;c;x) \sim \frac{\Gamma(c)}{\Gamma(a)} x^{a-c} e^x \sum_{n=0}^{\infty} \frac{(1-a)_n(c-a)_n}{n!x^n} \qquad x\to\infty$$

29. Using the integral representation (10.28), show that

$$U(a;c;x) \sim x^{-a} \qquad x\to\infty$$

30. Following the technique suggested in problem 28, derive the asymptotic series

$$U(a;c;x) \sim x^{-a} \sum_{n=0}^{\infty} \frac{(-1)^n(a)_n(1+a-c)_n}{n!x^n} \qquad x\to\infty$$

10.3 Relation to Other Functions

Specializations of either $M(a;c;x)$ or $U(a;c;x)$ lead to most of the other special functions introduced in earlier chapters. For example, it can readily be verified by comparing series or integral representations that

$$e^x = M(a;a;x) \tag{10.35}$$

$$\text{erfc}\, x = \frac{1}{\sqrt{\pi}} e^{-x^2} U\left(\frac{1}{2};\frac{1}{2};x^2\right) \tag{10.36}$$

$$H_{2n}(x) = (-1)^n \frac{(2n)!}{n!} M\left(-n;\frac{1}{2};x^2\right) \tag{10.37}$$

$$L_n(x) = M(-n;1;x) \tag{10.38}$$

$$\text{Ei}\,(x) = -e^x U(1;1;-x) \tag{10.39}$$

$$K_p(x) = \sqrt{\pi}\,(2x)^p e^{-x} U(p+\tfrac{1}{2};2p+1;2x) \tag{10.40}$$

among many other such relations (see the exercises).

The validity of (10.35) follows directly from

$$M(a;a;x) = \sum_{n=0}^{\infty} \frac{(a)_n x^n}{(a)_n n!} = \sum_{n=0}^{\infty} \frac{x^n}{n!} = e^x$$

while (10.36) and (10.37) are proved in Examples 1 and 2. Verifying (10.38), (10.39), and (10.40) is left to the exercises.

Example 1:   Show that $\mathrm{erfc}\, x = (1/\sqrt{\pi})e^{-x^2}U(\tfrac{1}{2};\tfrac{1}{2};x^2)$.

Solution: By introducing the substitution $t = x\sqrt{1+s}$, we find

$$\mathrm{erfc}\, x = \frac{2}{\sqrt{\pi}} \int_x^\infty e^{-t^2}\, dt$$

$$= \frac{1}{\sqrt{\pi}} x e^{-x^2} \int_0^\infty e^{-x^2 s}(1+s)^{-1/2}\, ds$$

Comparing this last integral with the integral representation (10.28) identifies the parameters $a = 1$ and $c = \tfrac{3}{2}$, and hence

$$\mathrm{erfc}\, x = \frac{1}{\sqrt{\pi}} x e^{-x^2} U\!\left(1; \frac{3}{2}; x^2\right)$$

However, using the identity $U(a;c;x) = x^{1-c}U(1+a-c;2-c;x)$, we can also write

$$\mathrm{erfc}\, x = \frac{1}{\sqrt{\pi}} e^{-x^2} U\!\left(\frac{1}{2}; \frac{1}{2}; x^2\right)$$

Example 2:   Show that

$$H_{2n}(x) = (-1)^n \frac{(2n)!}{n!} M\!\left(-n; \frac{1}{2}; x^2\right)$$

Solution: From the series definition of the Hermite polynomials, we have

$$H_{2n}(x) = \sum_{k=0}^{n} \frac{(-1)^k (2n)!}{k!(2n-2k)!} (2x)^{2n-2k}$$

$$= (-1)^n (2n)! \sum_{j=0}^{n} \frac{(-1)^j (2x)^{2j}}{(n-j)!(2j)!}$$

where the last step has resulted from the index change $j = n - k$. In terms of Pochhammer symbols, we write

$$(2j)! = 2^{2j} (\tfrac{1}{2})_j j!$$

$$(n-j)! = \frac{(-1)^j n!}{(-n)_j}$$

and therefore it follows that

$$H_{2n}(x) = (-1)^n \frac{(2n)!}{n!} \sum_{j=0}^{n} \frac{(-n)_j x^{2j}}{(1/2)_j \, j!} = (-1)^n \frac{(2n)!}{n!} M\left(-n; \frac{1}{2}; x^2\right)$$

Example 3: Use the relation (see problem 34 in Exercises 10.3)

$$(1-t)^{-a} M\left(a; 1-\frac{xt}{1-t}\right) = \sum_{n=0}^{\infty} (a)_n L_n(x) \frac{t^n}{n!}$$

to deduce that

$$\sum_{n=0}^{\infty} \frac{(-1)^n}{n!\Gamma(\mu-n)} L_n(x) = \frac{x^{\mu-1}}{[\Gamma(\mu)]^2} \qquad \mu > 0$$

Solution: Let $t = 1$ and $a = 1 - \mu$ in the given series to obtain

$$\sum_{n=0}^{\infty} (1-\mu)_n L_n(x) \frac{1}{n!} = \lim_{t \to 1} (1-t)^{\mu-1} M\left(1-\mu; 1; -\frac{xt}{1-t}\right)$$

$$= \frac{x^{\mu-1}}{\Gamma(\mu)}$$

the last step of which follows use of the asymptotic formula [Eq. (10.33)]

$$M\left(1-\mu; 1; -\frac{xt}{1-t}\right) \sim \frac{1}{\Gamma(\mu)} \left(\frac{xt}{1-t}\right)^\mu \qquad \frac{xt}{1-t} \to \infty$$

Finally, by recalling the identity (see problem 8 in Exercises 9.2)

$$(1-\mu)_n = \frac{(-1)^n \Gamma(\mu)}{\Gamma(\mu-n)}$$

we get the relation

$$\sum_{n=0}^{\infty} \frac{(-1)^n}{n!\Gamma(\mu-n)} L_n(x) = \frac{x^{\mu-1}}{[\Gamma(\mu)]^2} \qquad \mu > 0$$

10.3.1 Hermite functions
The DE

$$y'' - 2xy' + 2\nu y = 0 \qquad\qquad (10.41)$$

arises in the solution of Laplace's equation in parabolic coordinates. For $\nu = n \, (n = 0, 1, 2, \ldots)$, this is just the DE satisfied by the Hermite polynomials studied in Chap. 5. Therefore, in the general case where v is arbitrary we refer to the solutions of (10.41) as *Hermite functions*.

To find a general solution of (10.41), we start with the change of variable $t = x^2$, which converts the DE to the confluent hypergeometric form

$$t\frac{d^2y}{dt^2} + \left(\frac{1}{2} - t\right)\frac{dy}{dt} + \frac{v}{2}y = 0 \qquad (10.42)$$

Hence, (10.42) is just a special case of (10.17) for which $a = -v/2$ and $c = \frac{1}{2}$. Recalling Eq. (10.20), we see that a general solution of (10.42) is

$$y(t) = C_1 M\left(-\frac{v}{2};\frac{1}{2};t\right) + C_2 t^{1/2} M\left(\frac{1-v}{2},\frac{3}{2};t\right) \qquad (10.43)$$

and so the general solution of (10.41) is

$$y(x) = C_1 M\left(-\frac{v}{2};\frac{1}{2};x^2\right) + C_2 x M\left(\frac{1-v}{2},\frac{3}{2};x^2\right) \qquad (10.44)$$

It is customary to choose constants C_1 and C_2 to be

$$C_1 = \frac{2^v\sqrt{\pi}}{\Gamma[(1-v)/2]} \qquad C_2 = -\frac{2^{v+1}\sqrt{\pi}}{\Gamma(-v/2)}$$

and then define

$$H_v(x) = \frac{2^v\sqrt{\pi}}{\Gamma[(1-v)/2]} M\left(-\frac{v}{2};\frac{1}{2};x^2\right) - \frac{2^{v+1}\sqrt{\pi}}{\Gamma(-v/2)} x M\left(\frac{1-v}{2},\frac{3}{2};x^2\right)$$

$$(10.45)$$

which is called the **Hermite function** of degree v.

Various properties of the Hermite functions can be derived directly from the definition (10.45) in terms of confluent hypergeometric functions. For example, we can immediately deduce that

$$H_v(0) = \frac{2^v\sqrt{\pi}}{\Gamma[(1-v)/2]} \qquad H_v'(0) = -\frac{2^{v+1}\sqrt{\pi}}{\Gamma(-v/2)} \qquad (10.46)$$

Also by expressing the confluent hypergeometric functions in (10.45) in series form, a series for $H_v(x)$ can be derived for v not zero or a positive integer (see problem 49 in Exercises 10.3).

By comparing (10.45) with the definition of the confluent hypergeometric function of the second kind, it follows that the Hermite function can also be expressed as

$$H_\nu(x) = 2^\nu U\left(-\frac{\nu}{2}; \frac{1}{2}; x^2\right)$$
(10.47)

Hence, recalling the result of Example 1, we see that, for example,

$$H_{-1}(x) = \frac{\sqrt{\pi}}{2} e^{x^2} \operatorname{erfc} x$$
(10.48)

The basic recurrence formulas for the Hermite polynomials are satisfied as well by the Hermite functions; the proofs are left to the exercises.

10.3.2   Laguerre functions

The associated Laguerre polynomials are related to the confluent hypergeometric function by (see problem 6 in Exercises 10.3)

$$L_n^{(a)}(x) = \frac{(a+1)_n}{n!} M(-n; a+1; x)$$

$$= \frac{\Gamma(n+a+1)}{\Gamma(n+1)\Gamma(a+1)} M(-n; a+1; x) \qquad (10.49)$$

If we choose to replace the index n by the more general index ν (not restricted to nonnegative integer values), we have

$$L_\nu^{(a)}(x) = \frac{\Gamma(\nu+a+1)}{\Gamma(\nu+1)\Gamma(a+1)} M(-\nu; a+1; x) \qquad (10.50)$$

called the **Laguerre function** of degree ν.

For $\nu \neq n$ $(n = 0, 1, 2, \ldots)$, it is clear that the Laguerre function is not a polynomial, since the series for $M(-\nu; a+1; x)$ will be infinite in this case. Nonetheless, the basic recurrence formulas for the associated Laguerre polynomials (Sec. 5.3.3) continue to hold for the more general Laguerre function. In the case where ν is a negative integer, some immediate consequences of the defining relation (10.50) are

$$L_{-n}^{(a)}(x) = 0 \qquad n = 1, 2, 3, \ldots, \quad a > -1, \quad a \neq 0, 1, 2, \ldots \quad (10.51)$$

and (for $m = 1, 2, 3, \ldots$)

$$L_{-n}^{(m)}(x) = \begin{cases} 0 & n = 1, 2, 3, \ldots, m \\ \dfrac{(-1)^m (n)_m}{m!} & n = m + 1, m + 2, \ldots \end{cases} \qquad (10.52)$$

the proofs of which are left to the exercises.

Exercises   10.3

In problems 1 to 14, verify the given relation.

1. $\gamma(a, x) = \dfrac{x^a}{a} M(a; a + 1; -x)$

2. $\Gamma(a, x) = e^{-x} U(1 - a; 1 - a; x)$

3. $\mathrm{Ei}\,(x) = -e^x U(1; 1; -x)$

4. $\mathrm{li}\,(x) = -x U(1; 1; -\ln x)$

5. $H_{2n+1}(x) = (-1)^n \dfrac{(2n + 1)!}{n!} 2x M\left(-n; \dfrac{3}{2}; x^2\right)$

6. $L_n(x) = M(-n; 1; x)$

7. $L_n^{(a)}(x) = \dfrac{(a + 1)_n}{n!} M(-n; a + 1; x)$

8. $J_p(x) = \dfrac{(x/2)^p}{\Gamma(p + 1)} e^{-ix} M\left(p + \dfrac{1}{2}; 2p + 1; 2ix\right)$
   Hint: Start with the integral representation (6.32) in Sec. 6.3 and make the change of variable $t = 2s - 1$.

9. $I_p(x) = \dfrac{(x/2)^p}{\Gamma(p + 1)} e^{-x} M\left(p + \dfrac{1}{2}; 2p + 1; 2x\right)$

10. $K_p(x) = \sqrt{\pi}\,(2x)^p e^{-x} U\left(p + \dfrac{1}{2}; 2p + 1; 2x\right)$

11. $C(x) = \dfrac{x}{2}\left[M\left(\dfrac{1}{2}; \dfrac{3}{2}; \dfrac{1}{2} i\pi x^2\right) + M\left(\dfrac{1}{2}; \dfrac{3}{2}; -\dfrac{1}{2} i\pi x^2\right)\right]$

12. $S(x) = \dfrac{x}{2i}\left[M\left(\dfrac{1}{2}; \dfrac{3}{2}; \dfrac{1}{2} i\pi x^2\right) - M\left(\dfrac{1}{2}; \dfrac{3}{2}; -\dfrac{1}{2} i\pi x^2\right)\right]$

13. $\mathrm{Ci}\,(x) = -\dfrac{1}{2}[e^{-ix} U(1; 1; ix) + e^{ix} U(1; 1; -ix)]$

14. $\mathrm{Si}\,(x) = \dfrac{\pi}{2} + \dfrac{1}{2i}[e^{-ix} U(1; 1; ix) - e^{ix} U(1; 1; -ix)]$

In problems 15 to 30, verify the special cases.

15. $M(1; 2; -x) = \dfrac{1}{x}(1 - e^{-x})$

16. $M(2; 1; -x) = (1 - x)e^{-x}$

17. $M(1; 3; -x) = \dfrac{2}{x^2}(x + e^{-x} - 1)$

18. $M(3; 2; -x) = \left(1 - \dfrac{x}{2}\right)e^{-x}$

19. $M(\tfrac{1}{2}; \tfrac{1}{2}; -x^2) = e^{-x^2}$

20. $M\left(\dfrac{1}{2}; \dfrac{3}{2}; -x^2\right) = \dfrac{\sqrt{\pi}}{2x}\,\text{erf}\,x$

21. $M(-\tfrac{1}{2}; \tfrac{1}{2}; -x^2) = e^{-x^2} + \sqrt{\pi}\,x\,\text{erf}\,x$

22. $M\left(-\dfrac{1}{2}; \dfrac{3}{2}; -x^2\right) = \dfrac{1}{2}e^{-x^2} + \dfrac{\sqrt{\pi}}{4x}(1 + 2x^2)\,\text{erf}\,x$

23. $M(1; \tfrac{1}{2}; x^2) = 1 + \sqrt{\pi}\,xe^{x^2}\,\text{erf}\,x$

24. $M\left(1; \dfrac{3}{2}; x^2\right) = \dfrac{\sqrt{\pi}}{2x}e^{x^2}\,\text{erf}\,x$

25. $M\left(\dfrac{1}{2}; 1; -x\right) = e^{-x/2}I_0\left(\dfrac{x}{2}\right)$

26. $M\left(-\dfrac{1}{2}; 1; -x\right) = e^{-x/2}\left[(1 + x)I_0\left(\dfrac{x}{2}\right) + xI_1\left(\dfrac{x}{2}\right)\right]$

27. $M\left(\dfrac{1}{2}; 2; -x\right) = e^{-x/2}\left[I_0\left(\dfrac{x}{2}\right) + I_1\left(\dfrac{x}{2}\right)\right]$

28. $M\left(\dfrac{3}{2}; 2; -x\right) = e^{-x/2}\left[I_0\left(\dfrac{x}{2}\right) - I_1\left(\dfrac{x}{2}\right)\right]$

29. $M\left(\dfrac{3}{2}; 3; -x\right) = \dfrac{4}{x}e^{-x/2}I_1\left(\dfrac{x}{2}\right)$

30. $M\left(\dfrac{3}{2}; 4; -x\right) = \dfrac{4}{x}e^{-x/2}\left[I_0\left(\dfrac{x}{2}\right) + \left(1 - \dfrac{4}{x}\right)I_1\left(\dfrac{x}{2}\right)\right]$

31. Show that $(x > 0)$

(a) $\displaystyle\lim_{a \to \infty} M\left(a; c; -\dfrac{x}{a}\right) = \Gamma(c)x^{(1-c)/2}J_{c-1}(2\sqrt{x})$

(b) $\displaystyle\lim_{a \to \infty} M\left(a; c; \dfrac{x}{a}\right) = \Gamma(c)x^{(1-c)/2}I_{c-1}(2\sqrt{x})$

In problems 32 and 33, use properties of the confluent hyper-geometric function to sum the series.

32. $\displaystyle\sum_{n=0}^{\infty} \frac{n!}{(2n+1)!} x^{2n+1} = \sqrt{\pi}\, e^{x^2/4}\, \text{erf}\frac{x}{2}$

33. $\displaystyle\sum_{k=0}^{\infty} \frac{(n+k)!}{n!(k!)^2} (-1)^k x^k = e^{-x} L_n(x), \; n = 0, 1, 2, \ldots$

34. Use the Cauchy product to show that

$$(1-t)^{-a} M\left(a; 1; -\frac{xt}{1-t}\right) = \sum_{n=0}^{\infty} (a)_n L_n(x)\frac{t^n}{n!}$$

In problems 35 to 37, verify the integral relation.

35. $\displaystyle\int_0^{\infty} x^{2n+1} \cos bx \, e^{-ax^2/2}\, dx = \frac{2^n n!}{a^{n+1}} M\left(n+1; \frac{1}{2}; -\frac{b^2}{2a}\right), a > 0$

36. $\displaystyle\int_0^{\infty} x^{2n} \sin bx \, e^{-ax^2/2}\, dx = \frac{2^n n!}{a^{n+1}} bM\left(n+1; \frac{3}{2}; -\frac{b^2}{2a}\right), a > 0$

37. $\displaystyle\int_0^{\infty} J_0(ax) x^{2\mu-1} e^{-b^2 x^2}\, dx = \frac{\Gamma(\mu)}{2b^{2\mu}} M\left(\mu; 1; -\frac{a^2}{4b^2}\right), a \geq 0, b > 0,$

$\mu \neq 0, -1, -2, \ldots$

In problems 38 and 39, use the result of problem 28 in Exercises 10.2 to derive the asymptotic formula.

38. $\text{erf}\, x \sim 1, \, x \to \infty$

39. $\displaystyle I_p(x) \sim \frac{e^x}{\sqrt{2\pi x}} \sum_{n=0}^{\infty} \frac{(\frac{1}{2}-p)_n(\frac{1}{2}+p)_n}{n!(2x)^n}, \, x \to \infty$

In problems 40 to 43, use the result of problem 30 in Exercises 10.2 to derive the asymptotic formula.

40. $\displaystyle\text{erfc}\, x \sim \frac{e^{-x}}{x\sqrt{\pi}} \sum_{n=0}^{\infty} \frac{(-1)^n (\frac{1}{2})_n}{x^{2n}}, \, x \to \infty$

41. $\displaystyle\text{Ei}\,(x) \sim \frac{e^x}{x} \sum_{n=0}^{\infty} \frac{n!}{x^n}, \, x \to \infty$

42. $\displaystyle\text{li}\,(x) \sim \frac{x}{\ln x} \sum_{n=0}^{\infty} \frac{n!}{(\ln x)^n}, \, x \to \infty$

43. $\displaystyle K_p(x) \sim \sqrt{\frac{\pi}{2x}}\, e^{-x} \sum_{n=0}^{\infty} \frac{(-1)^n (\frac{1}{2}+p)_n(\frac{1}{2}-p)_n}{n!(2x)^n}, \, x \to \infty$

44. By expressing the confluent hypergeometric functions in (10.45) by their series representations, deduce that

$$H_v(x) = \frac{1}{2\Gamma(-v)} \sum_{k=0}^{\infty} (-1)^k \Gamma\left(\frac{k-v}{2}\right) \frac{(2x)^k}{k!}$$

$$v \neq n \quad (n = 0, 1, 2, \ldots)$$

In problems 45 to 47, use the result of problem 44 to deduce the given relation.

45. $H'_v(x) = 2vH_{v-1}(x)$

46. $2vH'_{v-1}(x) = 2xH'_v(x) - 2vH_v(x)$

47. $H_{v+1}(x) - 2xH_v(x) + 2vH_{v-1}(x) = 0$

48. For $v < 0$, show that

$$H_v(x) = \frac{1}{\Gamma(-v)} \int_0^\infty e^{-t^2 - 2xt} t^{-v-1} \, dt$$

49. Using the result of problem 48, deduce the asymptotic series (for $v < 0$)

$$H_v(x) \sim (2x)^v \sum_{n=0}^\infty \frac{(-1)^n (-v)_{2n}}{n!(2x)^{2n}} \qquad x \to \infty$$

50. Show that

$$H_{-1/2}(x) = \sqrt{\frac{x}{2\pi}} e^{x^2/2} K_{1/4}\left(\frac{x^2}{2}\right)$$

51. Derive the series representation $(a > -1)$

$$L_v^{(a)}(x) = \frac{\Gamma(v + a + 1)}{\Gamma(-v)\Gamma(v+1)} \sum_{k=0}^\infty \frac{\Gamma(k - v)}{\Gamma(k + a + 1)} \frac{x^k}{k!}$$

In problems 52 to 54, use the result of problem 51 to deduce the given relation.

52. $L_v^{(a)\prime}(x) = -L_{v-1}^{(a+1)}(x)$

53. $xL_v^{(a)\prime}(x) - vL_v^{(a)}(x) + (v + a)L_{v-1}^{(a)}(x) = 0$

54. $(v + 1)L_{v+1}^{(a)}(x) + (x - 1 - 2v - a)L_v^{(a)}(x) + (v + a)L_{v-1}^{(a)}(x) = 0$

55. Show that

$$L_{-n}^{(a)}(x) = 0 \qquad n = 1, 2, 3, \ldots, \qquad a > -1, \quad a \neq 0, 1, 2, \ldots$$

56. Show that (for $m = 1, 2, 3, \ldots$)

$$L_{-n}^{(m)}(x) = \begin{cases} 0 & n = 1, 2, 3, \ldots, m \\ \dfrac{(-1)^m (n)_m}{m!} & n = m + 1, m + 2, \ldots \end{cases}$$

10.4 Whittaker Functions

For purposes of developing certain theories concerning DEs, it is sometimes helpful to transform the equation to what is called the *Liouville standard form*.* To derive this form, we first write the DE

---

* Sometimes called the *normal form*, not to be confused with (10.53).

in *normal form*

$$y'' + A(x)y' + B(x)y = 0 \tag{10.53}$$

Next we set $y = u(x)v(x)$, for which

$$y' = uv' + u'v$$

$$y'' = uv'' + 2u'v' + u''v$$

and when these expressions are substituted into (10.53), we get

$$vu'' + (2v' + Av)u' + (v'' + Av' + Bv)u = 0 \tag{10.54}$$

By selecting $2v' + Av = 0$, the coefficient of u' can be made to vanish. A function v which gives this result is

$$v(x) = \exp\left[-\tfrac{1}{2}\int A(x)\,dx\right] \tag{10.55}$$

and thus (10.54) reduces to the *Liouville standard form*

$$u'' + Q(x)u = 0 \tag{10.56}$$

where
$$Q(x) = B(x) - \tfrac{1}{4}[A(x)]^2 - \tfrac{1}{2}A'(x) \tag{10.57}$$

If we rewrite the confluent hypergeometric DE

$$xy'' + (c - x)y' - ay = 0$$

in normal form, i.e.,

$$y'' + \left(\frac{c}{x} - 1\right)y' - \frac{a}{x}y = 0 \tag{10.58}$$

we can then identify the functions

$$A(x) = \frac{c}{x} - 1 \qquad B(x) = -\frac{a}{x} \tag{10.59}$$

Hence, the Liouville standard form of the confluent hypergeometric DE is

$$u'' + \left(-\frac{1}{4} + \frac{c - 2a}{2x} + \frac{2c - c^2}{4x^2}\right)u = 0 \tag{10.60}$$

From (10.55), we calculate

$$v(x) = \exp\left[-\frac{1}{2}\int\left(\frac{c}{x} - 1\right)dx\right] = e^{x/2}x^{-c/2} \tag{10.61}$$

and since $y = u(x)v(x)$ (or $u = y/v$), it follows that one solution of (10.60) is given by

$$u_1 = e^{-x/2}x^{c/2}M(a;c;x) \qquad c \neq 0, -1, -2, \ldots \qquad (10.62)$$

It is customary to introduce new parameters m and k by means of the transformations

$$\frac{c}{2} = m + \frac{1}{2} \qquad (c = 2m + 1)$$

$$\frac{c}{2} - a = k \qquad \left(a = \frac{1}{2} + m - k\right) \qquad (10.63)$$

so that in terms of these parameters, Eq. (10.60) becomes

$$u'' + \left(-\frac{1}{4} + \frac{k}{x} + \frac{\frac{1}{4} - m^2}{x^2}\right)u = 0 \qquad (10.64)$$

with solution $u_1 = M_{k,m}(x)$, where

$$M_{k,m}(x) = e^{-x/2}x^{m+1/2}M(\tfrac{1}{2} + m - k; 2m + 1; x)$$

$$2m \neq -1, -2, -3, \ldots \qquad (10.65)$$

We call $M_{k,m}(x)$ a **Whittaker function of the first kind.**

We have previously shown (Sec. 10.2.3) that when $c \neq 2, 3, 4, \ldots$, the function

$$y_2 = x^{1-c}M(1 + a - c; 2 - c; x)$$

is a second linearly independent solution of (10.58). Using the parameters m and k, and the relation $u = y/v$, it follows that when $2m$ is not an integer, the function

$$u_2 = e^{-x/2}x^{-m+1/2}M(\tfrac{1}{2} - m - k; -2m + 1; x) \qquad (10.66)$$

is a second linearly independent solution of (10.64). However, comparison of (10.66) with (10.65) identifies $u_2 = M_{k,-m}(x)$, and therefore a general solution of (10.64) is

$$u = C_1M_{k,m}(x) + C_2M_{k,-m}(x) \qquad 2m \neq 0, \pm 1, \pm 2, \ldots \qquad (10.67)$$

The solutions $M_{k,\pm m}(x)$ of (10.64) are not always the most convenient ones to use in forming a general solution, because of the restriction that $2m$ not be an integer. Therefore, in certain situations we find it preferable to introduce the **Whittaker function of the second kind**

$$W_{k,m}(x) = e^{-x/2}x^{m+1/2}U(\tfrac{1}{2} + m - k; 2m + 1; x) \qquad (10.68)$$

It can be shown that $W_{k,m}(x)$ is a solution of (10.64) that is linearly independent of $M_{k,m}(x)$, even when $2m = 0, 1, 2, \ldots$. That this is so follows from the linear independence of the confluent hypergeometric functions of the first and second kinds. In terms of $W_{k,m}(x)$, the general solution of (10.64) reads

$$u = C_1 M_{k,m}(x) + C_2 W_{k,m}(x) \qquad 2m \neq -1, -2, -3, \ldots \quad (10.69)$$

The Whittaker functions clearly have many properties which follow directly from those of Kummer's functions, some of which are discussed in the exercises. In most applications the choice of using Kummer's functions or Whittaker's functions is mostly a matter of convenience. Both sets of functions commonly occur in reference material, although the functions of Whittaker are somewhat less prominent.

Exercises 10.4

1. You are given Bessel's equation
$$x^2 y'' + xy' + (x^2 - p^2)y = 0$$
   (a) Find the Liouville standard form.
   (b) For $p = \tfrac{1}{2}$, use (a) to deduce that the general solution of Bessel's DE for this special case can be expressed as
$$y = C_1 x^{-1/2} \cos x + C_2 x^{-1/2} \sin x$$

2. It can be shown that oscillatory solutions of $u'' + Q(x)u = 0$ exist only if $Q(x) > 0$. Use this criterion to deduce that
$$my'' + cy' + ky = 0$$
   has oscillatory solutions only if $c^2 - 4mk < 0$.

3. Use the criterion stated in problem 2 to deduce that Bessel's modified equation
$$x^2 y'' + xy' - (x^2 + p^2)y = 0$$
   has no oscillatory solutions.

In problems 4 to 8, verify the given relation.

4. $\operatorname{erf} x = \dfrac{2}{\sqrt{\pi x}} e^{-x^2/2} M_{-1/4, 1/4}(x^2)$

5. $\operatorname{erfc} x = \dfrac{1}{\sqrt{\pi x}} e^{-x^2/2} W_{-1/4,1/4}(x^2)$

6. $\gamma(a, x) = \Gamma(a) - x^{(a-1)/2} e^{-x/2} W_{(a-1)/2, a/2}(x)$

7. $M_{0,m}(2x) = \Gamma(m + 1) 2^{2m+1/2} \sqrt{x} I_m(x)$

8. $W_{0,m}(2x) = \sqrt{\dfrac{2x}{\pi}} K_m(x)$

9. Show that
   (a) $W_{k,m}(x) = W_{k,-m}(x)$
   (b) $W_{-k,m}(-x) = W_{-k,-m}(-x)$

10. Show that $(n = 1, 2, 3, \ldots)$

   (a) $\dfrac{d^n}{dx^n} [e^{x/2} x^{m-1/2} M_{k,m}(x)]$

   $\qquad = (-1)^n (-2m)_n x^{m-n/2-1/2} e^{x/2} M_{k-n/2, m-n/2}(x)$

   (b) $\dfrac{d^n}{dx^n} [e^{x/2} x^{m-1/2} W_{k,m}(x)]$

   $\qquad = (-1)^n (1/2 - m - k)_n x^{m-n/2-1/2} e^{x/2} W_{k-n/2, m-n/2}(x)$

In problems 11 to 13, derive the asymptotic formula.

11. $M_{k,m}(x) \sim x^{m+1/2}, \ x \to 0^+$

12. $M_{k,m}(x) \sim \dfrac{\Gamma(2m + 1)}{\Gamma(1/2 + m - k)} x^{-k} e^{x/2}, \ x \to \infty$

13. $W_{k,m}(x) \sim x^k e^{-x/2}, \ x \to \infty$

14. Show that the *parabolic cylinder function* defined by

$$D_n(x) = 2^{n/2+1/4} x^{-1/2} W_{n/2+1/4, -1/4}\left(\frac{x^2}{2}\right) \qquad n = 0, \pm 1, \pm 2, \ldots$$

satisfies the DE

$$y'' + \left(n + \frac{1}{2} - \frac{x^2}{4}\right) y = 0$$

15. Verify that (see problem 14)

$$\int_{-\infty}^{\infty} [D_0(x)]^2 \, dx = \sqrt{2\pi}$$

16. Verify that (see problem 14)

   (a) $D_n(x) = 2^{-n/2} e^{-x^2/4} H_n\left(\dfrac{x}{\sqrt{2}}\right)$

   (b) $D_{-1}(x) = \sqrt{\dfrac{\pi}{2}} e^{x^2/4} \operatorname{erfc}\left(\dfrac{x}{\sqrt{2}}\right)$

17. Show that (for $v + \frac{1}{2} + m > 0$, $k - v > 0$)

$$\int_0^\infty e^{-bt/2} t^{v-1} M_{k,m}(bt)\, dt = \frac{\Gamma(k - v)\Gamma(\frac{1}{2} + m + v)\Gamma(2m + 1)}{\Gamma(\frac{1}{2} + m + k)\Gamma(\frac{1}{2} + m - v)} b^{-v}$$

18. Evaluate the integrals

(a) $\displaystyle\int_0^\infty [M_{k,m}(x)]^2\, dx$

(b) $\displaystyle\int_0^\infty x^{-1}[M_{k,m}(x)]^2\, dx$

19. Show that the wronskian of Whittaker's functions is

$$W(M_{k,m}, W_{k,m})(x) = -\frac{\Gamma(2m + 1)}{\Gamma(m - k + \frac{1}{2})} \qquad 2m \neq -1, -2, -3, \ldots$$

20. Using the integral representation for $U(a; c; x)$, show that

(a) $\displaystyle W_{k,m}(x) = \frac{x^k e^{-x/2}}{\Gamma(m - k + \frac{1}{2})} \int_0^\infty e^{-t} t^{m-k-1/2} \left(1 + \frac{t}{x}\right)^{m+k-1/2} dt$

$m - k + \frac{1}{2} > 0$

(b) From (a), deduce the asymptotic series

$$W_{k,m}(x) \sim e^{-x/2} x^k \sum_{n=0}^\infty \frac{(-1)^n (m - k + \frac{1}{2})_n (\frac{1}{2} - m - k)_n}{n! x^n} \qquad x \to \infty$$

21. Given the set of polynomials*

$$G_n^m(x) = \sum_{k=0}^n \binom{m + n - k}{m} \frac{x^k}{k!}$$

show that

(a) $\displaystyle G_n^m(x) = \frac{1}{n!} x^{m+n+1} U(m + 1; m + n + 2; x)$

(b) $\displaystyle G_n^m(x) = \frac{1}{n!} x^a e^{x/2} W_{b,a+1/2}(x), \quad a = \frac{m + n}{2}, \quad b = \frac{n - m}{2}$

22. In solving for the wave function associated with the hydrogen atom (see Sec. 5.3.4) we are led to the radial wave equation

$$\frac{1}{\rho^2} \frac{d}{d\rho}\left(\rho^2 \frac{d\chi}{d\rho}\right) + \left[\frac{\lambda}{\rho} - \frac{1}{4} - \frac{l(l+1)}{\rho^2}\right]\chi = 0$$

---

* The polynomials $G_n^m(x)$ arise in the problem of finding the probability density function for the output of a cross correlator. For example, see L. E. Miller and J. S. Lee, "The Probability Density Function for the Output of an Analog Cross Correlator," *IEEE Trans. Inform. Theory*, **IT-20**: 433–440, July 1974, and L. C. Andrews and C. S. Brice, "The PDF and CDF for the Sum of N Filtered Outputs of an Analog Cross Correlator with Bandpass Inputs," *IEEE Trans. Inform. Theory*, **IT-29**: 299–306, March 1983.

(a)   Show that this DE has the standard form

$$u'' + \left[\frac{\lambda}{\rho} - \frac{1}{4} - \frac{l(l+1)}{\rho^2}\right]u \quad 0$$

(b)   From (a), deduce that one solution for $\chi(\rho)$ is

$$\chi(\rho) = \frac{1}{\rho}M_{\lambda, l+1/2}(\rho)$$

(c)   Show that the solution in (b) can also be expressed in the form

$$\chi(\rho) = e^{-\rho/2}\rho^l L_{\lambda-l-1}^{(2l+1)}(\rho)$$

where $L_\nu^{(a)}(\rho)$ is the generalized Laguerre function.

11

Generalized
Hypergeometric Functions

11.1 Introduction

The special properties associated with the hypergeometric and confluent hypergeometric functions have spurred a number of investigations into developing functions even more general than these. Some of this work was done in the nineteenth century by Clausen, Appell, and Lauricella (among others), but much of it has occurred during the last 70 years. Even the most recent names are too numerous to mention, but MacRobert and Meijer are among the most famous.

The importance of working with generalized functions of any kind stems from the fact that most special functions are simply special cases of them, and thus each recurrence formula or identity developed for the generalized function becomes a master formula from which a large number of relations for other functions can be deduced. New relations for some of the special functions have been discovered in just this way. Also the use of generalized functions often facilitates the analysis by permitting complex expressions to be represented more simply in terms of some generalized function. Operations such as differentiation and integration can sometimes be performed more readily on the resulting generalized functions than on the original complex expression, even though the two are equivalent. Finally, in many situations we resort to expressing our results in terms of these generalized functions because there are no simpler functions that we can call upon.

Our treatment of generalized hypergeometric functions is brief. For a deeper discussion the interested reader should consult one of the many publications devoted entirely to functions of this nature.*

11.2 The Set of Functions $_pF_q$

In general, we say that a series $\sum u_n(x)$ is a *hypergeometric-type series* if the ratio $u_{n+1}(x)/u_n(x)$ is a rational function of n. A general series of this type is

$$_pF_q(a_1, \ldots, a_p; c_1, \ldots, c_q; x) = \sum_{n=0}^{\infty} \frac{(a_1)_n \cdots (a_p)_n x^n}{(c_1)_n \cdots (c_q)_n \, n!} \qquad (11.1)$$

where p and q are nonnegative integers and no c_k $(k = 1, 2, \ldots, q)$ is zero or a negative integer. The function defined by (11.1), which we denote simply $_pF_q$, is called a **generalized hypergeometric function**. Clearly, (11.1) includes the special cases $_2F_1$ and $_1F_1$, which are the *hypergeometric* and *confluent hypergeometric functions*, respectively.

Applying the ratio test to (11.1) leads to

$$\lim_{n \to \infty} \left| \frac{u_{n+1}(x)}{u_n(x)} \right| = |x| \lim_{n \to \infty} \left| \frac{(a_1 + n) \cdots (a_p + n)}{(c_1 + n) \cdots (c_q + n)(1 + n)} \right| \qquad (11.2)$$

Hence, provided the series does not terminate, we see that

1. If $p < q + 1$, the series *converges* for all (finite) x.

2. If $p = q + 1$, the series *converges* for $|x| < 1$ and diverges for $|x| > 1$.

3. If $p > q + 1$, the series *diverges* for all x except $x = 0$.

The series (11.1) is therefore meaningful when $p > q + 1$ only if it truncates [see (11.9) below].

Because of its generality, the function $_pF_q$ includes a great variety of functions as special cases. Some of these special cases are given by†

$$e^x = \sum_{n=0}^{\infty} \frac{x^n}{n!} = {}_0F_0(-; -; x) \qquad (11.3)$$

---

*For example, see A. M. Mathai and R. K. Saxena, *Generalized Hypergeometric Functions with Applications in Statistics and Physical Sciences,* Lecture Notes in Mathematics, Springer, New York, 1973.

† The absence of a parameter in $_pF_q$ is emphasized by a dash.

$$(1-x)^{-a} = \sum_{n=0}^{\infty} (a)_n \frac{x^n}{n!} = {}_1F_0(a; -; x) \tag{11.4}$$

$$\cos x = \sum_{n=0}^{\infty} \frac{(-1)^n}{(1/2)_n} \frac{(x^2/4)^n}{n!} = {}_0F_1\left(-; \frac{1}{2}; -\frac{x^2}{4}\right) \tag{11.5}$$

$$J_0(x) = \sum_{n=0}^{\infty} \frac{(-1)^n}{(1)_n} \frac{(x^2/4)^n}{n!} = {}_0F_1\left(-; 1; -\frac{x^2}{4}\right) \tag{11.6}$$

$$F(a, b; c; x) = \sum_{n=0}^{\infty} \frac{(a)_n(b)_n}{(c)_n} \frac{x^n}{n!} = {}_2F_1(a, b; c; x) \tag{11.7}$$

$$M(a; c; x) = \sum_{n=0}^{\infty} \frac{(a)_n}{(c)_n} \frac{x^n}{n!} = {}_1F_1(a; c; x) \tag{11.8}$$

An important terminating series for the case $p > q + 1$ is the Hermite polynomial

$$H_n(x) = (2x)^n \left[{}_2F_0\left(-\frac{n}{2}, \frac{1-n}{2}; -; -\frac{1}{x^2}\right) \right] \tag{11.9}$$

11.2.1 Hypergeometric-type series

The general series (11.1) has been extensively studied over the years, and many of the important properties associated with the related $_pF_q$ functions have been developed. For this reason it is often advantageous to express a given series whose sum is unknown in the form of (11.1), since it provides a standard form by which to classify the series. In this standard form one may then be able to identify the function that the series defines; if not, at least some general theory concerning the specific function $_pF_q$ is probably available.

Not all series, of course, are specializations of the generalized hypergeometric series (11.1). Only those series for which the ratio of successive terms is a rational function of n are hypergeometric-type series. Forming the ratio of successive terms is also helpful in identifying the numerator and denominator parameters of the hypergeometric-type series and the argument of the function. That is, by writing the ratio of successive terms in the form

$$\frac{u_{n+1}(x)}{u_n(x)} = \frac{(a_1 + n) \cdots (a_p + n)x^{\mu}}{(c_1 + n) \cdots (c_q + n)(1 + n)} \tag{11.10}$$

we can easily identify the parameters $a_1, \ldots, a_p, c_1, \ldots, c_q$, and the argument x^{μ}, where μ is a real number. If the particular factor $1 + n$ is not in the denominator, it can be introduced by multiplying both numerator and denominator by it. Note, however, that all hyper-geometric series of the form (11.1) are defined so that the first term of the series is unity. Thus, if $u_0(x) \neq 1$ in the given series, we need to multiply the resulting $_pF_q$ series by $u_0(x)$. Let us illustrate the technique with an example.

Example 1: Express $f(x)$ in terms of a $_pF_q$ function, where

$$f(x) = \sum_{n=0}^{\infty} \frac{(-1)^n (2n)! \, (x/2)^{2n}}{(n!)^2 [\Gamma(n+p+1)]^2}$$

Solution: We first observe that the $n = 0$ term of the series is

$$u_0(x) = \frac{1}{[\Gamma(p+1)]^2}$$

Next, using properties of factorials and the gamma function, we find

$$\frac{u_{n+1}(x)}{u_n(x)} = -\frac{(2n+2)! \, (x/2)^{2n+2}}{[(n+1)!]^2 [\Gamma(n+p+2)]^2} \cdot \frac{(n!)^2 [\Gamma(n+p+1)]^2}{(2n)! \, (x/2)^{2n}}$$

$$= \frac{(\tfrac{1}{2}+n)(-x^2)}{(p+1+n)^2(1+n)}$$

and thus deduce that

$$f(x) = \frac{1}{[\Gamma(p+1)]^2} \sum_{n=0}^{\infty} \frac{(-1)^n (\tfrac{1}{2})_n}{(p+1)_n (p+1)_n} \frac{x^{2n}}{n!}$$

or

$$f(x) = \frac{1}{[\Gamma(p+1)]^2} \left[{}_1F_2\left(\frac{1}{2}; p+1, p+1; -x^2\right) \right]$$

Example 2: Show that

$$\sum_{n=1}^{\infty} \frac{(4x)^n}{n(2n-1)!} = 4 \sinh^2 \sqrt{x} \qquad x > 0$$

Solution: Because the given series does not start with $n = 0$, we begin with an index shift to obtain

$$\sum_{n=1}^{\infty} \frac{(4x)^n}{n(2n-1)!} = 4x \sum_{n=0}^{\infty} \frac{(4x)^n}{(n+1)(2n+1)!}$$

$$= 4x \sum_{n=0}^{\infty} \frac{(1)_n}{(2)_n (\tfrac{3}{2})_n} \frac{x^n}{n!}$$

$$= 4x \, {}_1F_2(1; 2, \tfrac{3}{2}; x)$$

where we have used the identities (see Sec. 9.2)

$$n + 1 = \frac{(n + 1)!}{n!} = \frac{(2)_n}{(1)_n}$$

$$(2n + 1)! = 2^{2n} (^3/_2)_n n!$$

The special case of Ramanujan's theorem (see problem 4 in Exercises 11.2)

$$_2F_3\left(a, b - a; b, \frac{b}{2}, \frac{b+1}{2}; \frac{x^2}{4}\right) = _1F_1(a; b; x) \, _1F_1(a; b; -x)$$

with $a = 1$ and $b = 2$ leads to

$$_2F_3(1, 1; 2, 1, ^3/_2; x) = _1F_2(1; 2; ^3/_2; x)$$

$$= _1F_1(1; 2; 2\sqrt{x}) \, _1F_1(1; 2; -2\sqrt{x})$$

This relation, coupled with Kummer's transformation, yields

$$\sum_{n=1}^{\infty} \frac{(4x)^n}{n(2n - 1)!} = 4x \, _1F_1(1; 2; 2\sqrt{x}) \, _1F_1(1; 2; -2\sqrt{x})$$

$$= 4x [\, _1F_1(1; 2; 2\sqrt{x})]^2 e^{-2\sqrt{x}}$$

Finally, with the aid of the identities (see problem 9 in Exercises 10.2 and problem 10a in Exercises 7.2)

$$_1F_1(1; 2; 2x) = \Gamma\left(\frac{3}{2}\right)\left(\frac{2}{x}\right)^{1/2} e^x I_{1/2}(x)$$

$$I_{1/2}(x) = \sqrt{\frac{2}{\pi x}} \sinh x$$

it follows that

$$\sum_{n=1}^{\infty} \frac{(4x)^n}{n(2n - 1)!} = 4x \left(\frac{\pi}{4}\right)\left(\frac{2}{\sqrt{x}}\right)\left(\frac{2}{\pi\sqrt{x}}\right) \sinh^2 \sqrt{x} = 4 \sinh^2 \sqrt{x}$$

The above method of summing the given series in Example 2 may not represent the easiest approach to this particular problem. We are using it merely as an illustrative procedure that may prove useful in dealing with more complex series.

Exercises 11.2

1. Show that

$$\frac{d}{dx}[_pF_q(a_1,\ldots,a_p;c_1,\ldots,c_q;x)]$$

$$= \frac{\displaystyle\prod_{k=1}^{p} a_k}{\displaystyle\prod_{j=1}^{1} c_j} \, _pF_q(a_1+1,\ldots,a_p+1;c_1+1,\ldots,c_q+1;x)$$

2. For $a+b \neq 1$, show that

$$_0F_1(-;a;x) \, _0F_1(-;b;x)$$

$$= \, _2F_3\left(\frac{a+b-1}{2},\frac{a+b}{2};a+b-1,a,b;4x\right)$$

3. Use the result of problem 2 to deduce that

$$[J_0(x)]^2 = \, _1F_2(\tfrac{1}{2};1,1;-x^2)$$

4. Verify *Ramanujan's theorem*

$$_1F_1(a;b;x) \, _1F_1(a;b;-x) = \, _2F_3\left(a,b-a;b,\frac{b}{2},\frac{b+1}{2};\frac{x^2}{4}\right)$$

5. Show that (for $n = 0, 1, 2, \ldots$)

$$_2F_0(-n,a;-;x) = (a)_n(-1)^n x^n \, _1F_1\left(-n;1-a-n;-\frac{1}{x}\right)$$

6. Use the result of problem 5 to show that ($n = 0, 1, 2, \ldots$)

   (a) $\displaystyle L_n^{(a)}(x) = \frac{(-1)^n}{n!}x^n\left[_2F_0\left(-n,-n-a;-\frac{1}{x}\right)\right]$

   (b) $\displaystyle H_n(x) = (2x)^n\left[_2F_0\left(-\frac{n}{2},\frac{1-n}{2};-;-\frac{1}{x^2}\right)\right]$

7. Verify *Kummer's second formula* $[2a \neq -(2n+1), n = 0, 1, 2, \ldots]$

$$e^{-x}\, _1F_1(a;2a;2x) = \, _0F_1\left(-;a+\frac{1}{2};\frac{x^2}{4}\right)$$

8. Use the result of problem 7 to show that ($x > 0$)

$$_0F_1\left(-;a+\frac{1}{2};\frac{x^2}{4}\right) = \left(\frac{x}{2}\right)^{1/2-a}\Gamma\left(a+\frac{1}{2}\right)I_{a-1/2}(x)$$

9. Show that (for $n = 0, 1, 2, \dots$)

$$_3F_2(-n, a, b; c, 1 - c + a + b - n; 1) = \frac{(c - a)_n (c - b)_n}{(c)_n (c - a - b)_n}$$

Hint: Expand the relation

$$F(c - a, c - b; c; x) = (1 - x)^{a+b-c} F(a, b; c; x)$$

in series form and compare like coefficients.

10. Show that

$$\int_0^x {}_0F_1\left[-; 1; -\frac{1}{4} t(x - t)\right] dt = x\, {}_0F_1\left(-; \frac{3}{2}; -\frac{x^2}{16}\right) = 2 \sin \frac{x}{2}$$

11. Show that

$$\int_0^x t^{1/2}(x - t)^{-1/2}[1 - t^2(x - t)^2]^{-1/2}\, dt = \frac{\pi}{2} x\left[{}_2F_1\left(\frac{1}{4}, \frac{3}{4}; 1; \frac{x^4}{16}\right)\right]$$

12. Show that $(s > 1)$

$$\int_0^\infty e^{-st} t^\nu [{}_pF_q(a_1, \dots, a_p; c_1, \dots, c_q; t)]\, dt$$

$$= \frac{\Gamma(\nu + 1)}{s^{\nu+1}}\left[{}_{p+1}F_q\left(\nu + 1, a_1, \dots, a_p; c_1, \dots, c_q; \frac{1}{s}\right)\right]$$

In problems 13 to 16, express the series as a function ${}_pF_q$.

13. $\displaystyle\sum_{n=0}^{\infty} \frac{(2n)!\,(2n + 1)!}{2^{4n}(n!)^4} x^n$

15. $\displaystyle\sum_{k=0}^{n} \binom{2n}{2k} k!\, x^k$

14. $\displaystyle x + \sum_{n=1}^{\infty} \frac{1 \times 3 \times \cdots \times (2n - 1)x^{2n+1}}{(2n + 1)(2n + 3) \cdots (4n + 1)}$

16. $\displaystyle\sum_{k=0}^{[n/2]} \frac{(-1)^k n!\,(2x)^{n-2k}}{(1/2)_k (n - 2k)!}$

17. *Bessel polynomials* are defined by*

$$b_n(x) = {}_2F_0\left(-n, 1 + n; -; -\frac{x}{2}\right)$$

Show that

(a) $\displaystyle K_{n+1/2}(x) = \frac{\pi}{2x} e^{-x} b_n\left(\frac{1}{x}\right)$

(b) $\displaystyle G_n^n(x) = \frac{x^n}{n!} b_n\left(\frac{2}{x}\right)$, where the polynomials $G_n^m(x)$ are defined in problem 21, Exercises 10.4

---

* Bessel polynomials were first studied by H. L. Krall and O. Frink, "A New Class of Orthogonal Polynomials: The Bessel Polynomials," *Trans. Amer. Math. Soc.*, **65**: 100–115, 1949. See also E. Grosswald, *The Bessel Polynomials*, Lecture Notes in Mathematics, Springer, New York, 1978.

18. For the polynomials defined by*

$$Z_n(x) = {}_2F_2(-n, 1+n; 1, 1; x)$$

show that

(a) $nZ_n'(x) - nZ_n(x) = -nZ_{n-1}(x) - xZ_{n-1}'(x)$

(b) $(1-t)^{-1}\left[{}_1F_1\left(\dfrac{1}{2}; 1; \dfrac{-4xt}{(1-t)^2}\right)\right] = \sum\limits_{n=0}^{\infty} Z_n(x)t^n$

(c) $(1-t)^{-1}\exp\left[\dfrac{-2xt}{(1-t)^2}\right]I_0\left[\dfrac{-2xt}{(1-t)^2}\right] = \sum\limits_{n=0}^{\infty} Z_n(x)t^n$

In problems 19 to 22, verify the formulas for products of Bessel functions.

19. $J_p(x)J_\nu(x) = \dfrac{(x/2)^{p+\nu}}{\Gamma(p+1)\Gamma(\nu+1)}$

$$\times\left[{}_2F_3\left(\dfrac{p+\nu+1}{2}, \dfrac{p+\nu+2}{2}; p+1, \nu+1, p+\nu+1; -x^2\right)\right]$$

20. $[J_p(x)]^2 = \dfrac{(x/2)^{2p+1}}{[\Gamma(p+1)]^2}\left[{}_1F_2\left(p+\dfrac{1}{2}; p+1, 2p+1; -x^2\right)\right]$

21. $J_p(x)J_{p+1}(x) = \dfrac{(x/2)^{2p+1}}{\Gamma(p+1)\Gamma(p+2)}\left[{}_1F_2\left(p+\dfrac{3}{2}; p+2, 2p+2; -x^2\right)\right]$

22. $J_p(x)I_p(x) = \dfrac{(x/2)^{2p}}{[\Gamma(p+1)]^2}\left[{}_0F_3\left(-; \dfrac{p+1}{2}, \dfrac{p+2}{2}, p+1; -\dfrac{x^4}{64}\right)\right]$

23. Show that

$${}_0F_1(-; a; px)\,{}_0F_1(-; b; qx) = \sum\limits_{n=0}^{\infty}\dfrac{(px)^n}{(a)_n n!}\,{}_2F_1\left(-n, 1-a-n, b; \dfrac{p}{q}\right)$$

24. For $p = q = 1$, show that the result of problem 23 reduces to that of problem 2.

25. Show that

$${}_1F_1(a; c; px)\,{}_1F_1(b; d; qx) = \sum\limits_{n=0}^{\infty}\dfrac{(a)_n}{(c)_n}\dfrac{(px)^n}{n!}$$

$$\times {}_3F_2\left(-n, b, 1-c-n; d, 1-a-n; -\dfrac{q}{p}\right)$$

---

*These polynomials were introduced by H. Bateman, "Two Systems of Polynomials for the Solution of Laplace's Integral Equation," *Duke Math. J.*, **2**: 569–577, 1936.

11.3   Other Generalizations

In the first half of the twentieth century new theories concerning generalized functions began to flourish. Most of this work followed Barnes' use of the gamma function in 1907 to develop a new theory of the hypergeometric function $_2F_1$. In the 1930s, both the E *function* of MacRobert and the G *function* of Meijer were introduced in an attempt to give meaning to the symbol $_pF_q'$ for the case $p > q + 1$. The E function is actually a special case of the G function and for that reason is less prominent in the literature.

11.3.1   The Meijer G function

In 1936, C. S. Meijer introduced the **G function**[*]

$$
G_{p,q}^{m,n}\!\left(x \;\middle|\; \begin{matrix} a_1, \ldots, a_p \\ c_1, \ldots, c_q \end{matrix}\right) = \sum_{k=1}^{m} \frac{\prod_{j=1}^{m}{}' \Gamma(c_j - c_k) \prod_{j=1}^{n} \Gamma(1 + c_k - a_j) x^{c_k}}{\prod_{j=m+1}^{q} \Gamma(1 + c_k - c_j) \prod_{j=n+1}^{p} \Gamma(a_j - c_k)}
$$

$$
\times {}_pF_{q-1}[1 + c_k - a_1, \ldots, 1 + c_k - a_p; 1 + c_k - c_1, \ldots, *, \ldots,
$$

$$
1 + c_k - c_q; (-1)^{p-m-n} x] \quad (11.11)
$$

where $1 \le m \le q$, $0 \le n \le p \le q - 1$, no two of the c_k's ($k = 1, 2, \ldots, m$) differ by zero or an integer, and $a_j - c_k \ne 1, 2, 3, \ldots$ for $j = 1, 2, \ldots, n$ and $k = 1, 2, \ldots, m$. If $p = q$, we restrict $|x| < 1$. For notational convenience, we often write

$$
G_{p,q}^{m,n}\!\left(x \;\middle|\; \begin{matrix} a_p \\ c_p \end{matrix}\right) \equiv G_{p,q}^{m,n}\!\left(x \;\middle|\; \begin{matrix} a_1, \ldots, a_p \\ c_1, \ldots, c_q \end{matrix}\right) \quad (11.12)
$$

or if confusion is not likely, we simply write $G_{p,q}^{m,n}(x)$.

Because of its relation with the $_pF_q$ functions given by (11.11), it is clear that the G function incorporates a great many other functions as special cases. Some of these special cases are given by the

---

[*] C. S. Meijer, "Einige Integraldarstellungen aus der Theorie der Besselschen und der Whittaker Funktionen," *Akad. Wet. Amst. Proc.*, **39**: 394–403, 519–527, 1936. The prime in the product symbol $\prod'$ denotes the omission of the term when $j = k$. Also in the parameter set of $_pF_{q-1}$ the parameter corresponding to $1 + c_k - c_k$ (indicated by $*$) is to be omitted. Last, an empty product is interpreted as unity.

following:

$$G_{01}^{10}(x \mid a) = x^a e^{-x} \tag{11.13}$$

$$G_{11}^{11}\left(x \;\middle|\; \begin{matrix} 1-a \\ 0 \end{matrix}\right) = \Gamma(a)(1+x)^{-a} \tag{11.14}$$

$$G_{02}^{10}(x \mid a, b) = x^{(a+b)/2} J_{a-b}(2\sqrt{x}) \tag{11.15}$$

$$G_{12}^{11}\left(x \;\middle|\; \begin{matrix} ^{1}\!/_{2} \\ p, -p \end{matrix}\right) = \sqrt{\pi}\, e^{-x/2} I_p\left(\frac{x}{2}\right) \tag{11.16}$$

$$G_{02}^{20}(x \mid a, b) = 2x^{(a+b)/2} K_{a-b}(2\sqrt{x}) \tag{11.17}$$

$$G_{12}^{11}\left(x \;\middle|\; \begin{matrix} 1-a \\ 0, 1-c \end{matrix}\right) = \frac{\Gamma(a)}{\Gamma(c)}\,[{}_1F_1(a;c;-x)] \tag{11.18}$$

$$G_{22}^{12}\left(x \;\middle|\; \begin{matrix} 1-a, 1-b \\ 0, 1-c \end{matrix}\right) = \frac{\Gamma(a)\Gamma(b)}{\Gamma(c)}\,[{}_2F_1(a, b;c;-x)] \tag{11.19}$$

Meijer redefined the G function in 1941* in terms of a Barnes contour integral in the complex plane that ultimately led to an interpretation of the symbol ${}_pF_q$ when $p > q + 1$. In particular, as a consequence of his more general definition, we have the important property

$$G_{p,q}^{m,n}\left(\frac{1}{x} \;\middle|\; \begin{matrix} a_p \\ c_q \end{matrix}\right) = G_{q,p}^{n,m}\left(x \;\middle|\; \begin{matrix} 1-c_q \\ 1-a_p \end{matrix}\right) \tag{11.20}$$

which allows us to transform from a G function for which $p > q$ to one for which $p < q$ (and vice versa). This property, combined with (11.11), is particularly useful in developing asymptotic formulas for certain ${}_pF_q$ functions with large arguments. For instance, see the following example.

Example 3: Use the Meijer G function to derive the asymptotic formula

$$_1F_1(a;c;-x) \sim \frac{\Gamma(c)}{\Gamma(c-a)} x^{-a} \qquad x \to \infty$$

---

*C. S. Meijer, "Neue Integraldarstellungen für Whittakersche Funckionen," *Proc. Ned. Akad. v. Wetensch., Amsterdam*, **44:** 81–92, 1941.

Solution: Based on (11.18) and (11.20), we have

$$
{}_1F_1(a;c;-x) = \frac{\Gamma(c)}{\Gamma(a)} G_{12}^{11}\left(x \,\middle|\, \begin{matrix} 1-a \\ 0, 1-c \end{matrix}\right)
$$

$$
= \frac{\Gamma(c)}{\Gamma(a)} G_{21}^{11}\left(\frac{1}{x} \,\middle|\, \begin{matrix} 1, c \\ a \end{matrix}\right)
$$

Finally, we use (11.11) to obtain

$$
{}_1F_1(a;c;-x) = \frac{\Gamma(c)}{\Gamma(a)}\frac{\Gamma(a)}{\Gamma(c-a)}\left(\frac{1}{x}\right)^a\left[{}_2F_0\left(a;1+a-c;-;\frac{1}{x}\right)\right]
$$

$$
\sim \frac{\Gamma(c)}{\Gamma(c-a)} x^{-a} \qquad x \to \infty
$$

The basic properties of the G function are far too numerous for us to discuss in any detail. Also the proofs of many of these properties (and any real understanding of this function) require knowledge of complex variable theory. Hence, for our purposes, we will be content to merely list a few of the simplest properties without justification.

If one of the parameters in the numerator set coincides with one of the parameters in the denominator set, the order of the G function may decrease. For example, if $a_j = c_k$ for some $j = 1, 2, \ldots, n$ and some $k = m+1, m+2, \ldots, q$, then

$$
G_{p,q}^{m,n}\left(x \,\middle|\, \begin{matrix} a_p \\ c_q \end{matrix}\right) = G_{p-1,q-1}^{m,n-1}\left(x \,\middle|\, \begin{matrix} a_1, \ldots, a_{j-1}, a_{j+1}, \ldots, a_p \\ c_1, \ldots, c_{k-1}, c_{k+1}, \ldots, c_q \end{matrix}\right) \tag{11.21}
$$

An analogous relationship exists if $a_j = c_k$ for some $j = n+1, n+2, \ldots, p$ and some $k = 1, 2, \ldots, n$. In this case it is m, and not n, that decreases by one unit in addition to p and q decreasing by one unit.

Multiplication of the Meijer G function by powers of x leads to the simple relation

$$
x^r G_{p,q}^{m,n}\left(x \,\middle|\, \begin{matrix} a_p \\ c_q \end{matrix}\right) = G_{p,q}^{m,n}\left(x \,\middle|\, \begin{matrix} a_p + r \\ c_q + r \end{matrix}\right) \tag{11.22}
$$

where the implication is that each numerator and denominator parameter is increased by the power r. Differentiation of this function is also easily performed, although there are several varieties of formulas. A particularly simple differentiation formula is given by

$$
\frac{d}{dx}\left[x^{-c_1} G_{p,q}^{m,n}\left(x \,\middle|\, \begin{matrix} a_p \\ c_q \end{matrix}\right)\right] = -x^{-1-c_1} G_{p,q}^{m,n}\left(x \,\middle|\, \begin{matrix} a_1, \ldots, a_p \\ c_1 + 1, c_2, \ldots, c_q \end{matrix}\right) \tag{11.23}
$$

As an illustration of the use of the last two properties, consider the special example

$$\frac{d}{dx} G_{02}^{20}(x \mid a, 0) = -x^{-1}G_{02}^{20}(x \mid a, 1)$$

$$= -G_{02}^{20}(x \mid a - 1, 0) \qquad (11.24)$$

To check this result and to emphasize the efficiency of the G function notation, we note from (11.17) above that

$$G_{02}^{20}(x \mid a, 0) = 2x^{a/2}K_a(2x^{1/2})$$

Thus, (11.24) is equivalent to the formula

$$\frac{d}{dx} [2x^{a/2}K_a(2x^{1/2})] = -2x^{(a-1)/2}K_{a-1}(2x^{1/2}) \qquad (11.25)$$

Of course, we can derive this result directly through application of the product formula and chain rule, which yields

$$\frac{d}{dx} [2x^{a/2}K_a(2x^{1/2})] = 2x^{a/2}K_a'(2x^{1/2})x^{-1/2} + ax^{-1+a/2}K_a(2x^{1/2})$$

$$= 2x^{a/2}[-x^{-1/2}K_{a-1}(2x^{1/2})] \qquad (11.26)$$

where the last step is obtained through application of the identity

$$K_p'(x) = -K_{p-1}(x) - \frac{p}{x}K_p(x) \qquad (11.27)$$

Further simplification of (11.26) yields the same result as (11.25).

This last example gives some hint of the power and economy in using the G function as a tool of analysis. The difficulty that often exists in working with this function is the recognition of the particular $G_{p,q}^{m,n}(x)$ as one of the elementary or special functions. However, there are countless instances in which the G function of interest is not related to any known function.

Owing to the generality of the G function, any integral involving this function serves as a master integral formula for a whole class of special cases. Two such integrals, which are essentially Laplace and

Hankel transform relations, are given by

$$\int_0^\infty e^{-\beta x} x^{-\lambda} G_{p,q}^{m,n}\left(\alpha x \left| \begin{matrix} a_1, \ldots, a_p \\ c_1, \ldots, c_q \end{matrix} \right.\right) dx = \beta^{\lambda-1} G_{p+1,q}^{m,n+1}\left(\frac{\alpha}{\beta} \left| \begin{matrix} \lambda, a_1, \ldots, a_p \\ c_1, \ldots, c_q \end{matrix} \right.\right)$$

$$(11.28)$$

$$\beta > 0 \qquad p + q < 2(m + n) \qquad c_j - \lambda > -1 \qquad j = 1, \ldots, m$$

and

$$\int_0^\infty x^{2\lambda} J_\nu(\beta x) G_{p,q}^{m,n}\left(\alpha x^2 \left| \begin{matrix} a_1, \ldots a_p \\ c_1, \ldots, c_q \end{matrix} \right.\right) dx$$

$$= \frac{2^{2\lambda}}{\beta^{2\lambda+1}} G_{p+2,q}^{m,n+1}\left(\frac{4\alpha}{\beta^2} \left| \begin{matrix} \mu, a_1, \ldots, a_p, \sigma \\ c_1, \ldots, c_q \end{matrix} \right.\right)$$

$$(11.29)$$

$$\beta > 0 \qquad \mu = \frac{1-\nu}{2} - \lambda \qquad \sigma = \frac{1+\nu}{2} - \lambda$$

$$p + q < 2(m + n) \qquad c_j + \lambda + \frac{\upsilon}{2} > -\tfrac{1}{2} \qquad j = 1, \ldots, m$$

$$a_j + \lambda < \tfrac{3}{4} \qquad j = 1, \ldots, n$$

Example 4: Use the Meijer G function to evaluate

$$I = \int_0^\infty x^\alpha J_\ell(x) J_m(x) J_n(\beta x) \, dx$$

Solution: Integrals of this type are prominent in wave propagation problems. To start, we use the identity (recall problem 19 in Exercises 11.2)

$$J_\ell(x) J_m(x) = \frac{(x/2)^{\ell+m}}{\Gamma(\ell+1)\Gamma(m+1)}$$

$$\times {}_2F_3\left(\frac{\ell+m+1}{2}, \frac{\ell+m+2}{2}; \ell+1, m+1, \ell+m+1; -x^2\right)$$

which we can further write in terms of the Meijer G function as

$$J_\ell(x) J_m(x) =$$

$$\frac{(x/2)^{\ell+m}(\ell+m+1)}{\Gamma\left(\dfrac{\ell+m+1}{2}\right)\Gamma\left(\dfrac{\ell+m+2}{2}\right)} G_{24}^{12}\left(x^2 \left| \begin{matrix} \dfrac{\ell+m-1}{2}, \dfrac{-\ell-m}{2} \\ 0, -\ell, -m, -\ell-m \end{matrix} \right.\right)$$

Next, using the Hankel transform relation (11.29), we obtain:

$$I = \frac{1}{\beta}\left(\frac{2}{\beta}\right)^{\alpha+\ell+m} \frac{2^{-\ell-m}\Gamma(\ell+m+1)}{\Gamma\left(\dfrac{\ell+m+1}{2}\right)\Gamma\left(\dfrac{\ell+m+2}{2}\right)}$$

$$\times G_{44}^{13}\left(\frac{4}{\beta^2}\left|\begin{array}{l} \dfrac{1-\alpha-\ell-m-n}{2}, \dfrac{1-\ell-m}{2}, -\dfrac{\ell+m}{2}, \dfrac{1-\alpha-\ell-m+n}{2} \\ 0, -\ell, -m, -\ell-m \end{array}\right.\right)$$

For $\beta > 2$, we use (11.11) to obtain

$$I = a^{-\alpha-1}2^{-\ell-m}\left(\frac{2}{\beta}\right)^{\alpha+\ell+m} \frac{\Gamma\left(\dfrac{n+\alpha+\ell+m+1}{2}\right)}{\Gamma(\ell+1)\Gamma(m+1)\Gamma\left(\dfrac{n-\alpha-\ell-m+1}{2}\right)}$$

$$\times {}_4F_3\left(\frac{\ell+m+1}{2}, \frac{\ell+m+2}{2}, \frac{\alpha+\ell+m+n+1}{2}, \frac{\alpha+\ell+m-n+1}{2}; \right.$$

$$\left. \ell+1, m+1, \ell+m+1; \frac{4}{\beta^2}\right)$$

When $\beta < 2$, we use (11.20) to write

$$I = \frac{1}{\beta}\left(\frac{2}{\beta}\right)^{\alpha+\ell+m} \frac{2^{-\ell-m}\Gamma(\ell+m+1)}{\Gamma\left(\dfrac{\ell+m+1}{2}\right)\Gamma\left(\dfrac{\ell+m+2}{2}\right)}$$

$$\times G_{44}^{31}\left(\frac{\beta^2}{4}\left|\begin{array}{l} 1, \ell+1, m+1, \ell+m+1 \\ \dfrac{\alpha+\ell+m+n+1}{2}, \dfrac{\ell+m+1}{2}, \dfrac{\ell+m+2}{2}, \dfrac{\alpha+\ell+m-n+1}{2} \end{array}\right.\right)$$

We can, of course, rewrite this G function in terms of ${}_pF_q$ functions by use of (11.11) once again. Doing so leads to a sum of three ${}_4F_3$ functions in this case which we leave to the reader to finish.

One of the major areas of application where the G function has proved effective is probability theory. For example, the probability density function associated with the product of n random variables of the same distribution has been found in terms of G functions.* While

---

* M. D. Springer and W. E. Thompson, "The Distribution of Products of Beta, Gamma and Gaussian Random Variables," *SIAM J. Appl. Math.* **18**(4) (June 1970).

certain special cases of the G functions associated with such products can be expressed in terms of simpler functions, the general case most likely cannot. In such instances, the G functions must be dealt with directly for computational purposes.

11.3.2 The MacRobert E function

In the late 1930s, T. M. MacRobert also made an attempt to give meaning to the symbol $_pF_q$ when $p > q + 1$.[*] For the values $p \leq q + 1$, he introduced the function (called the **E function**)

$$E(a_1, \ldots, a_p; c_1, \ldots, c_q; x) = \frac{\prod\limits_{k=1}^{p} \Gamma(a_k)}{\prod\limits_{j=1}^{q} \Gamma(c_j)} \left[_pF_q\left(a_1, \ldots, a_p; c_1, \ldots, c_q; -\frac{1}{x}\right) \right]$$

$$(11.30)$$

where $x \neq 0$ if $p < q$ and $|x| > 1$ if $p = q + 1$, while for the values $p \geq q + 1$,[†]

$$E(a_1, \ldots, a_p; c_1, \ldots, c_q; x) = \sum_{n=1}^{p} \frac{\prod\limits_{k=1}^{p}{}' \Gamma(a_k - a_n)}{\prod\limits_{j=1}^{q} \Gamma(c_j - a_n)} \Gamma(a_n) x^{a_n}$$

$$\times {}_{q+1}F_{p-1}[a_n, a_n - c_1 + 1, \ldots, a_n - c_q + 1;$$

$$a_n - a_1, \ldots, *, \ldots, a_n - a_p + 1; (-1)^{p+q} x] \quad (11.31)$$

where $|x| < 1$ if $p = q + 1$.

MacRobert's E function never gained wide acceptance in the literature, mostly because it was found to be a special case of the Meijer G function,

$$E(a_1, \ldots, a_p; c_1, \ldots, c_q; x) = G_{q+1, p}^{p, 1}\left(x \left| \begin{array}{c} 1, c_1, \ldots, c_q \\ a_1, \ldots, a_p \end{array} \right.\right) \quad (11.32)$$

Hence, all properties of the E function are simple consequences of properties of the G function.

---

[*] T. M. MacRobert, "Induction Proofs of the Relations between Certain Asymptotic Expansions and Corresponding Generalized Hypergeometric Series," *Proc. Roy. Soc. Edinburgh*, **58**: 1–13, 1937–1938.

[†] See the footnote at the beginning of Sec. 11.3.1.

Exercises 11.3

1. From the definition (11.11), show that

$$_pF_q(a_1, \ldots, a_p; c_1, \ldots, c_q; x)$$

$$= \frac{\displaystyle\prod_{j=1}^{q} \Gamma(c_j)}{\displaystyle\prod_{j=1}^{p} \Gamma(a_j)} G_{p,q+1}^{1,p}\left(-x \;\middle|\; \begin{matrix} 1-a_1, \ldots, 1-a_p \\ 0, 1-c_1, \ldots, 1-c_q \end{matrix}\right)$$

$$= \frac{\displaystyle\prod_{j=1}^{q} \Gamma(c_j)}{\displaystyle\prod_{j=1}^{p} \Gamma(a_j)} G_{q+1,p}^{p,1}\left(-\frac{1}{x} \;\middle|\; \begin{matrix} 1, c_1, \ldots, c_q \\ a_1, \ldots, a_p \end{matrix}\right)$$

In problems 2 to 15, use the result of problem 1 and properties of the G function to deduce the given relation.

2. $G_{01}^{10}(x \mid 0) = e^{-x}$

3. $G_{01}^{10}(x \mid a) = x^a e^{-x}, \; x \geq 0$

4. $G_{11}^{10}\left(x \;\middle|\; \begin{matrix} a+b+1 \\ a \end{matrix}\right) = \dfrac{x^a(1-x)^b}{\Gamma(b+1)}, \; 0 < x < 1$

5. $G_{11}^{11}\left(x \;\middle|\; \begin{matrix} 1-a \\ 0 \end{matrix}\right) = \Gamma(a)(1+x)^{-a}, \; |x| < 1$

6. $G_{02}^{10}\left(\dfrac{x^2}{4} \;\middle|\; \begin{matrix} \frac{1}{2}, 0 \end{matrix}\right) = \dfrac{1}{\sqrt{\pi}} \sin x, \; x \geq 0$

7. $G_{02}^{10}\left(\dfrac{x^2}{4} \;\middle|\; \begin{matrix} 0, \frac{1}{2} \end{matrix}\right) = \dfrac{1}{\sqrt{\pi}} \cos x$

8. $G_{02}^{10}(x^2 \mid \frac{1}{2}p, -\frac{1}{2}p) = J_p(2x), \; x \geq 0$

9. $G_{02}^{10}(x^2 \mid a, b) = x^{a+b} J_{a-b}(2x), \; x \geq 0$

10. $G_{22}^{10}\left(x \;\middle|\; \begin{matrix} 1, 1 \\ 1, 0 \end{matrix}\right) = \ln(1+x), \; |x| < 1$

11. $G_{12}^{11}\left(x \;\middle|\; \begin{matrix} 1-a \\ 0, 1-c \end{matrix}\right) = \dfrac{\Gamma(a)}{\Gamma(c)} [_1F_1(a; c; -x)]$

12. $G_{22}^{12}\left(x \;\middle|\; \begin{matrix} 1-a, 1-b \\ 0, 1-c \end{matrix}\right) = \dfrac{\Gamma(a)\Gamma(b)}{\Gamma(c)} [_2F_1(a, b; c; -x)], \; |x| < 1$

13. $G_{12}^{11}\left(x \;\middle|\; \begin{matrix} \frac{1}{2} \\ p, -p \end{matrix}\right) = \sqrt{\pi} \, e^{-x/2} I_p\left(\dfrac{x}{2}\right), \; x \geq 0$

14. $G_{12}^{11}\left(x \left| \begin{matrix} a \\ b, c \end{matrix}\right.\right) = \dfrac{\Gamma(1-a+b)}{\Gamma(1-c+b)} x^b [{}_1F_1(1-a+b; 1+b-c; -x)]$

15. $G_{13}^{11}\left(x \left| \begin{matrix} {}^{1/2} \\ p, 0, -p \end{matrix}\right.\right) = \sqrt{\pi}\,[J_p(\sqrt{x})]^2, \; x \geq 0$

In problems 16 to 21, verify the relation.

16. $J_p'(x) - G_{13}^{11}\left(\dfrac{x^2}{4} \left| \begin{matrix} -{}^{1/2} \\ {}^{1/2}(p-1),\, -{}^{1/2}(p+1),\, {}^{1/2} \end{matrix}\right.\right)$

17. $K_p(2\sqrt{x}) = \tfrac{1}{2} x^{-p/2} G_{02}^{20}(x \,|\, p, 0)$

   Hint: Use problem 8 in Exercises 11.2.

18. $K_p(x) = \sqrt{\pi}\, e^x G_{12}^{20}\left(2x \left| \begin{matrix} {}^{1/2} \\ p, -p \end{matrix}\right.\right)$

19. $L_n^{(a)}(x) = \dfrac{(-1)^n}{n!} e^x G_{12}^{20}\left(x \left| \begin{matrix} -n-a \\ 0, -a \end{matrix}\right.\right)$

20. $M_{k,m}(x) = \dfrac{\Gamma(2m+1)}{\Gamma(m+k+{}^{1/2})} e^{x/2} G_{12}^{11}\left(x \left| \begin{matrix} 1-k \\ {}^{1/2}+m,\, {}^{1/2}-m \end{matrix}\right.\right)$

21. $W_{k,m}(x) = \dfrac{e^{-x/2}}{\Gamma({}^{1/2}-k+m)\Gamma({}^{1/2}-k-m)} G_{12}^{21}\left(x \left| \begin{matrix} 1+k \\ {}^{1/2}+m,\, {}^{1/2}-m \end{matrix}\right.\right)$

22. Use the results of problems 6 and 8 to deduce that $(x > 0)$

$$J_{1/2}(x) = \sqrt{\dfrac{2}{\pi x}} \sin x$$

23. Verify that $(p \leq q+1)$

$$\dfrac{d}{dx} E(a_1, \ldots, a_p; c_1, \ldots, c_q; x)$$

$$= x^{-2} E(a_1 + 1, \ldots, a_p + 1; c_1 + 1, \ldots, c_q + 1; x)$$

24. Show that

$${}_2F_2(a, b; c, d; -x) \sim \dfrac{\Gamma(c)\Gamma(d)\Gamma(b-a)}{\Gamma(b)\Gamma(c-a)\Gamma(d-a)} x^{-a}$$

$$+ \dfrac{\Gamma(c)\Gamma(d)\Gamma(a-b)}{\Gamma(a)\Gamma(c-b)\Gamma(d-b)} x^{-b} \qquad x \to \infty$$

25. Use Eqs. (11.11), (11.20), and the result of problem 13 to deduce that

$$I_p(x) \sim \dfrac{e^x}{\sqrt{2\pi x}} \qquad x \to \infty$$

26. (a)   Use problem 8 and Eq. (11.28) to show that

$$\int_0^\infty e^{-st}t^{p/2}J_p(2\sqrt{t})\,dt = s^{-(1+p/2)}G_{12}^{11}\left(\frac{1}{s}\left|\begin{array}{c} p/2 \\ p/2,\ -p/2 \end{array}\right.\right) \qquad s > 0$$

(b)   By use of Eq. (11.20) and problem 3, show that

$$\int_0^\infty e^{-st}t^{p/2}J_p(2\sqrt{t})\,dt = \frac{1}{s^{p+1}}e^{-1/s} \qquad p > -\tfrac{1}{2}, \qquad s > 0$$

27. (a)   Use problem 2 and Eq. (11.29) to deduce that

$$\int_0^\infty e^{-st}t^{p/2}J_p(2\sqrt{t})\,dt = G_{22}^{11}\left(s\left|\begin{array}{c} -p,\ 0 \\ 0 \end{array}\right.\right) \qquad s > 0$$

(b)   By use of Eq. (11.20) and problem 14, show that

$$\int_0^\infty e^{-st}t^{p/2}J_p(2\sqrt{t})\,dt = \frac{1}{s^{p+1}}e^{-1/s} \quad p > -\tfrac{1}{2}, \qquad s > 0$$

28. Use the technique of problem 27 to deduce that

$$\int_0^\infty x^\mu e^{-\alpha x^2}J_p(\beta x)\,dx = \frac{\beta^p \Gamma\left(\dfrac{p+\mu+1}{2}\right)}{2^{p+1}\alpha^{(p+\mu+1)/2}\Gamma(p+1)}$$

$$\times\ {}_1F_1\left(\frac{p+\mu+1}{2};p+1;-\frac{\beta^2}{4\alpha}\right) \qquad \alpha > 0, \qquad p+\mu > -1$$

In problems 29 to 32, use (11.28) or (11.29) to derive the given integral formula.

29. $\displaystyle\int_0^\infty e^{-st}L_n(t)\,dt = \frac{1}{s}\left(1-\frac{1}{s}\right)^n,\ s > 0,\ n = 0, 1, 2, \ldots$

30. $\displaystyle\int_0^\infty e^{-t^2}t^{2n}\cos 2xt\,dt = \frac{(-1)^n\sqrt{\pi}}{2^{2n+1}}e^{-x^2}H_{2n}(x),\ n = 0, 1, 2, \ldots$

31. $\displaystyle\int_0^\infty e^{-t^2}t^n H_n(xt)\,dt = \tfrac{1}{2}\sqrt{\pi}\,n!\,P_n(x),\ n = 0, 1, 2, \ldots$

32. $\displaystyle\int_0^\infty e^{-t}t^{n+m/2}J_m(2\sqrt{xt})\,dt = n!\,e^{-x}x^{m/2}L_n^{(m)}(x),\ m > 0,\ n = 0, 1, 2, \ldots$

12

Applications Involving Hypergeometric-Type Functions

12.1 Introduction

In this final chapter we illustrate the use of the general family of hypergeometric functions in various applications. Although we have chosen specific examples from the fields of *statistical communication theory, fluid mechanics,* and *random fields,* the techniques we use are sufficiently general that they apply to a wider range of applications. As before, we assume only a working knowledge of the subjects in order to follow the exposition.

12.2 Statistical Communication Theory

Communication systems may be broadly classified in terms of **linear operations,** such as *amplification* and *filtering,* and **nonlinear operations,** such as *modulation* and *detection.* Random noise, which appears at the input to any communications receiver, interferes with the reception of incoming radio and radar signals. When this noise is channeled through a passband linear filter whose bandwidth is narrow compared with the center frequency ω_0 of the filter, the output is called **narrowband noise** and has the representation (recall Sec. 8.3.1)

$$\mathbf{n}(t) = \mathbf{x}(t) \cos \omega_0 t - \mathbf{y}(t) \sin \omega_0 t \qquad (12.1)$$

where $\mathbf{x}(t)$ and $\mathbf{y}(t)$ are independent *gaussian* (or *normal*) random processes with zero means and equal mean-squared values N. If the incoming *signal* embedded in the noise process is a simple sinusoid $A \cos \omega_0 t$, the output of the linear filter then takes the form

$$\mathbf{v}(t) = A \cos \omega_0 t + \mathbf{n}(t)$$

$$= \mathbf{r}(t) \cos [\omega_0 t + \mathbf{\theta}(t)] \qquad (12.2)$$

where

$$\mathbf{r}(t) = \sqrt{[A + \mathbf{x}(t)]^2 + \mathbf{y}^2(t)}$$

$$\mathbf{\theta}(t) = \tan^{-1} \frac{\mathbf{y}(t)}{A + \mathbf{x}(t)} \qquad (12.3)$$

The joint distribution for the *envelope* $\mathbf{r}(t)$ and *phase* $\mathbf{\theta}(t)$ is given by [recall Eq. (8.30)]

$$p_{\mathbf{r}\mathbf{\theta}}(r, \theta) = \frac{r}{2\pi N} \exp \left[-\frac{(r^2 + A^2 - 2Ar \cos \theta)}{2N} \right] \qquad (12.4)$$

By integrating (12.4) over θ (modulo 2π) we previously found that the marginal density function of the envelope $\mathbf{r}(t)$ was the *rician distribution* [see Eq. (8.34)]

$$p_{\mathbf{r}}(r) = \frac{r}{N} e^{-(r^2 + A^2)/2N} I_0 \left(\frac{Ar}{N} \right) \qquad r > 0 \qquad (12.5)$$

where $I_0(x)$ is the *modified Bessel function of the first kind* and order zero. Similarly, the integration of (12.4) over $0 < r < \infty$ leads to the marginal phase distribution [see Eq. (8.35)]

$$p_{\mathbf{\theta}}(\theta) = \frac{1}{2\pi} e^{-s} \{ 1 + \sqrt{\pi s} \, e^{s \cos^2 \theta} \cos \theta [1 + \mathrm{erf} (\sqrt{s} \cos \theta)] \}$$

$$-\pi < \theta \leq \pi \qquad (12.6)$$

where $\mathrm{erf} x$ is the *error function* and $s = A^2/2N$ denotes the input signal-to-noise ratio (SNR). By relating the error function to the *confluent hypergeometric function* through use of (see problem 21 in Exercises 10.3)

$$e^{-x^2} + \sqrt{\pi} x \, \mathrm{erf} x = {}_1F_1(-\tfrac{1}{2}; \tfrac{1}{2}; -x^2) \qquad (12.7)$$

we can also express (12.6) in the form

$$p_{\mathbf{\theta}}(\theta) = \frac{1}{2\pi} e^{-s \sin^2 \theta} [\sqrt{\pi s} \cos \theta + {}_1F_1(-\tfrac{1}{2}; \tfrac{1}{2}; -s \cos^2 \theta)]$$

$$-\pi < \theta \leq \pi \qquad (12.8)$$

The form given by (12.8) is particularly useful for developing asymptotic relations. For instance, if the input SNR s is very small ($s \ll 1$), then clearly

$$_1F_1(-\tfrac{1}{2}; \tfrac{1}{2}; -s \cos^2 \theta) \sim 1 \quad s \ll 1$$

and (12.8) reduces to the *uniform distribution*

$$p_\theta(\theta) \sim \frac{1}{2\pi} \quad -\pi < \theta \le \pi \tag{12.9}$$

On the other hand, if $s \gg 1$, the density function (12.8) can be closely approximated by a gaussian or normal distribution. That is, under this condition the variance of $\theta(t)$ is necessarily small, and we can make the approximations

$$\sin \theta \sim \theta \quad \cos \theta \sim 1 \quad |\theta| \ll 1$$

and also

$$_1F_1(-\tfrac{1}{2}; \tfrac{1}{2}; -s \cos^2 \theta) \sim \sqrt{\pi s} \quad s \gg 1$$

Hence (12.8) reduces to the gaussian density function

$$p_\theta(\theta) \sim \sqrt{\frac{s}{\pi}} e^{-s\theta^2} \tag{12.10}$$

which has mean value zero and variance $1/2s$.

12.2.1 Nonlinear devices

We now wish to briefly discuss the output of certain *nonlinear devices* such as rectifiers, detectors, and nonlinear amplifiers. Let us suppose the narrowband waveform

$$\mathbf{v}(t) = A \cos \omega_0 t + \mathbf{n}(t)$$

$$= \mathbf{r}(t) \cos [\omega_0 t + \theta(t)] \tag{12.11}$$

representing a sinusoidal signal plus noise, is fed to a memoryless nonlinear device whose output $\mathbf{w}(t) = g[\mathbf{v}(t)]$ is a function of the input $\mathbf{v}(t)$ at the same instant of time, say, at $t = t_1$. The output function $g(r \cos \phi)$, where $\phi = \omega_0 t_1 + \theta$, is therefore *not* a function of t. It is an even periodic function of ϕ which has the *Fourier cosine series* representation (recall Sec. 1.4)

$$g(r \cos \phi) = \tfrac{1}{2}g_0(r) + \sum_{n=1}^{\infty} g_n(r) \cos n\phi \tag{12.12}$$

where the Fourier coefficients are defined by

$$g_n(r) = \frac{2}{\pi} \int_0^\pi g(r \cos \phi) \cos n\phi \, d\phi \qquad n = 0, 1, 2, \dots \quad (12.13)$$

Each term of the series (12.12) can be thought of as the output of a zonal filter centered at the nth harmonic of ω_0 (see Fig. 12.1). For example, the first term of the series $\frac{1}{2}g_0(r)$, which represents the low-pass component of the output called the *audio* or *video output*, is the only term used in *detection* devices. The next term of the series $g_1(r) \cos(\omega_0 t_1 + \theta)$, which has the same frequency as the input waveform, is the only significant term in the case of a *nonlinear amplifier* or *limiter*. Notice that the term $g_1(r) \cos(\omega_0 t_1 + \theta)$ reproduces the input phase exactly, but distorts the input amplitude.

The so-called **vth-law device,** which falls into the categories of *half-wave, full-wave even,* or *full-wave odd,* is a common nonlinear device. In the case of a full-wave even power-law device, the output $\mathbf{w}(t)$ is related to the input $\mathbf{r}(t) \cos \phi(t)$ by

$$\mathbf{w}(t) = |\mathbf{r}(t) \cos \phi(t)|^v \qquad (12.14)$$

where v is a nonnegative constant not necessarily restricted to integer values. The cases $v = 1$ and $v = 2$ represent a *linear rectifier* and *square-law rectifier*, respectively. If we follow the nonlinear device (12.14) by a low-pass zonal filter, the resulting configuration is known as an *envelope detector*. In this case we obtain

$$\frac{1}{2}g_0(r) = \frac{1}{\pi} \int_0^\pi r^v |\cos \phi|^v d\phi$$

$$= \frac{2}{\pi} r^v \int_0^{\pi/2} \cos^v \phi \, d\phi$$

which, from properties of the gamma function, reduces to

$$\frac{1}{2}g_0(r) = \frac{2\Gamma[(v+1)/2]}{\sqrt{\pi}\,\Gamma(1 + v/2)} r^v \qquad (12.15)$$

Hence, the random output $\mathbf{z}$ of the zonal filter (at a particular instant

Figure 12.1 Block diagram of nonlinear system.

of time) is

$$z = \frac{2\Gamma[(v+1)/2]}{\sqrt{\pi}\,\Gamma(1+v/2)} \mathbf{r}^v \tag{12.16}$$

The **output signal power** S_0 can be defined as the square of the mean output minus the mean value when the signal is absent. Thus, when the output signal is described by (12.16), we write

$$S_0 = [\langle \mathbf{z} \rangle - \langle \mathbf{z} \,|\, A = 0 \rangle]^2 \tag{12.17}$$

where $\langle\ \rangle$ denotes a statistical expectation. For example, let us consider the case where $\mathbf{z}$ is defined by (12.16) and the distribution of $\mathbf{r}$ is given by the rician density function

$$p_{\mathbf{r}}(r) = \frac{r}{N} e^{-(r^2 + A^2)/2N} I_0\left(\frac{Ar}{N}\right) \qquad r > 0 \tag{12.18}$$

In this case we find that (see problem 3 in Exercises 12.2)

$$\langle \mathbf{r}^v \rangle = \frac{1}{N} e^{-A^2/2N} \int_0^\infty r^{v+1} e^{-r^2/2N} I_0\left(\frac{Ar}{N}\right) dr$$

$$= (2N)^{v/2}\Gamma\left(1+\frac{v}{2}\right) {}_1F_1\left(1+\frac{v}{2}; 1; \frac{A^2}{2N}\right) e^{-A^2/2N}$$

$$= (2N)^{v/2}\Gamma\left(1+\frac{v}{2}\right) {}_1F_1\left(-\frac{v}{2}; 1; -\frac{A^2}{2N}\right) \tag{12.19}$$

where ${}_1F_1(a;c;x)$ is the *confluent hypergeometric function* and we have used Kummer's transformation (10.15) in the last step. Based on Eq. (12.17), therefore, the output signal power is

$$S_0 = \frac{4}{\pi} (2N)^v \Gamma^2\left(\frac{v+1}{2}\right) \left[{}_1F_1\left(-\frac{v}{2}; 1; -s\right) - 1 \right]^2 \tag{12.20}$$

where $s = A^2/2N$ is the *input* SNR.

In a similar manner we define the **output noise power** by

$$N_0 = \langle \mathbf{z}^2 \rangle - \langle \mathbf{z} \rangle^2 \tag{12.21}$$

Thus, for the case where $\mathbf{z}$ is defined by (12.16), we first obtain

$$\langle \mathbf{r}^{2v} \rangle = \frac{1}{N} e^{-A^2/2N} \int_0^\infty r^{2v+1} e^{-r^2/2N} I_0\left(\frac{Ar}{N}\right) dr$$

$$= (2N)^v \Gamma(1+v)\, {}_1F_1(-v; 1; -s) \tag{12.22}$$

where the integration follows from (12.19). The total noise power at the output is therefore given by

$$N_0 = \frac{4}{\pi}(2N)^\nu \Gamma^2\left(\frac{\nu+1}{2}\right)$$

$$\times \left[\frac{\Gamma(1+\nu)}{\Gamma^2(1+\nu/2)} \,{}_1F_1(-\nu; 1; -s) - {}_1F_1^2\left(-\frac{\nu}{2}; 1; -s\right)\right] \quad (12.23)$$

The *output* SNR as a function of the input SNR is obtained by forming the ratio S_0/N_0, which leads to the expression

$$\frac{S_0}{N_0} = \frac{\Gamma^2(1+\nu/2)[{}_1F_1(-\nu/2; 1; -s) - 1]^2}{\Gamma(1+\nu)\,{}_1F_1(-\nu; 1; -s) - \Gamma^2(1+\nu/2)\,{}_1F_1^2(-\nu/2; 1; -s)}$$

$$(12.24)$$

When the input SNR is either very small (weak-signal case) or very large (strong-signal case), we can simplify (12.24) by the use of *asymptotic formulas* for the confluent hypergeometric functions. For example, if $s \ll 1$, then (12.24) reduces to (see problem 8 in Exercises 12.2)

$$\frac{S_0}{N_0} \sim \frac{\Gamma^2(1+\nu/2)\nu^2 s^2}{4[\Gamma(1+\nu) - \Gamma^2(1+\nu/2)]} \qquad s \ll 1 \qquad (12.25)$$

whereas if $s \gg 1$, then

$$\frac{S_0}{N_0} \sim \frac{2s}{\nu^2} \qquad s \gg 1 \qquad (12.26)$$

Exercises   12.2

1. Expand e^{iur} in a Maclaurin series, where $i^2 = -1$.
   (a) Show that the *characteristic function* $\langle e^{iur} \rangle$ of the Rayleigh distribution $p_\mathbf{r}(r) = (r/N)e^{-r^2/2N}$, $r > 0$, leads to

   $$\langle e^{iu\mathbf{r}} \rangle = \sum_{n=0}^{\infty} \frac{(iu)^n}{n!} \langle \mathbf{r}^n \rangle$$

   $$= \sum_{n=0}^{\infty} \frac{(iu)^n}{n!} (2N)^{n/2} \Gamma\left(1 + \frac{n}{2}\right)$$

   (b) Splitting the series in (a) into even and odd terms, show that

   $$\langle e^{iu\mathbf{r}} \rangle = {}_1F_1\left(1; \frac{1}{2}; -\frac{Nu^2}{2}\right) + iu\sqrt{\frac{N\pi}{2}}\, e^{-Nu^2/2}$$

2. (a) For the joint distribution $p_2(r_1, r_2)$ given in problem 2 in Exercises 8.3, show that, for v, $\mu > -\frac{1}{2}$,

$$\langle r_1^v r_2^\mu \rangle = (2N)^{(v+\mu)/2} \Gamma\left(1 + \frac{v}{2}\right) \Gamma\left(1 + \frac{\mu}{2}\right)$$

$$\times (1 - \rho^2)^{(v+\mu)/2} \, _2F_1\left(1 + \frac{v}{2}, 1 + \frac{\mu}{2}; 1; \rho^2\right)$$

$$= (2N)^{(v+\mu)/2} \Gamma\left(1 + \frac{v}{2}\right) \Gamma\left(1 + \frac{\mu}{2}\right)$$

$$\times \, _2F_1\left(-\frac{v}{2}, -\frac{\mu}{2}; 1; \rho^2\right)$$

(b) For $\rho \to 1$, show that the answer in (a) reduces to

$$\langle r_1^v r_2^\mu \rangle = (2N)^{(v+\mu)/2} \Gamma\left(1 + \frac{v + \mu}{2}\right)$$

(c) For the special case $v = \mu = 2$, show that
$$\langle r_1^2 r_2^2 \rangle = 4N^2(1 + \rho^2)$$

3. The rician distribution (12.5) is given.
   (a) Show that the statistical moments are given by

$$\langle r^v \rangle = (2N)^{v/2} \Gamma\left(1 + \frac{v}{2}\right) \, _1F_1\left(-\frac{v}{2}; 1; -\frac{A^2}{2N}\right) \qquad v > -\frac{1}{2}$$

   (b) Show that the case $v = 1$ leads to the result

$$\langle r \rangle = \sqrt{\frac{\pi N}{2}} \, e^{-s/2}\left[(1 + s)I_0\left(\frac{s}{2}\right) + sI_1\left(\frac{s}{2}\right)\right]$$

   where $s = A^2/2N$.
   Hint: Use problem 26 in Exercises 10.3.

4. (a) Based on Eq. (12.22), show that
$$\langle r^{2k} \rangle = (2N)^k k! L_k(-s) \qquad k = 0, 1, 2, \ldots$$

   where $L_k(x)$ is the kth Laguerre polynomial.
   (b) For $k = 1$ and $k = 2$, show that (a) becomes
$$\langle r^2 \rangle = A^2 + 2N$$
$$\langle r^4 \rangle = A^2 + 8AN + 8N^2$$

5. For $rA/\sqrt{N} \gg 1$, show that the rician distribution (12.5) can be approximated by (for $r \cong A$)

$$p_r(r) \cong \sqrt{\frac{r}{2\pi AN}} \, e^{-(r-A)^2/2N} \cong \frac{1}{\sqrt{2\pi N}} e^{-(r-A)^2/2N}$$

6. Starting with the joint density function (12.4), show that the marginal phase distribution can be expressed in the form

$$p_\theta(\theta) = \frac{1}{2\pi} \sum_{n=0}^{\infty} \epsilon_n \frac{s^{n/2}}{n!} \Gamma\left(1 + \frac{n}{2}\right)$$

$$\times {}_1F_1\left(\frac{n}{2}; n+1; -s\right) \cos n\theta \qquad -\pi < \theta \le \pi$$

where $s = A^2/2N$ and $\epsilon_0 = 1$ and $\epsilon_n = 2$, $n \ge 1$.
Hint: Use problem 21a in Exercises 7.2.

7. For the joint distribution $p_2(r_1, r_2; \theta_1, \theta_2)$ given in problem 2 in Exercises 8.3, show that the joint distribution of the phases is given by

$$p_2(\theta_1, \theta_2) = \int_0^\infty \int_0^\infty p_2(r_1, r_2; \theta_1, \theta_2)\, dr_1\, dr_2$$

$$= \frac{1}{(2\pi)^2} \sum_{n=0}^{\infty} \frac{1}{n!} \epsilon_n \rho^n \Gamma^2\left(1 + \frac{n}{2}\right) {}_2F_1\left(\frac{n}{2}, \frac{n}{2}; n+1; \rho^2\right)$$

$$\times \cos\left[n(\theta_2 - \theta_1)\right] \qquad -\pi < \theta_1, \theta_2 \le \pi$$

where $\epsilon_0 = 1$ and $\epsilon_n = 2$, $n \ge 1$.
Hint: Use problem 21a in Exercises 7.2.

8. Show that
   (a) For $s \ll 1$, Eq. (12.24) reduces to (12.25).
   (b) For $s \gg 1$, Eq. (12.24) reduces to (12.26).

9. For the case of a *square-law envelope detector* $(v = 2)$, show that the output SNR given by (12.25) reduces to

$$\frac{S_0}{N_0} = \frac{s^2}{1 + 2s}$$

10. A *full-wave odd power-law device* is defined by

$$\mathbf{w}(t) = \begin{cases} \mathbf{v}(t)^v & \mathbf{v} \ge 0 \\ -|\mathbf{v}(t)|^v & \mathbf{v} < 0 \end{cases}$$

where $\mathbf{v} = \mathbf{r}\cos\phi$. Show that the output SNR is given by

$$\frac{S_0}{N_0} = \frac{\Gamma^2\left(\dfrac{v+3}{2}\right) {}_1F_1^2\left(\dfrac{1-v}{2}; 2; -s\right) s}{\Gamma(1+v)\, {}_1F_1(-v; 1; -s) - \Gamma^2\left(\dfrac{v+3}{2}\right) {}_1F_1^2\left(\dfrac{1-v}{2}; 2; -s\right) s}$$

11. The case $v = 0$ in problem 10 is called an *ideal limiter* (*clipper*). Show that when $v - 0$, the output SNR becomes

$$\frac{S_0}{N_0} = \frac{\pi s [I_0(s/2) + I_1(s/2)]^2}{4e^s - \pi s [I_0(s/2) + I_1(s/2)]^2}$$

Hint: Recall problem 27 in Exercises 10.3.

12. Referring to problem 11, derive the asymptotic cases

(a) $\dfrac{S_0}{N_0} \sim \dfrac{1}{4} \pi s, \; s \ll 1$

(b) $\dfrac{S_0}{N_0} \sim 2s, \; s \gg 1$

12.3 Fluid Mechanics

Fluid mechanics is the theory of the motion of liquids and gases. Although fluid motion is generally three-dimensional, a fluid flow in three-dimensional space is called *two-dimensional* if the velocity vector **V** is always parallel to a fixed plane (xy plane), and if the velocity components u and v parallel to this plane along with the pressure p and fluid density ρ are all constant along any normal to the plane. This situation permits us to confine our attention to just a single plane which we interpret as a cross section of the three-dimensional region under consideration. Our brief discussion here is limited to only two-dimensional flow problems.

An *ideal fluid* is one in which the stress on an element of area is wholly normal and independent of the orientation of the area (also called a *perfect* or *inviscid fluid*). In contrast, the stress on a small area is no longer normal to that area for a *viscous fluid* in motion. If the density is constant, we say the flow is *incompressible*. Of course, the notions of an ideal fluid or incompressible fluid are only idealizations that are valid when certain effects can be safely neglected in the analysis of a real fluid.

12.3.1 Unsteady hydrodynamic flow past an infinite plate

Viscous fluid flow past a continuous moving surface like a flat plate is a basic problem in the study of hydrodynamics. Suppose that x denotes the distance along a two-dimensional porous plate and y is the distance normal to the plate. The variables u and v correspond, respectively, to the velocity components in the x and y directions. Let us assume that decelerated fluid particles are removed from the boundary layer through the porous plate by *suction*. We further assume that the velocity component u will equal zero at the surface of the plate for all time, approach a function $f(t)$ as $y \to \infty$, and equal $f(0)$ at time $t = 0$. Under these conditions, it can be shown that the velocity component $u(y, t)$ is a solution of the boundary-value problem*

$$\frac{\partial u}{\partial t} - v_s \frac{\partial u}{\partial y} = \frac{\partial f}{\partial t} + v \frac{\partial^2 u}{\partial y^2} \qquad 0 < y < \infty, \qquad t > 0$$

$$u(y, 0) = f(0) \qquad 0 < y < \infty \tag{12.27}$$

$$u(0, t) = 0 \qquad u(y, t) \to f(t) \qquad \text{as} \qquad y \to \infty, \qquad t > 0$$

where v_s is the (constant) suction velocity and v is the kinematic viscosity of the fluid.

To solve (12.27), we use the Laplace transform defined by

$$\mathcal{L}\{u(y, t); t \to p\} = \int_0^\infty e^{-pt} u(y, t)\, dt = U(y, p) \tag{12.28}$$

Hence it follows that

$$\mathcal{L}\left\{\frac{\partial u}{\partial t}(y, t); t \to p\right\} = pU(y, p) - u(y, 0) = pU(y, p) - f(0)$$

$$\mathcal{L}\left\{\frac{\partial u}{\partial y}(y, t); t \to p\right\} = \frac{\partial U}{\partial y}(y, p)$$

$$\mathcal{L}\left\{\frac{\partial^2 u}{\partial y^2}(y, t); t \to p\right\} = \frac{\partial^2 U}{\partial y^2}(y, p)$$

and

$$\mathcal{L}\left\{\frac{\partial f}{\partial t}(t); p\right\} = pF(p) - f(0)$$

So by applying the Laplace transform termwise to the partial DE in (12.27), we arrive at the transformed problem

$$v \frac{\partial^2 U}{\partial y^2} + v_s \frac{\partial U}{\partial y} - pU = -pF(p) \tag{12.29}$$

$$U(0, p) = 0 \qquad U(y, p) \to F(p) \qquad \text{as} \qquad y \to \infty$$

---

*See J. T. Stuart, "A Solution of the Navier-Stokes and Energy Equations Illustrating the Response of Skin Friction and Temperature of an Infinite Plate Thermometer to Fluctuations in the Stream Velocity," *Proc. R. Soc. (Lond.)* **A231:** 116–130 (1955) and K. Vajravelu, L. C. Andrews, and R. N. Mohapatra, "Exact Solutions of the Unsteady Hydrodynamic and Hydromagnetic Flows Past an Infinite Plate," *Acta Mech.*, **74:** 185–193, 1988.

For a fixed value of p, we can treat (12.29) as an ordinary DE subject to prescribed boundary conditions. Its solution by elementary methods is readily found to be

$$U(y,p) = F(p)\{1 - \exp[-(a + \tfrac{1}{2}\sqrt{b + cp})y]\} \qquad (12.30)$$

where a, b, and c are constants defined by

$$a = \frac{v_s}{2v} \qquad b = \left(\frac{v_s}{v}\right)^2 \qquad c = \frac{4}{v}$$

To calculate the inverse transform of (12.30), we first write

$$\exp[-(a + \tfrac{1}{2}\sqrt{b + cp})y] = e^{-ay}e^{-(y/2)\sqrt{b+cp}}$$

and then through use of some standard properties of the Laplace transform given by*

$$\mathcal{L}^{-1}\{e^{-\alpha\sqrt{p}}; t\} = \frac{\alpha}{2\sqrt{\pi}}t^{-3/2}e^{-\alpha^2/4t}$$

and $$\mathcal{L}^{-1}\{G(b + cp); t\} = \frac{1}{c}e^{-bt/c}g\left(\frac{t}{c}\right) \qquad c > 0$$

where $g(t) = \mathcal{L}^{-1}\{G(p); t\}$, we find that

$$\mathcal{L}^{-1}\left\{\exp\left[-\left(a + \frac{1}{2}\sqrt{b + cp}\right)y\right]; p \to t\right\}$$

$$= \frac{y}{4}\sqrt{\frac{c}{\pi}}e^{-ay}t^{-3/2}e^{-bt/c}e^{-cy^2/16t} \qquad (12.31)$$

Then, by making use of the Laplace convolution theorem, we obtain the solution of (12.27) in the form

$$u(y, t) = f(t) - \frac{y}{4}\sqrt{\frac{c}{\pi}}e^{-ay}I \qquad (12.32)$$

where $$I = \int_0^t f(t - \tau)\tau^{-3/2}e^{-bt/c}e^{-cy^2/16\tau}\,d\tau \qquad (12.33)$$

---

*For example, see L. C. Andrews and B. K. Shivamoggi, *Integral Transforms for Engineers and Applied Mathematicians*, Macmillan, New York, 1988.

For the special case $f(t) = u_0$, it can be shown that I has the series representation (see problem 1 in Exercises 12.3)

$$I = \frac{u_0}{\sqrt{t}} e^{-cy^2/16t} \sum_{n=0}^{\infty} \frac{(-1)^n (bt/c)^n}{n!} U\left(1; \frac{3}{2} - n; \frac{cy^2}{16t}\right) \qquad (12.34)$$

where $U(a; c; x)$ is the *confluent hypergeometric function of the second kind* (see Sec. 10.2). By use of the Kummer transformation

$$U(a; c; x) = x^{1-c} U(1 + a - c; 2 - c; x)$$

we can write (12.34) in the equivalent form

$$I = \frac{4u_0}{y\sqrt{c}} e^{-cy^2/16t} \sum_{n=0}^{\infty} \frac{(-1)^n (by^2/16)^n}{n!} U\left(n + \frac{1}{2}; n + \frac{1}{2}; \frac{cy^2}{16t}\right) \qquad (12.35)$$

Thus the solution we seek now takes the form

$$u(y, t) = u_0 \left[1 - \frac{1}{\sqrt{\pi}} e^{-ay} e^{-cy^2/16t} \right.$$

$$\left. \times \sum_{n=0}^{\infty} \frac{(-1)^n (by^2/16)^n}{n!} U\left(n + \frac{1}{2}; n + \frac{1}{2}; \frac{cy^2}{16t}\right) \right] \qquad (12.36)$$

Asymptotic cases corresponding to $y^2 \ll 16/b$, $t \ll cy^2/16$, and $t \to \infty$ are taken up in the exercises.

12.3.2 Transonic flow and the Euler-Tricomi equation

A basic problem in aerodynamics involves the calculation of the air disturbance caused by an airfoil moving at a steady speed U through otherwise undisturbed air. Neglecting viscosity, the velocity $\mathbf{V}$ of the air is related to a *velocity potential function* ϕ by $\mathbf{V} = U\mathbf{i} + \nabla\phi$, where we are assuming the airfoil is moving in the x direction with $\mathbf{i}$ the unit vector along that axis. It has been shown that, for subsonic and supersonic flows, ϕ satisfies (approximately) the *linearized* partial differential equation

$$(1 - M^2) \frac{\partial^2 \phi}{\partial x^2} + \frac{\partial^2 \phi}{\partial y^2} = 0 \qquad (12.37)$$

where M, called the *Mach number*, is the ratio of U to the speed of sound. (*Subsonic flow* is characterized by $M < 1$ and *supersonic flow* by $M > 1$.) The transition between subsonic and supersonic flows for which $M \approx 1$ is called **transonic flow**. The investigation of properties of a two-dimensional flow corresponding to the transition from subsonic to supersonic is of fundamental interest. Unfortunately, Eq. (12.37) is not valid for transonic flow since in this case a nonlinear term, neglected in the derivation of (12.37), must be retained. That is, for transonic flows we must solve the *nonlinear* equation

$$2\alpha_* \frac{\partial \phi}{\partial x} \frac{\partial^2 \phi}{\partial x^2} = \frac{\partial^2 \phi}{\partial y^2} \qquad (12.38)$$

where α_* is a constant depending on properties of the air.

Through a *hodograph transformation*, which leads to taking the velocity components u and v as the independent variables, (12.38) can be written as the *linear* **Euler-Tricomi equation**

$$\frac{\partial^2 \Phi}{\partial \eta^2} - \eta \frac{\partial^2 \Phi}{\partial \theta^2} = 0 \qquad (12.39)$$

where η and θ are essentially polar coordinates in the uv plane and Φ is a function of the velocity components $u = V \cos \theta$ and $v = V \sin \theta$, related to the potential function ϕ by $\phi = -\Phi + V \partial\Phi/\partial V$. Equation (12.39) is classified as *hyperbolic* in the half-plane $\eta > 0$ and *elliptic* in the half-plane $\eta < 0$. Fundamentally these classifications reflect the fact that transonic flow is "mixed" in the original xy plane, subsonic in some regions, and supersonic in others.

In many cases of physical interest, the origin of the $\eta\theta$ plane has special significance since it is a singular point of the solution. Particular integrals of (12.39) possessing certain properties of homogeneity may be obtained by making the transformations

$$\xi = 1 - \frac{4\eta^3}{9\theta^2} \qquad \Phi = \theta^{2k} w(\xi)$$

where k is a constant equal to the degree of homogeneity of the function Φ. Making this transformation, we find that (12.39) becomes

$$\xi(1 - \xi)w'' + [\tfrac{5}{6} - 2k - (\tfrac{3}{2} - 2k)\xi]w' - k(k - \tfrac{1}{2})w = 0 \quad (12.40)$$

We recognize (12.40) as the *hypergeometric equation* [see Eq. (9.24)] for which $a = -k$, $b = \tfrac{1}{2} - k$, and $c = \tfrac{5}{6} - 2k$. Hence, if $2k + \tfrac{1}{6} \neq 0, \pm1, \pm2, \dots$, it follows that a general solution of (12.40) is given by

$$w = A[{}_2F_1(-k, \tfrac{1}{2} - k; \tfrac{5}{6} - 2k; \xi)]$$
$$+ B\xi^{2k + 1/6}[{}_2F_1(k + \tfrac{1}{6}, k + \tfrac{2}{3}; 2k + \tfrac{7}{6}; \xi)] \quad (12.41)$$

where A and B are arbitrary constants. Finally, in terms of the variables Φ, η, and θ, this solution takes the form

$$\Phi = \theta^{2k}\left\{A\left[\,_2F_1\left(-k, \frac{1}{2} - k; \frac{5}{6} - 2k; 1 - \frac{4\eta^3}{9\theta^2}\right)\right]\right.$$

$$\left. + B\left(1 - \frac{4\eta^3}{9\theta^2}\right)^{2k+1/6}\left[\,_2F_1\left(k + \frac{1}{6}, k + \frac{2}{3}; 2k + \frac{7}{6}; 1 - \frac{4\eta^3}{9\theta^2}\right)\right]\right\} \quad (12.42)$$

Other forms of the solution are also possible (e.g., see problem 11 in Exercises 12.3).

For additional details concerning transonic flow problems, the reader should consult chap. 12 in L. D. Landau and E. M. Lifshitz, *Fluid Mechanics*, 2d ed., Pergamon, New York, 1987.

Exercises 12.3

1. (a) By expressing $e^{bt/c}$ in a Maclaurin series followed by the change of variables $\tau = t/(s+1)$, show that when $f(t) = u_0$, the integral (12.33) assumes the form

$$I = \frac{u_0}{\sqrt{t}}e^{-cy^2/16t}\sum_{n=0}^{\infty}\frac{(-1)^n(bt/c)^n}{n!}\int_0^{\infty}\frac{e^{-cy^2s/16t}}{(s+1)^{n+1/2}}\,ds$$

(b) By recognizing the integral in (a), show further that

$$I = \frac{u_0}{\sqrt{t}}e^{-cy^2/16t}\sum_{n=0}^{\infty}\frac{(-1)^n(bt/c)^n}{n!}U\left(1; \frac{3}{2} - n; \frac{cy^2}{16t}\right)$$

2. Under the assumption $y^2 \ll 16/b$, use the $n = 0$ term of the series (12.36) to show that

$$u(y, t) \sim u_0\left[1 - e^{-ay}\operatorname{erfc}\left(\frac{y}{4}\sqrt{\frac{c}{t}}\right)\right] \qquad y^2 \ll \frac{16}{b}$$

3. Under the assumption $t \ll cy^2/16$, use the asymptotic relation

$$U(a; c; x) \sim x^{-a} \qquad x \to \infty$$

to show that (12.36) reduces to

$$u(y, t) \sim u_0\left(1 - \frac{4}{y}\sqrt{\frac{t}{\pi c}}e^{-ay - bt/c - cy^2/16t}\right) \qquad t \ll \frac{cy^2}{16}$$

4. To determine the steady-state solution for the problem described by (12.27), first show that

$$\lim_{t\to\infty} I = \int_0^{\infty}\tau^{-3/2}e^{-b\tau/c}e^{-cy^2/16\tau}\,d\tau$$

$$= \frac{4\sqrt{\pi}}{y}e^{-\sqrt{by}/2}$$

and use this result to obtain the classic expression

$$\lim_{t\to\infty} u(y, t) = u_0[1 - e^{(u_g/v)y}]$$

Hint: Recall problem 18 in Exercises 3.2.

5. Use the technique of Sec. 12.3.1 to show that the solution of the hydromagnetic flow problem

$$\frac{\partial u}{\partial t} - \frac{\partial u}{\partial y} = -mu + \frac{\partial^2 u}{\partial y^2} \qquad 0 < y < \infty, \qquad t > 0$$

$$u(0, t) = k \qquad u(y, t) \to 0 \qquad \text{as} \qquad y \to \infty, \qquad t > 0$$

$$u(y, 0) = 0 \qquad 0 < y < \infty$$

where m is a constant proportional to the magnetic field and k is another constant, is given by

$$u(y, t) = \frac{k}{\sqrt{\pi}} e^{-y/2 - y^2/4t}$$

$$\times \sum_{n=0}^{\infty} \frac{(-1)^n (1 + 4m)^n (y/4)^{2n}}{n!} U\left(n + \frac{1}{2}; n + \frac{1}{2}; \frac{y^2}{4t}\right)$$

6. For the case $t \ll y^2/4$, show that the solution in problem 5 reduces to

$$u(y, t) \sim \frac{2k}{y} \sqrt{\frac{t}{\pi}} \exp\left[-\frac{y}{2} - \frac{y^2}{4t} - \frac{(1 + 4m)t}{4}\right] \qquad y > 0$$

7. For the case $y \ll 4$, show that the solution in problem 5 reduces to

$$u(y, t) \sim k e^{-y/2} \operatorname{erfc}\left(\frac{y}{2\sqrt{t}}\right) \qquad t > 0$$

8. For the steady-state case $(t \to \infty)$, show that the solution in problem 5 reduces to

$$u(y, t) = k \exp\left[-(1 + \sqrt{1 + 4m})\frac{y}{2}\right]$$

9. For $\theta = \pm 2\eta^{3/2}/3$, show that the solution (12.42) reduces to

$$\Phi = A_1 \eta^{3k}$$

where A_1 is an arbitrary constant.

10. For $\eta = 0$, show that the solution (12.42) reduces to

$$\Phi = A_1 \theta^{2k} + B_1$$

where A_1 and B_1 are arbitrary constants.

11. By making the transformation $\xi = 4\eta^3/9\theta^2$, $\Phi = \theta^{2k}w(\xi)$, show that another solution of (12.39) is given by

$$\Phi = \theta^{2k}\left\{A\left[{}_2F_1\left(-k, \frac{1}{2}-k; \frac{2}{3}; \frac{4\eta^3}{9\theta^2}\right)\right]\right.$$

$$\left.+ B\left(\frac{\eta}{\theta^{2/3}}\right)\left[{}_2F_1\left(\frac{1}{2}-k, \frac{5}{6}-k; \frac{4}{3}; \frac{4\eta^3}{9\theta^2}\right)\right]\right\}$$

where A and B are arbitrary constants.

12. For the solution in problem 11, show that
   (a) $\Phi = A\theta^{2k}$ when $\eta = 0$
   (b) $\Phi = A_1\eta^{3k} + B_1$ when $\theta = 2\eta^{3/2}/3$, where A_1 and B_1 are arbitrary constants

12.4 Random Fields

Random functions of a vector spatial variable $\mathbf{r}$, such as velocity fluctuations, temperature fluctuations, or index of refraction fluctuations, are called **random fields.** In general, the study of random fields is completely analogous to that of *random processes* wherein the random function depends on time t. Because the probability distribution function associated with a particular random field $T(\mathbf{r})$ is often quite difficult to obtain in practice, we study the statistical characteristics of such a field through its average or expected value, or through its *autocovariance function* $B(\mathbf{r}_1, \mathbf{r}_2) = \langle T(\mathbf{r}_1)T(\mathbf{r}_2)\rangle$, where $\mathbf{r}_1$ and $\mathbf{r}_2$ are two points in space and $\langle \ \rangle$ denotes a statistical average.

We say a random field is *homogeneous* if its mean value is constant and if its autocovariance function depends only on the difference $\mathbf{r}_1 - \mathbf{r}_2$, that is, if $B(\mathbf{r}_1, \mathbf{r}_2) = B(\mathbf{r}_1 - \mathbf{r}_2)$. Further, if $B(\mathbf{r}_1 - \mathbf{r}_2)$ depends only on the scalar distance $r = |\mathbf{r}_1 - \mathbf{r}_2|$, we say the homogeneous random field is *isotropic,* and in this case we write simply $B(\mathbf{r}_1, \mathbf{r}_2) = B(r)$.

The three-dimensional Fourier transform of the autocovariance function of a random field is called the *power spectral density,* which for statistically homogeneous and isotropic fields reduces to the single integral

$$\Phi(\kappa) = \frac{1}{2\pi^2\kappa}\int_0^\infty B(r)r \sin \kappa r \, dr \qquad (12.43)$$

where κ is the magnitude of the wavenumber vector (spatial frequency) measured in radians per meter. By the properties of

Fourier transforms, it can also be shown that

$$B(r) = \frac{4\pi}{r} \int_0^\infty \Phi(\kappa)\kappa \sin \kappa r \, d\kappa \qquad (12.44)$$

In practice, it is often the related **structure function** $D(r)$ of the random field that is of greatest interest. In terms of the auto-covariance function, this function is defined by

$$D(r) = 2[B(0) - B(r)] \qquad (12.45)$$

12.4.1 Structure function of temperature

The *structure function* $D_T(r)$ associated with random temperature fluctuations in a turbulent medium, such as the atmosphere, is defined by

$$D_T(r) = 8\pi \int_0^\infty \left(1 - \frac{\sin \kappa r}{\kappa r}\right)\Phi_T(\kappa)\kappa^2 \, d\kappa \qquad (12.46)$$

where $\Phi_T(\kappa)$ is the power spectral density associated with the temperature fluctuations. A model for $\Phi_T(\kappa)$ commonly used in optical wave propagation is

$$\Phi_T(\kappa) = \frac{1}{4\pi} \beta\chi\epsilon^{-1/3} \frac{\exp\left(-\kappa^2/\kappa_m^2\right)}{(\kappa_0^2 + \kappa^2)^{11/6}} \qquad (12.47)$$

where β is the Obukhov-Corrsin constant, χ is the rate of dissipation of the mean-squared temperature fluctuations, and ϵ is the rate of viscous dissipation of the turbulent kinetic energy per unit of mass of the fluid. The remaining parameters κ_m and κ_0 are basically reciprocals, respectively, of the *inner scale l_0 and outer scale L_0* of the temperature fluctuations, that is, $\kappa_m = 5.92/l_0$ and $\kappa_0 = 2\pi/L_0$. The inner and outer scale parameters l_0 and L_0 are characteristic length scales associated with the smallest and largest scale sizes, respectively, in the turbulent flow that cause random fluctuations in the temperature field. The inner scale is generally on the order of millimeters while the outer scale is on the order of meters.

To evaluate the integral in (12.46), first we expand the sine function appearing in the integrand in a Maclaurin series, which leads to

$$D_T(r) = 2\beta\chi\epsilon^{-1/3} \sum_{n=1}^\infty \frac{(-1)^{n-1}r^{2n}}{(2n+1)!} \int_0^\infty \kappa^{2n+2} \frac{\exp\left(-\kappa^2/\kappa_m^2\right)}{(\kappa_0^2 + \kappa^2)^{11/6}} \, d\kappa \qquad (12.48)$$

Then, by making an appropriate change of variable, we deduce that

$$D_T(r) = \beta\chi\epsilon^{-1/3}\kappa_0^{-2/3}\sum_{n=1}^{\infty}\frac{(-1)^{n-1}}{(2n+1)!}\Gamma\left(n+\frac{3}{2}\right)(\kappa_0 r)^{2n}$$

$$\times U\left(n+\frac{3}{2}; n+\frac{2}{3}; \frac{\kappa_0^2}{\kappa_m^2}\right) \quad (12.49)$$

where we have used Eq. (10.28) to evaluate the integrals in terms of the *confluent hypergeometric function of the second kind* discussed in Sec. 10.2.2 (see problem 1 in Exercises 12.4). To simplify this last result, we consider separate cases.

For small separation distances $r \ll l_0$, we can approximate (12.49) by the first term of the series to obtain

$$D_T(r) \sim \frac{\sqrt{\pi}}{8}\beta\chi\epsilon^{-1/3}\kappa_0^{4/3}U\left(\frac{5}{2}; \frac{5}{3}; \frac{\kappa_0^2}{\kappa_m^2}\right)r^2 \quad r \ll l_0 \quad (12.50)$$

Under most conditions of atmospheric turbulence we find that $\kappa_0^2/\kappa_m^2 \ll 1$, so by making the further approximation [recall Eq. (10.30a)]

$$U(a;c;x) \sim \frac{\Gamma(c-1)}{\Gamma(a)}x^{1-c} \quad c>1, \quad 0<x\ll1 \quad (12.51)$$

we obtain

$$D_T(r) \sim C_T^2 l_0^{-4/3}r^2 \quad r \ll l_0 \quad (12.52)$$

where
$$C_T^2 = 2.41\beta\chi\epsilon^{-1/3} \quad (12.53)$$

is the *temperature structure constant*.

For $r > l_0$, a more useful form of the structure function (12.49) can be derived by expressing the confluent hypergeometric functions in (12.49) in terms of the confluent hypergeometric function of the first kind by use of the formula (recall problem 20 in Exercises 10.2)

$$U(a;c;x) = \frac{\Gamma(1-c)}{\Gamma(1+a-c)}[{}_1F_1(a;c;x)]$$

$$+ \frac{\Gamma(c-1)}{\Gamma(a)}x^{1-c}[{}_1F_1(1+a-c;2-c;x)] \quad (12.54)$$

Then, representing each of the confluent hypergeometric functions in (12.54) in a series and interchanging the order of summation of the resulting double series, we find

$$D_T(r) = -1.05 C_T^2 \kappa_0^{-2/3} \sum_{n=0}^{\infty} \frac{(\frac{3}{2})_n}{(\frac{2}{3})_n} \frac{(\kappa_0^4/\kappa_m^4)^n}{n!}$$

$$\times \sum_{k=1}^{\infty} \frac{(n+\frac{3}{2})_k}{(\frac{3}{2})_k (n+\frac{2}{3})_k} \frac{(\kappa_0^2 r^2/4)^k}{k!}$$

$$-1.685 C_T^2 \kappa_m^{-2/3} \sum_{n=0}^{\infty} \frac{(\frac{11}{6})_n}{(\frac{4}{3})_n} \frac{(\kappa_0^2/\kappa_m^2)^n}{n!}$$

$$\times \sum_{k=1}^{\infty} \frac{(-n-\frac{1}{3})_k}{(\frac{3}{2})_k} \frac{(-\kappa_m^2 r^2/4)^k}{k!}$$

Next, summing the innermost summations, we get

$$D_T(r) = 1.05 C_T^2 \kappa_0^{-2/3} \sum_{n=0}^{\infty} \frac{(\frac{3}{2})_n}{(\frac{2}{3})_n} \frac{(\kappa_0^2/\kappa_m^2)^n}{n!}$$

$$\times \left[1 - {}_1F_2\left(n+\frac{3}{2}; \frac{3}{2}, n+\frac{2}{3}; \frac{\kappa_0^2 r^2}{4}\right) \right]$$

$$+ 1.685 C_T^2 \kappa_m^{-2/3} \sum_{n=0}^{\infty} \frac{(\frac{11}{6})_n}{(\frac{4}{3})_n} \frac{(\kappa_0^2/\kappa_m^2)^n}{n!}$$

$$\times \left[{}_1F_1\left(-n-\frac{1}{3}; \frac{3}{2}; -\frac{\kappa_m^2 r^2}{4}\right) - 1 \right] \qquad (12.55)$$

where ${}_1F_2(a; b, c; x)$ is a *generalized hypergeometric function*. If we restrict $r < L_0$, then both remaining series in (12.55) can be approximated by their respective first terms ($n = 0$), which yields

$$D_T(r) \sim 1.05 C_T^2 \kappa_0^{-2/3} \left[1 - {}_0F_1\left(-; \frac{2}{3}; \frac{\kappa_0^2 r^2}{4}\right) \right]$$

$$+ 1.685 C_T^2 \kappa_m^{-2/3} \left[{}_1F_1\left(-\frac{1}{3}; \frac{3}{2}; \frac{-\kappa_m^2 r^2}{4}\right) - 1 \right] \qquad r < L_0 \quad (12.56)$$

Still further simplification in the form of the structure function is achieved if we restrict our attention to the range of values $l_0 \ll r \ll L_0$, called the *inertial convective subrange*. Here we use the asymptotic formulas

$${}_0F_1\left(-; \frac{2}{3}; \frac{\kappa_0^2 r^2}{4}\right) \sim 1 \qquad r \ll L_0 \qquad (12.57)$$

$${}_1F_1\left(-\frac{1}{3}; \frac{3}{2}; \frac{-\kappa_m^2 r^2}{4}\right) \sim \frac{\Gamma(\frac{3}{2})}{\Gamma(\frac{11}{6})} \left(\frac{\kappa_m^2 r^2}{4}\right)^{1/3} \qquad r \gg l_0 \qquad (12.58)$$

from which we obtain the classic result

$$D_T(r) \sim C_T^2 r^{2/3} \qquad l_0 \ll r \ll L_0 \qquad (12.59)$$

When $r > L_0$, we approximate the second series in (12.55) by a different method. In this case we start with the asymptotic formula

$$_1F_1\left(-n - \frac{1}{3}; \frac{3}{2}; \frac{-\kappa_m^2 r^2}{4}\right) \sim \frac{\Gamma(3/2)}{\Gamma(n + {}^{11}/_6)}\left(\frac{\kappa_m^2 r^2}{4}\right)^{n+1/3} \qquad r \gg l_0 \quad (12.60)$$

and sum the resulting series to obtain

$$D_T(r) \sim 1.05 C_T^2 \kappa_0^{-2/3}\left[1 - {}_0F_1\left(-; \frac{2}{3}; \frac{\kappa_0^2 r^2}{4}\right)\right]$$

$$+ 1.685 C_T^2 \kappa_m^{-2/3}\left[\frac{\Gamma(3/2)}{\Gamma(n + {}^{11}/_6)}\left(\frac{\kappa_m r}{2}\right)^{2/3} {}_0F_1\left(-; \frac{4}{3}; \frac{\kappa_0^2 r^2}{4}\right) - 1\right] \qquad r > L_0$$

$$(12.61)$$

Finally, by observing that (recall problem 8 in Exercises 11.2)

$$_0F_1\left(-; \frac{2}{3}; \frac{\kappa_0^2 r^2}{4}\right) = \Gamma\left(\frac{2}{3}\right)\left(\frac{\kappa_0 r}{2}\right)^{1/3} I_{-1/3}(\kappa_0 r) \qquad (12.62)$$

$$_0F_1\left(-; \frac{4}{3}; \frac{\kappa_0^2 r^2}{4}\right) = \Gamma\left(\frac{4}{3}\right)\left(\frac{\kappa_0 r}{2}\right)^{-1/3} I_{1/3}(\kappa_0 r) \qquad (12.63)$$

we find that (12.61) can be expressed in the form

$$D_T(r) \sim 1.05 C_T^2 \kappa_0^{-2/3}\left[1 - \frac{(4\kappa_0 r)^{1/3}}{\Gamma(1/3)} K_{1/3}(\kappa_0 r)\right] \qquad r > L_0 \quad (12.64)$$

where $I_p(x)$ and $K_p(x)$ are *modified Bessel functions* of the first and second kinds, respectively.

Exercises 12.4

1. Make the change of variable $t = (\kappa/\kappa_0)^2$ in the integral (12.48) to show that

$$\int_0^\infty \kappa^{2n+2} \frac{\exp(-\kappa^2/\kappa_m^2)}{(\kappa_0^2 + \kappa^2)^{11/6}} \, d\kappa = \frac{1}{2}\kappa_0^{2n-2/3}\Gamma\left(n + \frac{3}{2}\right)$$

$$\times U\left(n + \frac{3}{2}; n + \frac{2}{3}; \frac{\kappa_0^2}{\kappa_m^2}\right)$$

2. Under the assumption $\kappa_0 \to 0$ (infinite outer scale), show that Eq. (12.19) reduces to

$$D_T(r) \sim 1.685 C_T^2 \kappa_m^{-2/3} \left[{}_1F_1\left(-\frac{1}{3};\frac{3}{2};\frac{-\kappa_m^2 r^2}{4}\right) - 1 \right]$$

3. Under the condition $\kappa_0 r \ll 1$, show that Eq. (12.64) reduces to the inertial convective range form

$$D_T(r) \sim C_T^2 r^{2/3} \qquad l_0 \ll r \ll L_0$$

4. For $\sqrt{\lambda L} \ll l_0$, the *phase structure function* associated with an optical plane wave propagating through locally isotropic and homogeneous atmospheric turbulence is computed by evaluating

$$D_S(\rho) = 8\pi^2 k^2 L \int_0^\infty [1 - J_0(\kappa\rho)]\Phi_n(\kappa)\kappa\,d\kappa$$

where k is the wavenumber of the optical wave, $\lambda = 2\pi/k$ is the wavelength, L is the propagation path length, and

$$\Phi_n(\kappa) = 0.033 C_n^2 \frac{\exp\left(-\kappa^2/\kappa_m^2\right)}{(\kappa_0^2 + \kappa^2)^{11/6}}$$

Show that

$$D_S(\rho) = 1.303 k^2 L C_n^2 \kappa_0^{-5/3} \sum_{n=1}^\infty \frac{(-1)^{n-1}(\kappa_0\rho/2)^{2n}}{n!} U\left(n+1; n+\frac{1}{6}; \frac{\kappa_0^2}{\kappa_m^2}\right)$$

5. For $\kappa_0 \to 0$, show that the first term of the series in problem 4 reduces $D_S(\rho)$ to the asymptotic result

$$D_S(\rho) \sim 0.33\Gamma(\tfrac{1}{6})k^2 L C_n^2 \kappa_m^{1/3}\rho^2 \qquad \rho \ll l_0$$

6. (a) For $\kappa_0 \to 0$, use the asymptotic formula for $U(a;c;x)$ to show that the series in problem 4 sums to

$$D_S(\rho) \sim 1.56\Gamma\left(\frac{1}{6}\right)k^2 L C_n^2 \kappa_m^{-5/3}\left[{}_1F_1\left(-\frac{5}{6};1;\frac{-\kappa_m^2\rho^2}{4}\right) - 1 \right]$$

(b) From (a), deduce that

$$D_S(\rho) \sim 2.91 k^2 L C_n^2 \rho^{5/3} \qquad l_0 \ll \rho$$

7. For a plane wave propagating through locally isotropic and homogeneous turbulence, the *variance of log amplitude* of the wave is defined by the integral

$$\sigma_\chi^2 = \frac{1}{3}\pi^2 L^2 \int_0^\infty \Phi_n(\kappa)\kappa^5\,d\kappa \qquad \frac{\sqrt{\lambda L}}{l_0} \ll 1$$

where all parameters are defined in problem 4. Under the additional assumption $\kappa_0^2/\kappa_m^2 \ll 1$, show that

$$\sigma_\chi^2 \sim 0.054\Gamma(\tfrac{7}{6})C_n^2 L^3 \kappa_m^{7/3}$$

8. Under the same conditions as stated in problem 7, the *covariance function of log amplitude* of an optical plane wave is defined by the integral

$$B_\chi(\rho) = \tfrac{1}{3}\pi^2 L^3 \int_0^\infty J_0(\kappa\rho)\Phi_n(\kappa)\kappa^5\,d\kappa$$

Show that

$$b_\chi(\rho) = \frac{B_\chi(\rho)}{B_\chi(0)} = {}_1F_1\!\left(\frac{7}{6}; 1; \frac{-\kappa_m^2\rho^2}{4}\right)$$

Bibliography

The available literature on special functions is vast, both in textbooks and in research papers. Rather than attempt to list any substantial part of it, we have opted to give a short list of some of the classical books as well as some more recent references. However, each of these references in turn supplies numerous additional references, including many of the early research papers on special functions.

Abramowitz, M., and I. A. Stegun (eds.): *Handbook of Mathematical Functions,* Dover, New York, 1965.

Arfken, G.: *Mathematical Methods for Physicists,* 3d ed., Academic, New York, 1985.

Artin, E.: *The Gamma Function,* Holt, Rinehart and Winston, New York, 1964.

Askey, R. A. (ed.): *Theory and Application of Special Functions,* Academic, New York, 1975.

Bell, W. W.: *Special Functions for Scientists and Engineers,* Van Nostrand, London, 1968.

Carlson, B. C.: *Special Functions of Applied Mathematics,* Academic, New York, 1977.

Erdelyi, A., et al.: *Higher Transcendental Functions,* 3 vols., Bateman Manuscript Project, McGraw-Hill, New York, 1953, 1955.

Gradshteyn, I. S., and I. M. Ryzhik: *Table of Integrals, Series, and Products,* Academic, New York, 1980.

Hochstadt, H.: *The Functions of Mathematical Physics,* Wiley, New York, 1971.

Jackson, D.: *Fourier Series and Orthogonal Polynomials,* Carus Math. Monogr. **6,** Mathematical Association America, Menasha, Wis., 1941.

Jahnke, E., et al.: *Tables of Higher Functions,* 6th ed., McGraw-Hill, New York, 1960.

Lebedev, N. N.: *Special Functions and Their Applications* (R. A. Silverman, transl. and ed.), Dover, New York, 1972.

Luke, Y. L.: *The Special Functions and Their Approximations,* 2 vols., Academic, New York, 1969.

Magnus, W., and F. Oberhettinger: *Formulas and Theorems for the Special Functions of Mathematical Physics,* Springer, New York, 1969.

Olver, F. W. J.: *Asymptotics and Special Functions,* Academic, New York, 1974.

Rainville, E. D.: *Special Functions,* Chelsea, New York, 1960.

Slater, L. J.: *Confluent Hypergeometric Functions,* Cambridge University Press, London, 1966.

Sneddon, I. N.: *Special Functions of Mathematical Physics and Chemistry,* 2d ed., Oliver and Boyd, Edinburgh, 1961.

Watson, G. N.: *A Treatise on the Theory of Bessel Functions,* 2d ed., Cambridge University Press, London, 1952.

A List of Special Function Formulas

For easy reference, the following is a selected list of formulas for many of the special functions discussed in the text.

Gamma Function

1. $\Gamma(x) = \displaystyle\int_0^\infty e^{-t}t^{x-1}\,dt,\ x>0$

2. $\Gamma(x) = 2\displaystyle\int_0^\infty e^{-t^2}t^{2x-1}\,dt,\ x<0$

3. $\Gamma(x) = \displaystyle\int_0^1 \left(\ln\frac{1}{t}\right)^{x-1}\,dt,\ x>0$

4. $\Gamma(x) = \displaystyle\lim_{n\to\infty}\frac{n!\,n^x}{x(x+1)(x+2)\cdots(x+n)}$

5. $\dfrac{1}{\Gamma(x)} = xe^{\gamma x}\displaystyle\prod_{n=1}^\infty\left(1+\frac{x}{n}\right)e^{-x/n}$

6. $\Gamma(x+1) = x\Gamma(x)$

7. $\Gamma(x)\Gamma(1-x) = \dfrac{\pi}{\sin \pi x}$

8. $\Gamma\left(\dfrac{1}{2}+x\right)\Gamma\left(\dfrac{1}{2}-x\right) = \dfrac{\pi}{\cos \pi x}$

9. $\sqrt{\pi}\,\Gamma(2x) = 2^{2x-1}\Gamma(x)\Gamma(x+\tfrac{1}{2})$

10. $\Gamma(1) = 1$

11. $\Gamma(\tfrac{1}{2}) = \sqrt{\pi}$

12. $\Gamma(n+1) = n!,\ n = 0, 1, 2, \ldots$

13. $\Gamma\left(n + \dfrac{1}{2}\right) = \dfrac{(2n)!}{2^{2n}n!}\sqrt{\pi}, \; n = 0, 1, 2, \ldots$

14. $\displaystyle\int_0^{\pi/2} \cos^{2x-1}\theta \, \sin^{2y-1}\theta \, d\theta = \dfrac{\Gamma(x)\Gamma(y)}{2\Gamma(x+y)}, \; x > y, \, y > 0$

15. $\Gamma(x+1) \sim \sqrt{2\pi x}\, x^x e^{-x}\left(1 + \dfrac{1}{12x} + \cdots\right), \; x \to \infty$

16. $n! \sim \sqrt{2\pi n}\, n^n e^{-n}, \; n \gg 1$

Beta Function

17. $B(x, y) = \displaystyle\int_0^1 t^{x-1}(1-t)^{y-1}\, dt, \; x > 0, \, y > 0$

18. $B(x, y) = \displaystyle\int_0^\infty \dfrac{t^{x-1}}{(1+t)^{x+y}}\, dt, \; x > 0, \, y > 0$

19. $B(x, y) = 2\displaystyle\int_0^{\pi/2} \cos^{2x-1}\theta \, \sin^{2y-1}\theta \, d\theta, \; x > 0, \, y > 0$

20. $B(x, y) = B(y, x)$

21. $B(x, y) = \dfrac{\Gamma(x)\Gamma(y)}{\Gamma(x+y)}, \; x > 0, \, y > 0$

Incomplete Gamma Function

22. $\gamma(a, x) = \displaystyle\int_0^x e^{-t}t^{a-1}\, dt, \; a > 0$

23. $\Gamma(a, x) = \displaystyle\int_x^\infty e^{-t}t^{a-1}\, dt, \; a > 0$

24. $\gamma(a, x) + \Gamma(a, x) = \Gamma(a)$

25. $\gamma(a, x) = x^a \displaystyle\sum_{n=0}^\infty \dfrac{(-1)^n x^n}{n!\,(n+a)}$

26. $\Gamma(a, x) = \Gamma(a) - x^a \displaystyle\sum_{n=0}^\infty \dfrac{(-1)^n x^n}{n!\,(n+a)}$

27. $\gamma(a+1, x) = a\gamma(a, x) - x^a e^{-x}$

28. $\Gamma(a+1, x) = a\Gamma(a, x) + x^a e^{-x}$

29. $\gamma(n+1, x) = n!\,[1 - e^{-x}e_n(x)], \; n = 0, 1, 2, \ldots$

30. $\Gamma(n+1, x) = n!\,e^{-x}e_n(x), \; n = 0, 1, 2, \ldots$

31. $\Gamma(a, x) \sim \Gamma(a)x^{a-1}e^{-x} \sum\limits_{n=0}^{\infty} \dfrac{x^{-n}}{\Gamma(a-n)}$, $a > 0$, $x \to \infty$

Digamma and Polygamma Functions

32. $\psi(x) = \dfrac{d}{dx} \ln \Gamma(x) = \dfrac{\Gamma'(x)}{\Gamma(x)}$

33. $\psi(x) = -\gamma + \sum\limits_{n=0}^{\infty} \left(\dfrac{1}{n+1} - \dfrac{1}{n+x} \right)$

34. $\psi(x+1) = \psi(x) + \dfrac{1}{x}$

35. $\psi(1-x) - \psi(x) = \pi \cot \pi x$

36. $\psi(x) + \psi(x + \tfrac{1}{2}) - 2 \ln 2 = 2\psi(2x)$

37. $\psi(1) = -\gamma$

38. $\psi(n+1) = -\gamma + \sum\limits_{k=1}^{n} \dfrac{1}{k}$, $n = 1, 2, 3, \ldots$

39. $\psi(x+1) = -\gamma + \sum\limits_{n=1}^{\infty} (-1)^{n+1}\zeta(n+1)x^n$, $-1 < x < 1$

40. $\psi(x+1) \sim \ln x + \dfrac{1}{2x} - \dfrac{1}{2} \sum\limits_{n=1}^{\infty} \dfrac{B_{2n}}{n} x^{-2n}$, $x \to \infty$

41. $\psi^{(m)}(x) = \dfrac{d^{m+1}}{dx^{m+1}} \ln \Gamma(x)$, $m = 1, 2, 3, \ldots$

42. $\psi^{(m)}(x) = (-1)^{m+1}m! \sum\limits_{n=0}^{\infty} \dfrac{1}{(n+x)^{m+1}}$

43. $\psi^{(m)}(1) = (-1)^{m+1}m! \, \zeta(m+1)$

44. $\psi^{(m)}(x+1) = (-1)^{m+1} \sum\limits_{n=0}^{\infty} (-1)^n \dfrac{(m+n)!}{n!} \zeta(m+n+1)x^n$,

    $-1 < x < 1$

Error Functions

45. $\operatorname{erf} x = \dfrac{2}{\sqrt{\pi}} \int_0^x e^{-t^2} \, dt$

46. $\operatorname{erf} x = \dfrac{2}{\sqrt{\pi}} \sum\limits_{n=0}^{\infty} \dfrac{(-1)^n x^{2n+1}}{n!(2n+1)}$, $-\infty < x < \infty$

47. erf $0 = 0$

48. erf $\infty = 1$

49. erfc $x = \dfrac{2}{\sqrt{\pi}} \displaystyle\int_x^\infty e^{-t^2}\, dt$

50. erfc $x = 1 - \text{erf}\, x$

51. erfc $x \sim \dfrac{e^{-x^2}}{\sqrt{\pi}\, x} \left[1 + \displaystyle\sum_{n=1}^\infty (-1)^n \dfrac{1 \times 3 \times \cdots \times (2n-1)}{(2x^2)^n} \right]$,

$x \to \infty$

Exponential Integrals

52. $\text{Ei}\,(x) = \displaystyle\int_{-\infty}^x \dfrac{e^t}{t}\, dt,\ x \neq 0$

53. $E_1(x) = -\text{Ei}(-x) = \displaystyle\int_x^\infty \dfrac{e^{-t}}{t}\, dt,\ x > 0$

54. $E_1(x) = \Gamma(0, x)$

55. $E_1(x) = -\gamma - \ln x - \displaystyle\sum_{n=1}^\infty \dfrac{(-1)^n x^n}{n!\, n},\ x > 0$

56. $\text{Ei}(x) \sim \dfrac{e^x}{x} \displaystyle\sum_{n=0}^\infty \dfrac{n!}{x^n},\ x \to \infty$

57. $E_1(x) \sim \dfrac{e^{-x}}{x} \displaystyle\sum_{n=0}^\infty \dfrac{(-1)^n n!}{x^n},\ x \to \infty$

Elliptic Integrals

58. $F(m, \phi) = \displaystyle\int_0^\phi \dfrac{d\theta}{\sqrt{1 - m^2 \sin^2 \theta}},\ 0 < m < 1$

59. $E(m, \phi) = \displaystyle\int_0^\phi \sqrt{1 - m^2 \sin^2 \theta}\, d\theta,\ 0 < m < 1$

60. $\Pi(m, \phi, a) = \displaystyle\int_0^\phi \dfrac{d\theta}{\sqrt{1 - m^2 \sin^2 \theta}\,(1 + a^2 \sin^2 \theta)},\ 0 < m < 1,$

$a \neq m, 0$

61. $K(m) = \displaystyle\int_0^{\pi/2} \dfrac{d\theta}{\sqrt{1 - m^2 \sin^2 \theta}},\ 0 < m < 1$

62. $E(m) = \displaystyle\int_0^{\pi/2} \sqrt{1 - m^2 \sin^2 \theta}\, d\theta,\ 0 < m < 1$

63. $\Pi(m, a) = \int_0^{\pi/2} \dfrac{d\theta}{\sqrt{1 - m^2 \sin^2 \theta}\,(1 + a^2 \sin^2 \theta)}$, $0 < m < 1$,

$a \neq m, 0$

64. $K(m) = \dfrac{\pi}{2} \sum\limits_{n=0}^{\infty} \binom{-1/2}{n}^2 m^{2n}$

65. $E(m) = \dfrac{\pi}{2} \sum\limits_{n=0}^{\infty} \binom{1/2}{n}\binom{-1/2}{n} m^{2n}$

Legendre Polynomials

66. $P_n(x) = \sum\limits_{k=0}^{[n/2]} \dfrac{(-1)^k (2n - 2k)!\, x^{n-2k}}{2^n k!\,(n-k)!\,(n-2k)!}$

67. $P_n(x) = \dfrac{1}{2^n n!} \dfrac{d^n}{dx^n} [(x^2 - 1)^n]$

68. $(1 - 2xt + t^2)^{-1/2} = \sum\limits_{n=0}^{\infty} P_n(x) t^n$

69. $P_n(1) = 1;\ P_n(-1) = (-1)^n$

70. $P_n'(1) = \frac{1}{2}n(n + 1);\ P_n'(-1) = (-1)^{n-1}[\frac{1}{2}n(n+1)]$

71. $P_{2n}(0) = \dfrac{(-1)^n (2n)!}{2^{2n} (n!)^2};\ P_{2n+1}(0) = 0$

72. $(n + 1)P_{n+1}(x) - (2n + 1)xP_n(x) + nP_{n-1}(x) = 0$

73. $P_{n+1}'(x) - 2xP_n'(x) + P_{n-1}'(x) - P_n(x) = 0$

74. $P_{n+1}'(x) - xP_n'(x) - (n + 1)P_n(x) = 0$

75. $xP_n'(x) - P_{n-1}'(x) - nP_n(x) = 0$

76. $P_{n+1}'(x) - P_{n-1}'(x) = (2n + 1)P_n(x)$

77. $(1 - x^2)P_n'(x) = nP_{n-1}(x) - nxP_n(x)$

78. $(1 - x^2)P_n''(x) - 2xP_n'(x) + n(n + 1)P_n(x) = 0$

79. $P_n(x) = \dfrac{1}{\pi} \int_0^{\pi} [x + (x^2 - 1)^{1/2} \cos \phi]^n \, d\phi$

80. $|P_n(x)| \leq 1,\ |x| \leq 1$

81. $|P_n(x)| < \left[\dfrac{\pi}{2n(1 - x^2)}\right]^{1/2},\ |x| < 1,\ n = 1, 2, 3, \ldots$

82. $\displaystyle\int_{-1}^{1} P_n(x)P_k(x)\, dx = 0,\ k \neq n$

83. $\displaystyle\int_{-1}^{1} [P_n(x)]^2\,dx = \frac{2}{2n+1}$

84. $\displaystyle\sum_{k=0}^{n} (2k+1)P_k(t)P_k(x) = \frac{n+1}{t-x}\,[P_{n+1}(t)P_n(x) - P_n(t)P_{n+1}(x)]$

Associated Legendre Functions

85. $P_n^m(x) = (1-x^2)^{m/2}\dfrac{d^m}{dx^m}\,[P_n(x)],\ m = 1, 2, 3, \ldots$

86. $P_n^0(x) = P_n(x)$

87. $P_n^{-m}(x) = (-1)^m\dfrac{(n-m)!}{(n+m)!}\,P_n^m(x)$

88. $(n-m+1)P_{n+1}^m(x) - (2n+1)xP_n^m(x) + (n+m)P_{n-1}^m(x) = 0$

89. $(1-x^2)P_n^{m\prime\prime}(x) - 2xP_n^{m\prime}(x) + \left[n(n+1) - \dfrac{m^2}{1-x^2}\right]P_n^m(x) = 0$

90. $\displaystyle\int_{-1}^{1} P_n^m(x)P_k^m(x)\,dx = 0,\ k \neq n$

91. $\displaystyle\int_{-1}^{1} [P_n^m(x)]^2\,dx = \frac{2(n+m)!}{(2n+1)(n-m)!}$

Hermite Polynomials

92. $H_n(x) = \displaystyle\sum_{k=0}^{[n/2]} \frac{(-1)^k n!}{k!\,(n-2k)!}\,(2x)^{n-2k}$

93. $H_n(x) = (-1)^n e^{x^2}\dfrac{d^n}{dx^n}\,(e^{-x^2})$

94. $\exp(2xt - t^2) = \displaystyle\sum_{n=0}^{\infty} H_n(x)\,\frac{t^n}{n!}$

95. $H_{2n}(0) = (-1)^n\dfrac{(2n)!}{n!};\ H_{2n+1}(0) = 0$

96. $H_{n+1}(x) - 2xH_n(x) + 2nH_{n-1}(x) = 0$

97. $H_n'(x) = 2nH_{n-1}(x)$

98. $H_n''(x) - 2xH_n'(x) + 2nH_n(x) = 0$

99. $\displaystyle\int_{-\infty}^{\infty} e^{-x^2}H_n(x)H_k(x)\,dx = 0,\ k \neq n$

100. $\int_{-\infty}^{\infty} e^{-x^2}[H_n(x)]^2\,dx = 2^n n!\sqrt{\pi}$

Laguerre Polynomials

101. $L_n(x) = \sum_{k=0}^{n} \dfrac{(-1)^k n!\,x^k}{(k!)^2(n-k)!}$

102. $L_n(x) = \dfrac{e^x}{n!}\dfrac{d^n}{dx^n}(x^n e^{-x})$

103. $(1-t)^{-1}\exp\left(-\dfrac{xt}{1-t}\right) = \sum_{n=0}^{\infty} L_n(x)t^n$

104. $L_n(0) = 1$

105. $(n+1)L_{n+1}(x) + (x-1-2n)L_n(x) + nL_{n-1}(x) = 0$

106. $L_n'(x) - L_{n-1}'(x) + L_{n-1}(x) = 0$

107. $xL_n'(x) = nL_n(x) - nL_{n-1}(x)$

108. $xL_n''(x) + (1-x)L_n'(x) + nL_n(x) = 0$

109. $\int_0^{\infty} e^{-x}L_n(x)L_k(x)\,dx = 0,\ k \ne n$

110. $\int_0^{\infty} e^{-x}[L_n(x)]^2\,dx = 1$

Associated Laguerre Polynomials

111. $L_n^{(m)}(x) = (-1)^m\dfrac{d^m}{dx^m}[L_{n+m}(x)],\ m = 1, 2, 3, \ldots$

112. $L_n^{(m)}(x) = \sum_{k=0}^{n}\dfrac{(-1)^k(m+n)!\,x^k}{(n-k)!\,(m+k)!\,k!}$

113. $L_n^{(m)}(x) = \dfrac{e^x}{n!}x^{-m}\dfrac{d^n}{dx^n}(x^{n+m}e^{-x})$

114. $(1-t)^{-m-1}\exp\left(-\dfrac{xt}{1-t}\right) = \sum_{n=0}^{\infty} L_n^{(m)}(x)t^n$

115. $L_n^{(m)}(0) = \dfrac{(n+m)!}{n!\,m!}$

116. $(n+1)L_{n+1}^{(m)}(x) + (x-1-2n-m)L_n^{(m)}(x) + (n+m)L_{n-1}^{(m)}(x) = 0$

117. $L_{n-1}^{(m)}(x) + L_n^{(m-1)}(x) - L_n^{(m)}(x) = 0$

118. $L_n^{(m)'}(x) = -L_{n-1}^{(m+1)}(x)$

119. $xL_n^{(m)''}(x) + (m + 1 - x)L_n^{(m)'}(x) + nL_n^{(m)}(x) = 0$

120. $\displaystyle\int_0^\infty e^{-x}x^m L_n^{(m)}(x)L_k^{(m)}(x)\,dx = 0,\ k \neq n$

121. $\displaystyle\int_0^\infty e^{-x}x^m [L_n^{(m)}(x)]^2\,dx = \dfrac{\Gamma(n + m + 1)}{n!}$

Gegenbauer Polynomials

122. $C_n^\lambda(x) = (-1)^n \displaystyle\sum_{k=0}^{[n/2]} \binom{-\lambda}{n-k}\binom{n-k}{k}(2x)^{n-2k}$

123. $(1 - 2xt + t^2)^{-\lambda} = \displaystyle\sum_{n=0}^\infty C_n^\lambda(x)t^n$

124. $C_n^\lambda(1) = (-1)^n \dbinom{-2\lambda}{n};\ C_n^\lambda(-1) = \dbinom{-2\lambda}{n}$

125. $C_{2n}^\lambda(0) = \dbinom{-\lambda}{n};\ C_{2n+1}^\lambda(0) = 0$

126. $(n + 1)C_{n+1}^\lambda(x) - 2(\lambda + n)xC_n^\lambda(x) + (2\lambda + n - 1)C_{n-1}^\lambda(x) = 0$

127. $C_n^{\lambda'}(x) = 2\lambda C_{n+1}^{\lambda+1}(x)$

128. $(1 - x^2)C_n^{\lambda''}(x) - (2\lambda + 1)xC_n^{\lambda'}(x) + n(n + 2\lambda)C_n^\lambda(x) = 0$

129. $\displaystyle\int_{-1}^1 (1 - x^2)^{\lambda-1/2}C_n^\lambda(x)C_k^\lambda(x)\,dx = 0,\ k \neq n$

130. $\displaystyle\int_{-1}^1 (1 - x^2)^{\lambda-1/2}[C_n^\lambda(x)]^2\,dx = \dfrac{2^{1-2\lambda}\pi}{(n + \lambda)}\dfrac{\Gamma(n + 2\lambda)}{[\Gamma(\lambda)]^2 n!}$

Chebyshev Polynomials

131. $T_n(x) = \dfrac{n}{2}\displaystyle\sum_{k=0}^{[n/2]} \dfrac{(-1)^k(n - k - 1)!}{k!\,(n - 2k)!}(2x)^{n-2k},\ n = 1, 2, 3, \ldots$

132. $U_n(x) = \displaystyle\sum_{k=0}^{[n/2]} \binom{n-k}{k}(-1)^k(2x)^{n-2k}$

133. $\dfrac{1 - xt}{1 - 2xt + t^2} = \displaystyle\sum_{n=0}^\infty T_n(x)t^n$

134. $(1 - 2xt + t^2)^{-1} = \displaystyle\sum_{n=0}^\infty U_n(x)t^n$

135. $T_n(1) = 1;\ U_n(1) = n + 1$

136. $T_{2n}(0) = (-1)^n;\ T'_{2n+1}(0) = 0$

137. $U_{2n}(0) = (-1)^n;\ U_{2n+1}(0) = 0$

138. $T_{n+1}(x) - 2xT_n(x) + T_{n-1}(x) = 0$

139. $U_{n+1}(x) - 2xU_n(x) + U_{n-1}(x) = 0$

140. $T'_n(x) = U_n(x) - xU_{n-1}(x)$

141. $(1 - x^2)U_{n-1}(x) = xT_n(x) - T_{n+1}(x)$

142. $(1 - x^2)T''_n(x) - xT'_n(x) + n^2 T_n(x) = 0$

143. $(1 - x^2)U''_n(x) - 3xU'_n(x) + n(n + 2)U_n(x) = 0$

144. $\displaystyle\int_{-1}^{1} (1 - x^2)^{-1/2}T_n(x)T_k(x)\,dx = 0,\ k \neq n$

145. $\displaystyle\int_{-1}^{1} (1 - x^2)^{1/2}U_n(x)U_k(x)\,dx = 0,\ k \neq n$

146. $\displaystyle\int_{-1}^{1} (1 - x^2)^{-1/2}[T_n(x)]^2\,dx = \begin{cases} \pi & n = 0 \\ \dfrac{\pi}{2} & n \geq 1 \end{cases}$

147. $\displaystyle\int_{-1}^{1} (1 - x^2)^{1/2}[U_n(x)]^2\,dx = \dfrac{\pi}{2}$

Bessel Functions of the First Kind

148. $\displaystyle J_p(x) = \sum_{k=0}^{\infty} \frac{(-1)^k (x/2)^{2k+p}}{k!\,\Gamma(k + p + 1)}$

149. $\displaystyle \exp\left[\frac{1}{2}x\left(t - \frac{1}{t}\right)\right] = \sum_{n=-\infty}^{\infty} J_n(x)t^n,\ t \neq 0$

150. $J_{-n}(x) = (-1)^n J_n(x),\ n = 0, 1, 2, \ldots$

151. $J_0(0) = 1;\ J_p(0) = 0,\ p > 0$

152. $\displaystyle J_{1/2}(x) = \sqrt{\frac{2}{\pi x}}\,\sin x$

153. $\displaystyle J_{-1/2}(x) = \sqrt{\frac{2}{\pi x}}\,\cos x$

154. $\displaystyle J_n(x + y) = \sum_{k=-\infty}^{\infty} J_k(x)J_{n-k}(x)$

155. $\dfrac{d}{dx}\,[x^p J_p(x)] = x^p J_{p-1}(x)$

156. $\dfrac{d}{dx}\,[x^{-p} J_p(x)] = -x^{-p} J_{p+1}(x)$

157. $\left(\dfrac{d}{x\,dx}\right)^m [x^p J_p(x)] = x^{p-m} J_{p-m}(x)$

158. $\left(\dfrac{d}{x\,dx}\right)^m [x^{-p} J_p(x)] = (-1)^m x^{-p-m} J_{p+m}(x)$

159. $J_p'(x) + \dfrac{p}{x} J_p(x) = J_{p-1}(x)$

160. $J_p'(x) - \dfrac{p}{x} J_p(x) = -J_{p+1}(x)$

161. $J_{p-1}(x) + J_{p+1}(x) = \dfrac{2p}{x} J_p(x)$

162. $J_{p-1}(x) - J_{p+1}(x) = 2J_p'(x)$

163. $x^2 J_p''(x) + x J_p'(x) + (x^2 - p^2) J_p(x) = 0$

164. $W(J_p, J_{-p})(x) = -\dfrac{2\sin p\pi}{\pi x}$

165. $J_p(x) J_{1-p}(x) + J_{-p}(x) J_{p-1}(x) = \dfrac{2\sin p\pi}{\pi x}$

166. $J_n(x) = \dfrac{1}{\pi} \displaystyle\int_0^\pi \cos\,(n\phi - x\sin\phi)\,d\phi,\ n = 0, 1, 2, \ldots$

167. $J_p(x) = \dfrac{(x/2)^p}{\sqrt{\pi}\,\Gamma(p + \frac{1}{2})} \displaystyle\int_{-1}^1 e^{ixt}(1 - t^2)^{p-1/2}\,dt,\ p > -\dfrac{1}{2},\ x > 0$

168. $\displaystyle\int x^p J_{p-1}(x)\,dx = x^p J_p(x) + C$

169. $\displaystyle\int x^{-p} J_{p+1}(x)\,dx = -x^{-p} J_p(x) + C$

170. $\displaystyle\int_0^b x J_p(k_m x) J_p(k_n x)\,dx = 0,\ m \neq n;\ J_p(k_n b) = 0,\ n = 1, 2, 3, \ldots$

171. $\displaystyle\int_0^b x [J_p(k_n x)]^2\,dx = \frac{1}{2} b^2 [J_{p+1}(k_n b)]^2;\ J_p(k_n b) = 0,\ n = 1, 2, 3, \ldots$

172. $J_p(x) \sim \dfrac{(x/2)^p}{\Gamma(p + 1)},\ p \neq -1, -2, -3, \ldots,\ x \to 0^+$

173. $J_p(x) \sim \sqrt{\dfrac{2}{\pi x}} \cos\left[x - \dfrac{(p + \frac{1}{2})\pi}{2}\right], \ x \to \infty$

Bessel Functions of the Second Kind

174. $Y_p(x) = \dfrac{(\cos p\pi)J_p(x) - J_{-p}(x)}{\sin p\pi}$

175. $Y_0(x) = \dfrac{2}{\pi} J_0(x)\left(\ln\dfrac{x}{2} + \gamma\right)$

$\qquad - \dfrac{2}{\pi} \displaystyle\sum_{k=1}^{\infty} \dfrac{(-1)^k(x/2)^{2k}}{(k!)^2}\left(1 + \dfrac{1}{2} + \cdots + \dfrac{1}{k}\right)$

176. $Y_n(x) = \dfrac{2}{\pi} J_n(x) \ln\dfrac{x}{2} - \displaystyle\sum_{k=0}^{n-1} \dfrac{(n - k - 1)!}{k!}\left(\dfrac{x}{2}\right)^{2k-n}$

$\qquad - \dfrac{1}{\pi} \displaystyle\sum_{k=0}^{\infty} \dfrac{(-1)^k(x/2)^{2k+n}}{k!\,(k + n)!}\,[\psi(k + n + 1) + \psi(k + 1)],$

$\qquad n = 1, 2, 3, \ldots$

177. $Y_{-n}(x) = (-1)^n Y_n(x), \ n = 0, 1, 2, \ldots$

178. $\dfrac{d}{dx}\,[x^p Y_p(x)] = x^p Y_{p-1}(x)$

179. $\dfrac{d}{dx}\,[x^{-p} Y_p(x)] = -x^{-p} Y_{p+1}(x)$

180. $Y_p'(x) + \dfrac{p}{x} Y_p(x) = Y_{p-1}(x)$

181. $Y_p'(x) - \dfrac{p}{x} Y_p(x) = -Y_{p+1}(x)$

182. $Y_{p-1}(x) + Y_{p+1}(x) = \dfrac{2p}{x} Y_p(x)$

183. $Y_{p-1}(x) - Y_{p+1}(x) = 2Y_p'(x)$

184. $x^2 Y_p''(x) + x Y_p'(x) + (x^2 - p^2)Y_p(x) = 0$

185. $W(J_p, Y_p)(x) = \dfrac{2}{\pi x}$

186. $J_p(x)Y_{p+1}(x) + Y_p(x)J_{p+1}(x) = -\dfrac{2}{\pi x}$

187. $Y_0(x) \sim \dfrac{2}{\pi} \ln x, \ x \to 0^+$

188. $Y_p(x) \sim -\dfrac{\Gamma(p)}{\pi}\left(\dfrac{2}{x}\right)^p,\ p>0,\ x\to 0^+$

189. $Y_p(x) \sim \sqrt{\dfrac{2}{\pi x}}\sin\left[x - \dfrac{(p+\frac{1}{2})\pi}{2}\right],\ x\to\infty$

Modified Bessel Functions of the First Kind

190. $I_p(x) = \displaystyle\sum_{k=0}^{\infty}\dfrac{(x/2)^{2k+p}}{k!\,\Gamma(k+p+1)}$

191. $\exp\left[\dfrac{1}{2}x\left(t+\dfrac{1}{t}\right)\right] = \displaystyle\sum_{n=-\infty}^{\infty} I_n(x)t^n,\ t\neq 0$

192. $I_{-n}(x) = I_n(x),\ n=0,1,2,\dots$

193. $I_0(0)=1;\ I_p(0)=0,\ p>0$

194. $I_{1/2}(x) = \sqrt{\dfrac{2}{\pi x}}\sinh x$

195. $I_{-1/2}(x) = \sqrt{\dfrac{2}{\pi x}}\cosh x$

196. $I_n(x+y) = \displaystyle\sum_{k=-\infty}^{\infty} I_k(x)I_{n-k}(x)$

197. $\dfrac{d}{dx}[x^p I_p(x)] = x^p I_{p-1}(x)$

198. $\dfrac{d}{dx}[x^{-p} I_p(x)] = x^{-p} I_{p+1}(x)$

199. $I_p'(x) + \dfrac{p}{x}I_p(x) = I_{p-1}(x)$

200. $I_p'(x) - \dfrac{p}{x}I_p(x) = I_{p+1}(x)$

201. $I_{p-1}(x) + I_{p+1}(x) = 2I_p'(x)$

202. $I_{p-1}(x) - I_{p+1}(x) = \dfrac{2p}{x}I_p(x)$

203. $x^2 I_p''(x) + x I_p'(x) - (x^2+p^2)I_p(x) = 0$

204. $W(I_p, I_{-p})(x) = -\dfrac{2\sin p\pi}{\pi x}$

205. $I_0(x) = \dfrac{1}{\pi} \displaystyle\int_0^\pi e^{\pm x \cos\theta}\, d\theta$

206. $I_p(x) = \dfrac{(x/2)^p}{\sqrt{\pi}\,\Gamma(p + \frac{1}{2})} \displaystyle\int_{-1}^{1} e^{-xt}(1 - t^2)^{p - 1/2}\, dt, \; p > -\frac{1}{2},\; x > 0$

207. $I_p(x) \sim \dfrac{(x/2)^p}{\Gamma(p + 1)}, \; p \ne -1, -2, -3, \ldots,\; x \to 0^+$

208. $I_p(x) \sim \dfrac{e^x}{\sqrt{2\pi x}}, \; x \to \infty$

Modified Bessel Functions of the Second Kind

209. $K_p(x) = \dfrac{\pi}{2}\dfrac{I_{-p}(x) - I_p(x)}{\sin p\pi}$

210. $K_0(x) = -I_0(x)\left(\ln\dfrac{x}{2} + \gamma\right)$

$+ \displaystyle\sum_{k=1}^{\infty} \dfrac{(x/2)^{2k}}{(k!)^2}\left(1 + \dfrac{1}{2} + \cdots + \dfrac{1}{k}\right)$

211. $K_n(x) = (-1)^{n-1}I_n(x)\ln\dfrac{x}{2}$

$+ \dfrac{1}{2}\displaystyle\sum_{k=0}^{n-1}\dfrac{(-1)^k(n - k - 1)!}{k!}\left(\dfrac{x}{2}\right)^{2k-n}$

$+ \dfrac{(-1)^n}{2}\displaystyle\sum_{k=0}^{\infty}\dfrac{(x/2)^{2k+n}}{k!\,(k+n)!}[\psi(k + n + 1) + \psi(k + 1)],$

$n = 1, 2, 3, \ldots$

212. $K_{-p}(x) = K_p(x)$

213. $\dfrac{d}{dx}[x^p K_p(x)] = -x^p K_{p-1}(x)$

214. $\dfrac{d}{dx}[x^{-p} K_p(x)] = -x^{-p} K_{p+1}(x)$

215. $K_p'(x) + \dfrac{p}{x}K_p(x) = -K_{p-1}(x)$

216. $K_p'(x) - \dfrac{p}{x}K_p(x) = -K_{p+1}(x)$

217. $K_{p-1}(x) + K_{p+1}(x) = -2K_p'(x)$

218. $K_{p-1}(x) - K_{p+1}(x) = -\dfrac{2p}{x}K_p(x)$

219. $x^2K_p''(x) + xK_p'(x) - (x^2 + p^2)K_p(x) = 0$

220. $W(I_p, K_p)(x) = -\dfrac{1}{x}$

221. $I_p(x)K_{p+1}(x) + K_p(x)I_{p+1}(x) = \dfrac{1}{x}$

222. $K_p(x) = \dfrac{\sqrt{\pi}\,(x/2)^p}{\Gamma(p + \frac{1}{2})} \displaystyle\int_1^\infty e^{-xt}(t^2 - 1)^{p-1/2}\,dt, \; p > -\frac{1}{2}, \; x > 0$

223. $K_0(x) \sim -\ln x, \; x \to 0^+$

224. $K_p(x) \sim \dfrac{\Gamma(p)}{2}\left(\dfrac{2}{x}\right)^p, \; p > 0, \; x \to 0^+$

225. $K_p(x) \sim \sqrt{\dfrac{\pi}{2x}}\,e^{-x}, \; x \to \infty$

Hypergeometric Function

226. $F(a, b; c; x) = \displaystyle\sum_{n=0}^{\infty} \dfrac{(a)_n(b)_n}{(c)_n}\dfrac{x^n}{n!}, \; |x| < 1$

227. $F(a, b; c; x) = F(b, a; c; x)$

228. $\dfrac{d^k}{dx^k}F(a, b; c; x) = \dfrac{(a)_k(b)_k}{(c)_k}F(a + k, b + k; c + k; x),$

$\quad k = 1, 2, 3, \ldots$

229. $F(a, b; c; x) = \dfrac{\Gamma(c)}{\Gamma(b)\Gamma(c - b)} \displaystyle\int_0^1 t^{b-1}(1 - t)^{c-b-1}(1 - xt)^{-a}\,dt,$

$\quad c > b > 0$

230. $F(a, b; c; 1) = \dfrac{\Gamma(c)\Gamma(c - a - b)}{\Gamma(c - a)\Gamma(c - b)}$

Confluent Hypergeometric Function

231. $M(a; c; x) = \displaystyle\sum_{n=0}^{\infty} \dfrac{(a)_n x^n}{(c)_n\,n!}, \; -\infty < x < \infty$

232. $M(a; c; x) = e^x M(c - a; c; -x)$

233. $\dfrac{d^k}{dx^k}M(a; c; x) = \dfrac{(a)_k}{(c)_k}M(a + k; c + k; x), \; k = 1, 2, 3, \ldots$

234. $M(a;c;x) = \dfrac{\Gamma(c)}{\Gamma(a)\Gamma(c-a)} \displaystyle\int_0^1 e^{xt} t^{a-1}(1-t)^{c-a-1}\, dt,\ c > a > 0$

235. $M(a;c;x) \sim 1,\ x \to 0$

236. $M(a;c;x) \sim \dfrac{\Gamma(c)}{\Gamma(a)} x^{a-c} e^x \displaystyle\sum_{n=0}^{\infty} \dfrac{(1-a)_n(c-a)_n}{n!\,x^n},\ x \to \infty$

Pochhammer Symbol

237. $(a)_0 = 1,\ (a)_n = a(a+1)\cdots(a+n-1) = \dfrac{\Gamma(a+n)}{\Gamma(a)}$

238. $(0)_0 = 1$

239. $(1)_n = n!$

240. $(a)_{n+1} = a(a+1)_n$

241. $(a)_{n+k} = (a)_k(a+k)_n$

242. $(a)_{n-k} = \dfrac{(-1)^k (a)_n}{(1-a-n)_k}$

243. $(a)_{-n} = \dfrac{(-1)^n}{(1-a)_n}$

244. $(-k)_n = \begin{cases} \dfrac{(-1)^n k!}{(k-n)!} & 0 \le n \le k \\ 0 & n > k \end{cases}$

245. $(n+k)! = n!\,(n+1)_k = k!\,(k+1)_n$

246. $\Gamma(a+1-n) = \dfrac{(-1)^n \Gamma(a+1)}{(-a)_n}$

247. $\dbinom{-a}{n} = \dfrac{(-1)^n}{n!}(a)_n$

248. $(2n)! = 2^{2n}(\tfrac{1}{2})_n n!$

249. $(2n+1)! = 2^{2n}(\tfrac{3}{2})_n n!$

250. $(a)_{2n} = 2^{2n}(\tfrac{1}{2}a)_n(\tfrac{1}{2}+\tfrac{1}{2}a)_n$

Selected Answers
to Exercises

Chapter 1

Exercises 1.2

3. -1

7. Converges absolutely

9. Diverges

10. Converges conditionally

Exercises 1.3

3. Converges only for $x = 0$

6. Converges uniformly

15. (a) $(1 - x^2)^{-1/2} = \sum_{n=0}^{\infty} \binom{-1/2}{n} (-1)^n x^{2n}$

    (b) $\sin^{-1} x = \sum_{n=0}^{\infty} \frac{(2n)! \, x^{2n+1}}{2^{2n} (n!)^2 (2n+1)}$

18. $\sum_{n=1}^{\infty} (-1)^{n-1} \dfrac{2^{2n-1} x^{2n}}{(2n)!}$

20. $\frac{3}{4}(1 - 2x^2 + \frac{7}{6}x^4 - \frac{61}{180}x^6 + \cdots)$

Exercises 1.4

3. $\dfrac{\pi}{2} - \dfrac{4}{\pi} \sum_{n=1}^{\infty} \dfrac{\cos(2n+1)x}{(2n+1)^2}$

11. (a) $\frac{3}{4}$
    (b) $\frac{3}{4}$
    (c) $\frac{3}{4}$

Exercises 1.5

4. π

7. Converges absolutely

Exercises 1.6

1. $f(x) \sim bx^2$ as $x \to 0$, $f(x) \sim bx/a$ as $x \to \infty$

Exercises 1.7

5. $\frac{3}{2}$

7. 1

Chapter 2

Exercises 2.2

3. 60

31. $\pi \dfrac{\sqrt{2}}{4}$

7. $^{10}/_9$

35. $\dfrac{\Gamma(x/p)}{ps^{x/p}}$

Exercises 2.3

1. $\dfrac{[\Gamma(^1/_4)]^2}{2\sqrt{\pi}}$

9. $2\sigma^4$

Exercises 2.4

1. $\dfrac{2\pi}{\sqrt{3}}$

14. π

11. $\dfrac{\pi}{8}$

17. $\dfrac{\pi a^6}{32}$

Chapter 3

Exercises 3.3

1. (a) $\operatorname{erf}(\sqrt{x})$ (b) $\operatorname{erf}(\sqrt{x}) - 2\sqrt{\dfrac{x}{\pi}}\,e^{-x}$

Exercises 3.4

15. $\operatorname{Si}(b) - \operatorname{Si}(a)$

19. $\operatorname{Si}(ab) - {}^1/_2\{\operatorname{Si}[b(a+1)] + \operatorname{Si}[b(a-1)]\}$

Exercises 3.5

11. $12E(1/\sqrt{3})$

Chapter 4

Exercises 4.2

15. (a) $n = 1, 3, 5, \ldots$ (b) $n = 0, 2, 4, \ldots$

Exercises 4.4

27. $x^3 = \tfrac{3}{5}P_1(x) + \tfrac{2}{5}P_3(x)$

36. $f(x) = \dfrac{1}{2} + \displaystyle\sum_{n=1}^{\infty} \dfrac{(-1)^{n-1}(4n+1)(2n-2)!}{2^{2n}(n+1)!\,(n-1)!} P_{2n}(x)$

Exercises 4.5

1. None

5. Smooth

Exercises 4.6

3. $y = C_1'P_3(x) + C_2'Q_3(x)$

Exercises 4.8

1. (a) $u(r, \phi) = 1$

   (b) $u(r, \phi) = r \cos \phi$

   (c) $u(r, \phi) = \tfrac{2}{3}r^2 P_2(\cos \phi) + \tfrac{1}{3}$

   (d) $u(r, \phi) = \tfrac{4}{3}r^2 P_2(\cos \phi) - \tfrac{1}{3}$

8. $u(r, \phi) = \dfrac{2}{3} T_0 \left[1 - \left(\dfrac{r}{c}\right)^2 P_2(\cos \phi) \right]$

Chapter 6

Exercises 6.5

4. $f(x) = 2 \displaystyle\sum_{n=1}^{\infty} \dfrac{J_0(k_n x)}{k_n J_1(k_n)}$

Exercises 6.6

3. $y = C_1 J_{1/4}(x) + C_2 J_{-1/4}(x)$

Exercises 6.7

3. $y = C_1 J_k(x^2) + C_2 Y_k(x^2)$

5. $y = x^{-n}[C_1 J_n(x) + C_2 Y_n(x)]$

7. $y = \sqrt{x}\,[C_1 J_{1/2}(x) + C_2 J_{-1/2}(x)]$

11. $y = \sqrt{x}\,[C_1 J_0(2\sqrt{x}) + C_2 Y_0(2\sqrt{x})]$

13. (b) $y(t) = C_1 J_0\left(\dfrac{2}{m}\sqrt{at}\right) + C_2 Y_0\left(\dfrac{2}{m}\sqrt{at}\right)$

   (c) $y(x) = C_1 J_0\left(\dfrac{2}{m}\sqrt{a}\,e^{mx/2}\right) + C_2 Y_0\left(\dfrac{2}{m}\sqrt{a}\,e^{mx/2}\right)$

Chapter 7

Exercises 7.2

2. $y = C_1 I_1(x) + C_2 K_1(x)$

25. $y = \sqrt{x}[C_1 I_{1/3}(\tfrac{2}{3}x^{3/2}) + C_2 K_{1/3}(\tfrac{2}{3}x^{3/2})]$

27. $y = x^2[C_1 I_{2/3}(x^3) + C_2 K_{2/3}(x^3)]$

Chapter 8

Exercises 8.4

9. $u(r, \phi) = \dfrac{I_0(3r)}{I_0(3)} \sin 3z$

Index

ABOUT THE AUTHOR

Larry C. Andrews is a professor of mathematics at the
University of Central Florida where he is also a member of
the Department of Electrical Engineering. Dr. Andrews
received a doctoral degree in theoretical mechanics in 1970
from Michigan State University. His current research
interests include the propagation of laser beams through
turbulent media, detection theory, signal processing,
special functions, integral transforms, and differential
equations.